D1597360

MECHATRONICS

License, Disclaimer of Liability, and Limited Warranty

The CD-ROM that accompanies this book may only be used on a single PC. This license does not permit its use on the Internet or on a network (of any kind). By purchasing or using this book/CD-ROM package (the "Work"), you agree that this license grants permission to use the products contained herein, but does not give you the right of ownership to any of the textual content in the book or ownership to any of the information or products contained on the CD-ROM. Use of third party software contained herein is limited to and subject to licensing terms for the respective products, and permission must be obtained from the publisher or the owner of the software in order to reproduce or network any portion of the textual material or software (in any media) that is contained in the Work.

Jones and Bartlett Publishers, LLC ("the Publisher") and anyone involved in the creation, writing, or production of the accompanying algorithms, code, or computer programs ("the software") or any of the third party software contained on the CD-ROM or any of the textual material in the book, cannot and do not warrant the performance or results that might be obtained by using the software or contents of the book. The authors, developers, and the publisher have used their best efforts to insure the accuracy and functionality of the textual material and programs contained in this package; we, however, make no warranty of any kind, express or implied, regarding the performance of these contents or programs. The Work is sold "as is" without warranty (except for defective materials used in manufacturing the disc or due to faulty workmanship).

The authors, developers, and the publisher of any third party software, and anyone involved in the composition, production, and manufacturing of this work will not be liable for damages of any kind arising out of the use of (or the inability to use) the algorithms, source code, computer programs, or textual material contained in this publication. This includes, but is not limited to, loss of revenue or profit, or other incidental, physical, or consequential damages arising out of the use of this Work.

The sole remedy in the event of a claim of any kind is expressly limited to replacement of the book and/or the CD-ROM, and only at the discretion of the Publisher.

The use of "implied warranty" and certain "exclusions" vary from state to state, and might not apply to the purchaser of this product.

MECHATRONICS

GANESH S. HEGDE

JONES AND BARTLETT PUBLISHERS

Sudbury, Massachusetts

BOSTON TORONTO LONDON SINGAPORE

World Headquarters
Jones and Bartlett Publishers
40 Tall Pine Drive
Sudbury, MA 01776
978-443-5000
info@jbpub.com
www.jbpub.com

Jones and Bartlett Publishers
Canada
6339 Ormindale Way
Mississauga, Ontario L5V 1J2
Canada

Jones and Bartlett Publishers
International
Barb House, Barb Mews
London W6 7PA
United Kingdom

Jones and Bartlett's books and products are available through most bookstores and online booksellers. To contact Jones and Bartlett Publishers directly, call 800-832-0034, fax 978-443-8000, or visit our website www.jbpub.com.

Substantial discounts on bulk quantities of Jones and Bartlett's publications are available to corporations, professional associations, and other qualified organizations. For details and specific discount information, contact the special sales department at Jones and Bartlett via the above contact information or send an email to specialsales@jbpub.com.

Copyright © 2010 by Jones and Bartlett Publishers, LLC
Original Copyright © 2008 by Laxmi Publications Pvt. Ltd.

All rights reserved. No part of the material protected by this copyright may be reproduced or utilized in any form, electronic or mechanical, including photocopying, recording, or by any information storage and retrieval system, without written permission from the copyright owner.

The publisher recognizes and respects all marks used by companies, manufacturers, and developers as a means to distinguish their products. All brand names and product names mentioned in this book are trademarked or service marks of their respective companies. Any omission or misuse (of any kind) of service marks or trademarks, etc., is not an attempt to infringe on the property of others.

ISBN: 978-1-934015-29-2

Cover Design: Tyler Creative

Library of Congress Cataloging-in-Publication Data
Hegde, G. (Ganesh)
Mechatronics / G. Hegde.
p. cm.
Includes index.
ISBN 978-1-934015-29-2
1. Mechatronics. I. Title.
TJ163.12.H44 2009
621--dc22

2009004512

6048

Printed in the United States of America
13 12 11 10 09   10 9 8 7 6 5 4 3 2 1

This book is dedicated to

my son,

Anil Hegde

TABLE OF CONTENTS

Preface xxi

Chapter 1. Introduction to Mechatronics **1**

1.1 Origin and Evolution 2
1.2 Definition of Mechatronics 3
1.2.1 Sequential Integration 4
1.2.2 Concurrent Integration 5
1.3 Multidisciplinary Scenarios 6
1.3.1 Mechanical Design and Modeling 6
1.3.2 Actuators and Sensors 6
1.3.3 Vibration and Noise Control 6
1.3.4 Manufacturing 7
1.3.5 Motion Control 7
1.3.6 Microdevices and Optoelectronic Systems 7
1.3.7 Intelligent Control 7
1.3.8 System Integration 7
1.3.9 Automation Systems 7
1.4 Need for Mechatronics in Industry 8
1.4.1 Changing Market Conditions 8
1.4.2 Variety in Product Ranges 8
1.4.3 Short Production Runs 9
1.4.4 Good Product Quality and Consistency 9
1.4.5 Ease of Reconfiguration of the Process 9
1.4.6 Enhancement in Process Capabilities 9
1.4.7 Demand for Increased Flexibility 9
1.4.8 Distributed Control and Local Decision Making 9
1.5 Objectives of Mechatronics 10
1.5.1 Design Objective 10
1.5.2 Data Extraction Objective 10
1.5.3 Output Generation 10
1.5.4 Processing Objective 11
1.5.5 Control Objective 11

1.5.6 Communication Objective 11
1.5.7 Automation Objective 11
1.5.8 Display Objective 11
1.5.9 Performance Objective 11
1.6 Design of Mechatronic Systems 12
1.7 Modules in Mechatronic Systems 13
1.7.1 Functional Description of Each Module 14
1.8 Mechatronics Technology 15
1.9 Mechatronics and Engineering Skills 16
1.10 Overview of Mechatronics 17
1.10.1 What Is Mechatronics? 18
1.10.2 Why Mechatronics? 18
1.10.3 How Do Mechatronic Systems Work? 19
1.11 Systems and Mechatronics 19
1.11.1 System 19
1.11.2 Measurement System 20
1.11.3 Actuation System 20
1.11.4 Control System 20
1.11.5 Microprocessor System 21
1.12 Measurement System 22
1.13 Microprocessor-Based Controllers 22
1.14 Engine-Management System 23
1.15 Automatic Camera 26
1.15.1 Auto Focusing 26
1.15.2 Aperture Control 28
1.16 Automatic Washing Machine 29
1.16.1 Pre-wash Cycle 30
1.16.2 Main Wash Cycle 30
1.16.3 Rinse Cycle 31
1.16.4 Spin Cycle 31
1.17 Automatic Bathroom Scale 31
1.17.1 Working Principle 32
1.18 Advantages/Disadvantages of Mechatronics 32
Exercises 33

Chapter 2. Sensors and Transducers 35

2.1 Definition of Sensors 36
2.2 Classification of Sensors 38
2.3 Definition of Performance Parameters 39
2.4 Pressure Sensors 41
2.5 Flow Sensors 43
2.6 Transducers 45
2.6.1 Definition 45
2.6.2 Parameters Sensed 46
2.7 Classification of Transducers 46
2.8 Hall Effect Sensors 46
2.8.1 Principle of Hall Effect 47
2.8.2 Hall Effect MOSFET (Metal Oxide Semiconductor Field Effect Transistor) 47
2.8.3 Types of Hall Effect Sensors 48
2.8.4 Advantages of Hall Effect Sensors 49
2.8.5 Disadvantages of Hall Effect Sensors 49
2.8.6 Applications of Hall Effect Sensors 49
2.8.7 Application of Hall Effect 50
2.9 Light Sensors 50
2.9.1 Principle 50
2.9.2 Types of Light Sensors 51
2.10 Proximity Sensors 53
2.10.1 Inductive Proximity Sensors 53
2.10.2 Ultrasonic Sensors 54
2.11 Desirable Features for Sensors and Transducers 55
2.12 Optical Sensors 55
Exercises 56

Chapter 3. Hydraulic Systems 59

3.1 Introduction to Hydraulic Systems 60
3.2 Definition of Actuators and Actuator Systems 61
3.3 Classification of Actuators 61
3.4 Classification of Hydraulic Cylinders 63

3.5 Hydraulic Cylinders 64
3.5.1 Constructional Features 64
3.6 Configurations in Hydraulic Cylinders Based on Rod and Piston Styles 64
3.7 Configurations Based on Cylinder Style 66
3.8 Configurations Based on Mounting Style 67
3.9 Applications of Hydraulic Cylinders 68
3.10 Hydraulic Motors 68
3.11 Classification of Hydraulic Motors 69
3.12 Swash Plate Motors 69
3.12.1 Construction and Workings 69
3.12.2 Features 69
3.12.3 Application and Selection Parameters 70
3.13 Bent Axis Motors 70
3.13.1 Construction and Workings 70
3.13.2 Features 71
3.13.3 Application and Selection Parameters 71
3.14 Radial Piston Motors 71
3.14.1 Construction and Workings 71
3.14.2 Features 72
3.14.3 Application and Selection Parameters 72
3.15 Vane Motors 72
3.15.1 Construction and Workings 72
3.15.2 Features 73
3.15.3 Application and Selection Parameters 73
3.16 Gear Motors 74
3.16.1 Construction and Workings 74
3.16.2 Features 74
3.16.3 Application and Selection Parameters 75
3.17 Annular Gear Motors 75
3.17.1 Construction and Workings 75
3.17.2 Features 76
3.17.3 Application and Selection Parameters 76
3.18 Valves 76
3.19 Classification of Valves 77

3.20 Pressure Control Valves 77
3.20.1 Construction 77
3.20.2 Workings 78
3.21 Reducing/Regulating Valves 78
3.21.1 Construction 78
3.21.2 Workings 79
3.22 Sequence Valves 79
3.23 Flow Control Valves 80
3.23.1 Construction 81
3.23.2 Workings 81
3.23.3 Applications 81
3.24 Direction Control Valves 81
3.24.1 Construction 81
3.24.2 Workings 81
3.24.3 Application Features 82
3.24.4 Requirements of a Direction Control Valve 83
3.25 Check Valves 83
3.25.1 Construction 83
3.26 Symbols for Hydraulic System Components 84
3.26.1 Pumps 84
3.26.2 Linear Actuators 84
3.26.3 Rotary Actuators 84
3.26.4 Pressure Control Valves 85
3.26.5 Flow Control Valves 85
3.26.6 Direction Control Valves 85
3.26.7 Miscellaneous Elements 86
3.26.8 Operations 87
3.27 General Hydraulic Circuit 87
3.28 Hydraulic Circuits 89
3.29 Relieving Circuits 91
3.29.1 Functioning of Circuit 91
3.30 Reducing/Regulating Circuits 91
3.30.1 Functioning of Circuit 91
3.31 Counter-Balance Circuits 92
3.31.1 Functioning of Circuit 92

3.32 Sequence Circuits 93
3.32.1 Functioning of Circuit 93
3.33 Meter-In Circuits 94
3.33.1 Functioning of Circuit 94
3.34 Meter-Out Circuits 94
3.34.1 Functioning of Circuit 94
3.35 Direction Control Circuits 95
3.35.1 Functioning of Circuit 95
3.36 Hydrostatic Transmissions 96
Exercises 96

Chapter 4. Electrical Actuation Systems 99

4.1 Mechanical Switches 100
4.2 Design Varieties of Mechanical Switches 101
4.3 Limit Switches 103
4.4 Contact Bounce 103
4.5 Tips for Minimizing Contact Bounce 105
4.6 Methods to Prevent Bouncing (Hardware Solution) 105
4.7 Solid-State Switches 106
4.7.1 Diodes 106
4.7.2 Thyristors 107
4.7.3 Triacs 109
4.7.4 Bipolar Transistors 109
4.7.5 Darlington Pairs 111
4.8 Solenoids 111
4.8.1 Construction 111
4.8.2 Principle of Workings 112
4.8.3 Characteristics of Solenoids 112
4.9 Pulse-Latching Solenoids 113
4.9.1 Construction 113
4.9.2 Applications 114
4.10 Relays 114
4.10.1 Applications 115
4.11 Electric Motors 115

4.12 Classification of Electric Motors 116
4.13 Four-Pole D.C. Motors 117
4.13.1 Construction 117
4.13.2 Principle of Workings 117
4.13.3 Analysis 117
4.14 Self-Commutated D.C. Motors 118
4.15 Permanent Magnet D.C. Motors 119
4.15.1 Construction 119
4.15.2 Workings 120
4.15.3 Analysis 120
4.15.4 Characteristics 121
4.15.5 Control 121
4.16 Feature Comparison of D.C. Motors 122
4.17 Brushless D.C. Motors (BLDC) 122
4.17.1 Construction 122
4.17.2 Workings 123
4.17.3 Features 123
4.18 Disc-Type BLDC Motors 123
4.18.1 Applications of BLDC Motors 124
4.19 Asynchronous Motors 124
4.20 Single-Phase Induction Motors 125
4.20.1 Construction 125
4.20.2 Workings 126
4.20.3 Features 126
4.20.4 Construction of Four-Pole Asynchronous Motors 126
4.21 Synchronous Induction Motors 127
4.21.1 Construction 127
4.21.2 Workings 127
4.21.3 Applications 128
4.22 Comparison of D.C. and A.C. Commutator Motors 129
4.23 Stepper Motors 129
4.23.1 Construction 130
4.23.2 Workings 130
4.24 Single-Phase Stepper Motors 131
4.24.1 Specifications of Stepper Motors 132

4.25 Servomotors: Definition 133
4.26 Servo-Drive Control 134
4.26.1 Workings of a Servomotor 134
4.27 Control of BLDC Servomotor 135
4.28 A.C. Servomotor Control 135
Exercises 138

Chapter 5. System Models **141**

5.1 Elements of Mechanical Systems 142
5.1.1 Springs 142
5.1.2 Transfer Function 143
5.1.3 Position Feedback 144
5.1.4 Damping Element 144
5.1.5 Transfer Function 145
5.1.6 Velocity Feedback 146
5.1.7 Mass 147
5.1.8 Transfer Function 147
5.1.9 Acceleration Feedback 148
5.2 Spring-Mass-Damper System 148
5.3 An Unconventional Approach 150
5.3.1 Reduction of Figure 5.15 151
5.4 Arrangement of Mechanical Elements 151
5.4.1 Series Arrangement 151
5.4.2 Parallel Arrangement 152
5.5 Application: Rack-and-Pinion Arrangement 152
5.6 Elements of an Electrical System 154
5.6.1 R-L-C Circuit 156
5.7 Unconventional Solution to the R-L-C Circuit 157
5.8 Application to D.C. Servomotor 158
5.9 Hydraulic System Modeling 160
5.9.1 Resistance to Motion 160
5.9.2 Resistance to Acceleration 161
5.9.3 Resistance to Deformation 162
5.10 Modeling of Actuators 163
5.10.1 Linear Actuator 163

5.10.2 Rotary Actuator 164
5.11 Modeling of Control Valves 165
5.11.1 Flow Control Valves 166
5.11.2 Relief Valves 167
5.11.3 Direction Control Valves 168
5.12 Thermal Systems 169
5.12.1 Thermal Resistance 169
5.12.2 Thermal Capacitance 170
5.13 Modeling of Thermal Systems 171
5.13.1 Temperature Controller 172
Exercises 173

Chapter 6. Elements of Machine Tools 175

6.1 Structures 176
6.2 Design Considerations of Structures 177
6.3 Loads on Structures 178
6.4 Guide Ways 178
6.5 Characteristics of Good Guide Ways 179
6.6 Classification of Guide Ways 179
6.7 Principle of Slide Ways 180
6.8 Stick-Slip Phenomena 181
6.9 Principle of Anti-Friction Ways 182
6.10 Design Shapes of Slide Ways with Applications 183
6.11 Shapes of Anti-Friction Ways 184
6.12 Recirculating Type of Anti-Friction Ways 185
6.13 Hydrostatic Slide Ways 185
6.14 Hydrodynamic Slide Ways 186
6.15 Slide Ways and Anti-Friction Ways 188
6.16 Recirculating Ball Screw-and-Nut Arrangement 189
6.17 Advantages and Disadvantages of Anti-Friction Power Screws 190
6.18 Pre-Loading Ball Nuts 190
6.19 Planetary Roller Screws 192
6.20 Spindles and Spindle Bearings 193
6.21 Types of Loads on Spindles 196

6.22 Selection of Spindles 197
6.23 Types of Bearings 198
6.24 Sliding Bearings 199
6.25 Hydrodynamic Journal Bearings 200
6.26 Hydrostatic Journal Bearings 201
6.27 Bearing Material Selection 202
6.28 Anti-Friction Bearings 203
6.29 Comparison of Sliding Bearings and Anti-Friction Bearings 204
6.29.1 Design Considerations of Bearing Supports 205
6.30 Pre-Loading of Anti-Friction Bearings 205
6.31 Pre-Loading Methods 206
6.32 Advantages and Disadvantages of Anti-Friction Bearings 209
6.32.1 Advantages 209
6.32.2 Disadvantages 209
6.33 Selection of Anti-Friction Bearings 209
6.34 Frictionless Bearings 210
Exercises 212

Chapter 7. Signal Conditioning 215

7.1 Introduction to Signal Processing 216
7.2 Concept of Signal Conditioning 216
7.3 Need for Signal Conditioning 217
7.4 Operational Amplifiers 218
7.5 Voltage-to-Current Converter: Inverting Type Operational Amplifier 220
7.6 Current-to-Voltage Converter: Inverting Type Operational Amplifier 220
7.7 Non-Inverting Operational Amplifier 221
7.8 Summing Amplifier 221
7.9 Integrating Amplifier 222
7.10 Differential Amplifier 222
7.11 Logarithmic Amplifier 223
7.12 Schmitt Trigger Amplifier 223
7.13 Amplifier Errors 224

7.13.1 Input Bias Current 224
7.13.2 Drift 224
7.13.3 Frequency Response 225
7.13.4 Slew Rate 225
7.13.5 Gain Variation 226
7.13.6 Comparator 226
7.14 Protection 227
7.14.1 Zener Diode Protection 227
7.15 Filtering 228
7.15.1 Types of Filters 228
7.15.2 High-pass Filters 230
7.16 Multiplexers 230
7.17 Wheatstone Bridge 232
7.17.1 Temperature Compensation 233
7.17.2 Load Cell 234
7.18 Signal Processing 234
7.18.1 Linearization 234
7.18.2 Compensation 235
7.18.3 Signal Averaging 235
7.18.4 Fourier Analysis 235
7.19 Digital-to-Analog Converter (DAC) 236
7.19.1 Graded Resistor DAC 236
7.19.2 Ladder Type DAC 236
7.19.3 Specifications of DACs 237
7.20 Analog-to-Digital Converter (ADC) 237
7.20.1 Successive Approximation ADC 238
7.20.2 Integrating ADC 238
7.20.3 Sampling Theorem 239
7.21 Data Acquisition (DAQ) 240
7.21.1 Specifications of DAQ Board 240
Exercises 241

Chapter 8. Microprocessors and Microcontrollers 243

8.1 Introduction to Microprocessors 244
8.1.1 Stored Program Control 244

8.1.2 Digital Processing 244
8.1.3 Speed of Operations 244
8.1.4 Design Flexibility 244
8.1.5 Integration 244
8.1.6 Cost 245
8.1.7 Definition 245
8.2 Microprocessor-Based Digital Control 245
8.2.1 Advantages of Digital Control 246
8.3 Digital Number Systems 247
8.3.1 Binary System 247
8.3.2 Binary Addition 247
8.3.3 Binary Subtraction 248
8.3.4 Hexadecimal System 248
8.4 Logic Functions 249
8.5 Microprocessor Architecture Terminology 250
8.5.1 Central Processing Unit (CPU) 250
8.5.2 Registers 257
8.5.3 Memory 257
8.5.4 ALU 259
8.5.5 Arithmetic 259
8.5.6 Logical 259
8.5.7 Address 260
8.5.8 Buses 260
8.5.9 Data 262
8.5.10 Interrupts 263
8.5.11 Assembler 263
8.5.12 Read Cycle 265
8.5.13 Write Cycle 266
8.5.14 Explanation of Terminology 266
8.5.15 Fetch Cycle 267
8.6 Microcontrollers 268
8.6.1 Instruction Sets 269
8.7 Requirements for Control 270
8.8 Differences Between Microprocessors and Microcontrollers 271
8.9 Classification of Microcontrollers 272

8.10 Implementation of Control Requirements 272
8.10.1 Intel 8048 Family 272
8.10.2 Intel 8051 Single-Chip Family 273
8.10.3 Universal Peripheral Interface (UPI): 8041 Family 274
8.10.4 Analog Signal Processor: 2920 Family 275
Exercises 275

Index **277**

PREFACE

This book on mechatronics is different from other books in style and presentation. It has been structured with the interests of undergraduate students in mechanical engineering at various universities in mind. The book is not only useful as a textbook for students but also as a reference for practicing engineers. The concepts, definitions, and illustrations on the subject are dealt with in simple and lucid language so that readers will have no difficulty understanding the material. The book, being interdisciplinary in nature, incorporates the material to the extent needed by students of mechanical engineering.

Chapter 1 gives an introduction to mechatronics. Chapter 2 deals with input devices such as sensors and transducers. Hydraulic actuation systems are covered in Chapter 3. Electrical actuation is covered in detail in Chapter 4. Mechanical, electrical, and thermal systems are modeled in Chapter 5. The elements and structural aspects of CNC machines are tackled in Chapter 6. The basics of electronics and electrical streams in the form of signal conditioning form the content in Chapter 7. The computer fundamentals of microprocessors and microcontrollers are the subjects covered in Chapter 8.

The subjective contribution of several texts and reference books on mechatronics have helped my presentation and style, and I owe thanks to all of them.

Behind the scenes is the support of my colleagues, students, and especially, Dr. S.C. Sharma, Principal of the R.V. College of Engineering. My hearty acknowledgments to all those who helped me directly and indirectly.

I thank particularly my son, Anil Hegde, and my wife, Geeta Ganesh, for extending constant cooperation.

If the book satisfies students and readers, it would give great pleasure to the author.

—G. Hegde

CHAPTER 1

INTRODUCTION TO MECHATRONICS

This chapter, Introduction to Mechatronics, is an introductory delibration covering the definition, needs, objectives, and advantages/disadvantages of mechatronics starting with brief historical developments. Readers will get a broad overview of related disciplines, modules, design concepts, various systems in mechatronics, and some applications based on microprocessor-based controllers, such as engine-management systems, automatic cameras, automatic washing machines, and bathroom scales are discussed.

This chapter covers the following issues to a considerable extent:

- Origin and evolution of mechatronic with historical developments.
- Definition of mechatronics with two approaches illustrated with block diagrams.
- Multidisciplinary scenarios discussing various streams to be integrated in mechatronics.
- Need for mechatronics in industries highlighting the needs of customers and industries of the present era and the role played by mechatronics in satisfying them.
- Objectives of mechatronics revealing the purpose of its developments.
- Design of mechatronic systems comparing conventional designs with mechatronic designs.

- Descriptions of various modules in mechatronics describing the connection between them with a block diagram.
- Mechatronic technology covering technological aspects.
- Skillsets needed for mechatronic engineers.
- Overview of mechatronics describing the what, why, and how of mechatronics.
- Systems in mechatronics.
- Measurement systems and the function of elements.
- Microprocessor-based controllers.
- Examples of applications of mechatronic systems.
- Advantages and disadvantages of mechatronics.

1.1 ORIGIN AND EVOLUTION

A Japanese engineer from the Yasukawa Electric Company coined the term mechatronics in 1969 to reflect the merging of mechanical and electronics disciplines. In late 1970 an elective course entitled "Electro-mechanical Interfaces" linking electrical and mechanical aspects in the area of drive technology was introduced in the UK.

The engineering degree at Lancaster University, introduced by Professor Michael French, has a totally different concept of engineering design with a theme to connect different streams in mechanical and electrical engineering. This led to the development of a generation of engineers who could work in an inter-disciplinary environment. At Cambridge University a clear development of such a course with the coupling of fields of common need established successful links with industry and the institute to work together in design and manufacturing automation.

Until the early 1980s mechatronics was understood to mean mechanisms that are electrified or driven by electric energy to produce motion. But in mid-1980, the concept of mechatronics came to mean concurrent engineering between mechanics and electronics with controls concepts. In the mid-1980s a subject for post-graduates in mechatronics was established in the UK which was eventually introduced to undergraduates in 1988. Mechatronics gained legitimacy in academic circles in 1996 with the publication of the journal IEEE/ASME Transactions on Mechatronics.

The mechatronics stream has grown leaps and bounds in industrial settings, projecting its importance in automating technology, products, and processes to improve accuracy, control, and repeatability.

Historical Developments

- 1969 – Coining of term by Yasukawa Electric Company
- Late 1970 – "Electro-mechanical Interface" course introduced
- Early 1980 – Mechatronics meant mechanisms with electrified drives
- Mid-1980 – Mechatronics meant to be concurrent between mechanics and electronics
- Mid-1980 – Course introduced at PG level in the UK
- 1988 – Course introduced at UG level

1.2 DEFINITION OF MECHATRONICS

A mechanical system designed to execute a desired function works under input parameters such as force, torque, pressures, heat, etc., to produce respective outputs such as translation, rotation, expansion, and deformation. When the extraction of information inputs to a system is carried out by sensors and output information is extracted by the actuators to be processed by a controller (preferably microprocessor based) so that the desired function is executed with specified accuracy by an action of the feedback process, this is a mechatronic system. Such an arrangement is generally illustrated by the block diagram given in Figure 1.1. In the present representation the mechanism is a mechanical system. The sensors are electrical/electronic. Actuators can be mechanical/electrical. Microprocessor-based controllers are electronic. Hence, the total system is the integration of mechanical, electrical, and electronic sub-systems arranged to produce an accurate process or function automating the output. With the addition of computers, the mechatronic system becomes more sophisticated and flexible.

Defined succinctly: "Mechatronics is the synergistic integration of mechanical engineering with electronics and electrical systems with intelligent computer control in the design and manufacture of industrial products, processes, and operations."

The integration of mechanical, electrical, electronic, and computer systems into a mechatronic system is one of two approaches based on the design philosophy and flow. The approaches are:

1. Sequential integration
2. Concurrent integration

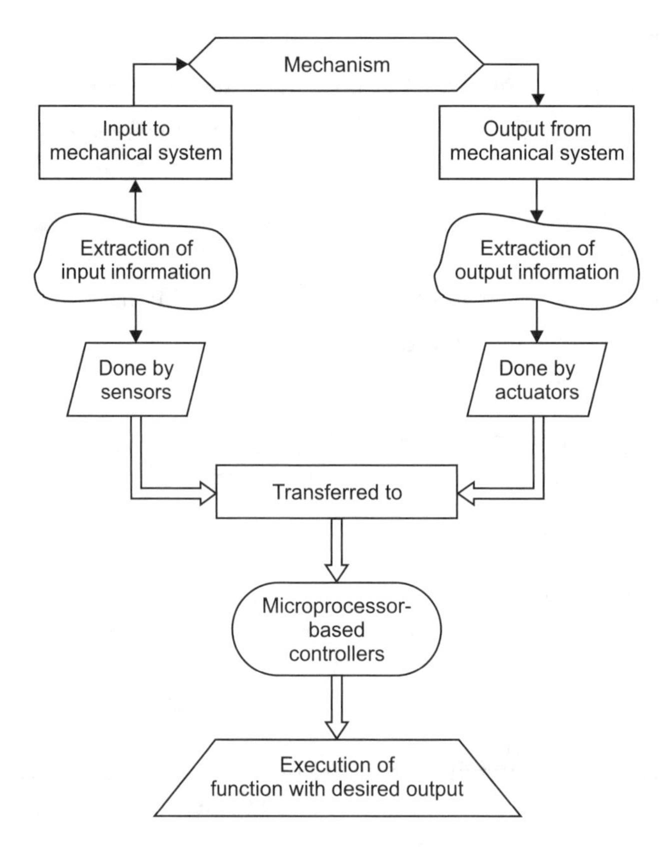

FIGURE 1.1 Definition of mechatronics.

1.2.1 Sequential Integration

Sequential integration is based on the design and accommodation of sub-systems following a serial sequence. In this approach the mechanical system is designed/ selected and developed first in all respects. The selection and design of the electrical component is carried out subsequently. Furthermore, microprocessor controls are added to complete the system. The block diagram is depicted in Figure 1.2. This has the drawback of lack of coordination between the groups of respective specializations, and lack of feedback between the teams may lead to reworks and rejections before arriving at the final outcome. The matching of specifications can increase the design cycle time. The conflicting situation would require managerial intervention and meetings for the exchange of information. This approach needs manpower and documentation and delibration before arriving at the final product. The repair and maintenance also gets decentralized and the owning of responsibility for malfunctions and failures cannot be attributed to a single team.

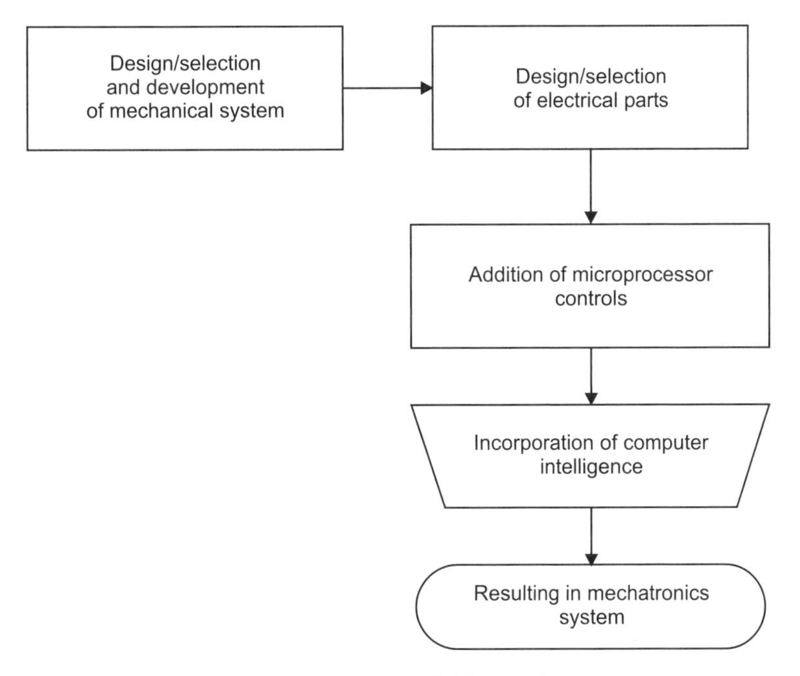

FIGURE 1.2 Sequential integration.

1.2.2 Concurrent Integration

Concurrent integration is characterized by parallel carryout of the design and development by a profile study of all the concerned streams. Mechatronic design is the sub-set of mechanical, electrical, and electronic design. Hence, mechatronics forms the common region of integration of the three sub-systems as shown in Figure 1.3. For concurrent integration designers need not know everything about mechanical, electrical, or electronics controls to design a mechatronic system. Most of the disadvantages of sequential integration are eliminated by the adaption of concurrent integration. However, this does not allow for mastery over any one stream but allows for a selective study of the needed features common to different specializations. In successfully integrating it combines the core disciplines of electronic/electrical engineering and computing and mechanical engineering with links into areas as diverse as manufacturing technology and management and working practices. The mechatronics approach to engineering design is featured by the resulting system that is simpler with fewer components and moving parts than the complete mechanical counterparts. The simplification is achieved through replacement of complex mechanical functions by electronic parts. In concurrent integration of the mechatronic approach the conceptual design process is established at the earliest stages. Here, the possibilities are kept open prior to the determination of the form of embodiment.

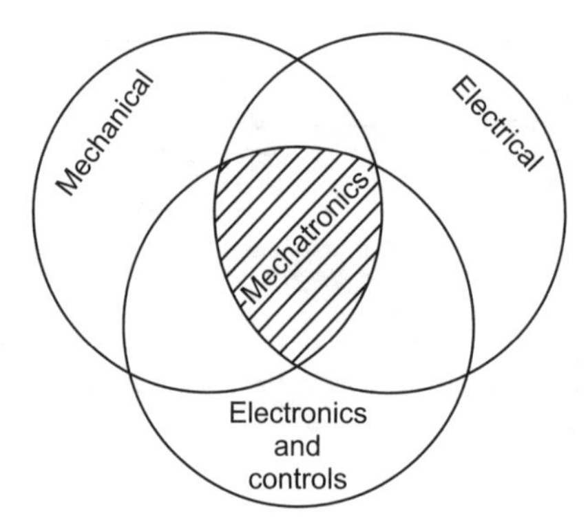

FIGURE 1.3 Concurrent integration.

1.3 MULTIDISCIPLINARY SCENARIOS

The following disciplines of engineering, at least in part, generally contribute to the development of mechatronic systems with the stages of formulation, selection, and standardization.

1.3.1 Mechanical Design and Modeling

The parts of the mechanisms are to be designed satisfactorily to accommodate the seen and unseen effects of forces and deformations. The processes are to be modeled for the control of parameters such as temperature and pressure to obtain the desired output from the system.

1.3.2 Actuators and Sensors

Actuators help in producing linear and rotary motions with desired force and torque. The sensors gather the input and output information to be processed by the controllers. Actuators can be electrical, hydraulic, or pneumatic which is determined by the need of the application and is generally selected based on the output requirements.

1.3.3 Vibration and Noise Control

The moving parts have mass and are in motion with the acceleration produced by the dynamic effects which are to be analyzed for natural frequency and amplitude of vibration. The vibrations are arrested by resilient dampers in the system. The moving parts produce noise due to friction which has to be reduced by proper lubrication and insulation.

1.3.4 Manufacturing

The process plan, material selection, facility plan, and communication between the system with proper quality control are the main considerations in manufacturing which also form mechatronics needs.

1.3.5 Motion Control

The conveyors and transfer lines aiding the handling of materials during processing are streamlined by the motion control derived from the mechatronics application. The motion control in robotics used to prevent collision with the hurdles in the work cell is also the application of mechatronics.

1.3.6 Microdevices and Optoelectronic Systems

Peizo-electric devices and microcontrollers are the miniature devices that impart and control micromotion, keeping the mechatronic systems compact in size and accurate in outcome. Optoelectronic devices such as optical encoders and sensors have a role in information extraction and measurement leading to the collection of data to be input to the processors.

1.3.7 Intelligent Control

Partial intelligent behaviors of a mechatronic system can be inducted through computer utilization operated by expert systems and artificial intelligent software. Computers aid in interfacing the other sub-systems of mechatronics by means of databases and knowledgebases driven through inference engines. Computer capacity is enhanced by the advent in electronic hardware capable of handling large data.

1.3.8 System Integration

The outputs from the systems of different streams of mechatronics have to be integrated on a common platform that will interpret the signal from the various systems. In electronics it is achieved by a VLSI (Very Large Scale Integration) design that includes self-calibration, monitoring, and testing of the systems under consideration.

1.3.9 Automation Systems

The input panels, display monitors, and output peripherals used in the system automate the proceedings by providing information. Computer networks such as LANs and WANs automate communication between the systems of mechatronics.

Automation not only eliminates overhead but also improves productivity through faster and accurate details needed by the engineer.

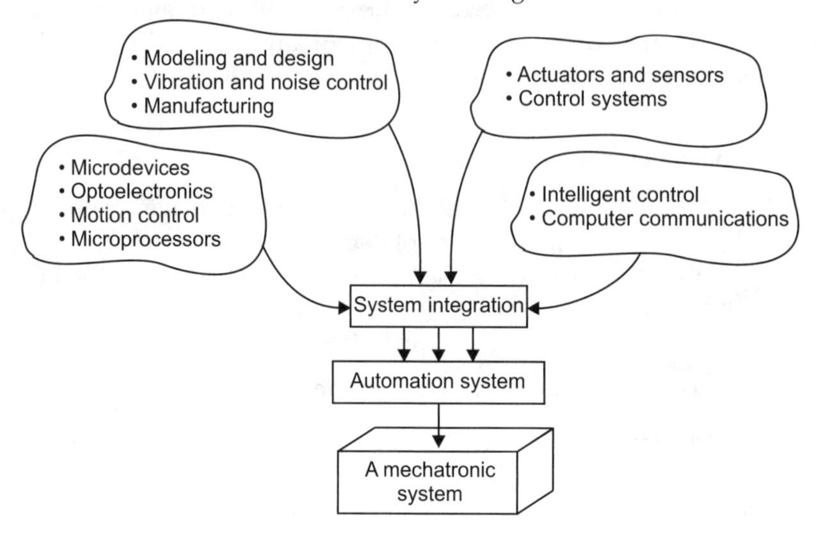

FIGURE 1.4 Interdisciplinary scenarios in mechatronics.

1.4 NEED FOR MECHATRONICS IN INDUSTRY

Industry is a plant organized to produce products and goods needed by people or industries. To produce parts machines are needed and to make machines parts are required. Accuracy with which the parts are produced depends on machine capability. With the improvement in part quality there is improvement in machine capabilities. With the help of mechatronics the products and process quality of competitive interests are produced to meet the demands. The need for mechatronics in industry can be summarized in the following paragraphs.

1.4.1 Changing Market Conditions

Market conditions are so volatile that often products become obsolete very fast because of the changing perceptions of consumers. Competition is so stringent that the sellers' market is turning into a buyers' market. To satisfy and attract customers the use of mechatronics in industry (manufacturing) and in products is an inevitability for enterpreneurs.

1.4.2 Variety in Product Ranges

Variations in size, shape, feature, facility, performance, and aesthetics are governed by customer likes, dislikes, and needs. Hence, manufacturers are compelled to produce a variety of products with a wide range. This is made easy by taking advantage of mechatronics.

1.4.3 Short Production Runs

Short product cycles, batch production, and job changeovers frequently influence the possibility of short production runs, market demand, and obsoletion of features. Batch production in an industry producing products of diversified specifications is not avoidable. Job changeovers on a machine with a variety of capabilities have the option of short production runs, and have the advantage of efficient machine utilization. The answer is mechatronics adaption.

1.4.4 Good Product Quality and Consistency

For a better reputation, surviving the competition, and better export turnovers maintaining product quality and producing the same quality repeatedly is the achiever's philosophy. Again, to reach this goal, mechatronics is the answer.

1.4.5 Ease of Reconfiguration of the Process

The processing flexibility inherent in a system is decided by the ease with which the facility can be reconfigured between the runs. Resorting to mechatronics can suitably favor the reconfiguration of the processes carried out on the machine.

1.4.6 Enhancement in Process Capabilities

Process capability is enhanced by a decreased deviation in the mean of the desired parameters. This is possible by good repeatability, accuracy, and resolution of the production system. The capability of reproducing the same quality (repeatability) with minimum error (accuracy) depends on the least possible capability (resolution) of a machine that is enhanced by mechatronic features.

1.4.7 Demand for Increased Flexibility

Flexible Manufacturing Systems (FMS), Computer Numerical Control (CNC) machine tools, robots, and Automatically Guided Vehicles (AGVs) are some of the advanced applications of mechatronics that render flexibility in manufacturing.

1.4.8 Distributed Control and Local Decision Making

Industries need centralized locally decision-making capacities and decentralized action control which is possible by automation and mechatronics.

Plant optimization, diagnostic-based maintenance, and large data handling, the developments in VLSI, have lead to self-calibration, self-testing, and self-monitoring capabilities in manufacturing systems.

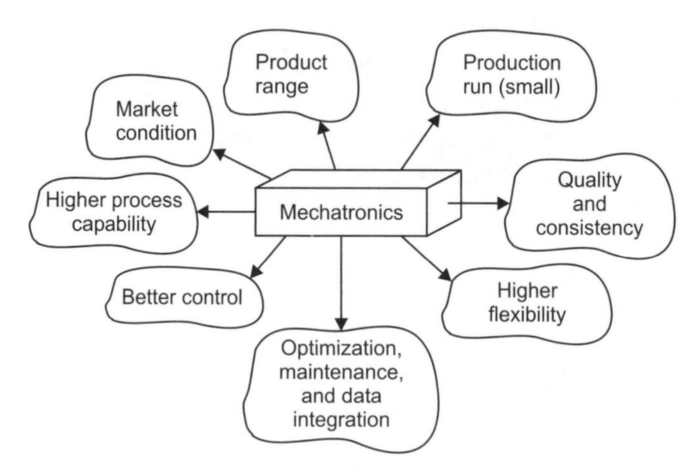

FIGURE 1.5 Need diagram of mechatronics.

1.5 OBJECTIVES OF MECHATRONICS

The replacement of the traditional approach with mechatronics is derived using the following objectives.

1.5.1 Design Objective

Design from the mechanical point-of-view should be exercised keeping in mind the incorporation and selection of mechatronic features and components. Design should have the objectives of adopting a minimum number of components, a minimum number of moving parts, and should be compact in size.

1.5.2 Data Extraction Objective

The input data for processing has to be extracted using sensors. Proper sensors should be selected for data such as force, torque, pressure, or position.

1.5.3 Output Generation

The development of movements such as translation and rotation has to be accomplished using electric actuators such as motors, solenoids, and piezo-electric elements for small outputs and by hydraulic or pneumatic actuators when the force and displacement requirement is large.

1.5.4 Processing Objective

The processing of data has to be done with the aid of microprocessors which can process multiple inputs and outputs received from sensors and transferred to actuators.

1.5.5 Control Objective

The error in output compared to the desired output is minimized using a feedback control system. The use of microcontrollers make the system compact and sophisticated.

1.5.6 Communication Objective

The need for communication is satisfied with the induction of a Manufacturing Automation Protocol (MAP), Technical Office Protocol (TOP), Open System Interconnection (OSI), and a Control Area Network (CAN). One output is used to control different functional needs at decentralized locations by a proper communication network.

1.5.7 Automation Objective

The purposes of system integration, data integration, functional flexibility, and reconfiguration of a program are accomplished by automation objectives.

1.5.8 Display Objective

Conventional analog displays with many moving parts are replaced by monitors and LED digital displays for convenience and improvement.

1.5.9 Performance Objective

The refinement and accuracy of outputs from mechatronic systems will induct better quality in products manufactured, functions executed, and processes carried out with an improvement in performance.

The objectives of mechatronics are not only to develop a sophisticated product, but also to automate the processes and operations of the manufacturer. Certain objectives concentrate on conversion into mechatronic products with advanced features, sensors, actuators, controllers, and processors. But concepts such as total manufacturing solutions call for MAP with integrated communication providing distributed output control with centralized input information. Robotics, CNC, FMS, and CAN significantly contribute to factory automation, highlighting the importance of the objectives of mechatronics in a wide variety of applications.

1.6 DESIGN OF MECHATRONIC SYSTEMS

The concurrent interaction between many skills and streams of engineering establishes the complex process of mechatronic design to replace successfully traditionally designed mechanical systems. Other supplementing disciplines in mechatronics are electronics, control theory, and electrical and computer technology.

Mechatronic design includes the following philosophy and examples:

- Replacement of mechanical parts with electronic components such as piezo-electric actuators.
- Replacing analog indicators with a digital Light Emitting Diode (LED) displays.
- Replacement of analog measurement system for inputs with sensors.
- Instead of using bimetallic temperature switches, microprocessor-based thermodiode sensors can perform the on/off function for a heating system leading to accurate temperature control under varying conditions.
- Microprocessors and microcontrollers are embedded in the system and control programs are installed to produce different motions for a particular operation; the manufacturer takes the responsibility of programming.
- Using an on/off timer with a time delay can replace the cam and follower-operated mechanical switch with movements, friction, and bouncing. The PLC solution makes it possible in mechatronic designs.
- A stepper motor with a microcontroller for motion control can be a substitute for a mechanically linked mechanism of rotation and oscillation as is the case of a windscreen wiper.
- Deflection of a spring (leaf) connected to a rack-and-pinion arrangement to convert deflection to rotation and further transferring the movements to a pointer of an indicator were followed in the conventional mechanical designs of a weighing machine. The use of load cells with strain gauges, an analog-to-digital converter, a microprocessor to process the signals, and an LED display of the weights in digital form is the mechatronic approach to the design of a bathroom scale.
- The mechatronic approach has refined accurate result outputs compared to crude and inaccurate readings of traditional analog systems with mechanical designs.
- Remarkable improvement in flexibility, precision, and performance by programmed control of the parameters giving faster outputs are the

common characteristics of mechatronically designed systems. Handling the complexity of disturbances and deviations is effective with the mechatronic design approach.

1.7 MODULES IN MECHATRONIC SYSTEMS

A mechatronic system is identified by basic modules categorized based on functions carried out to form a total system. A block diagram schematically showing the flow of information between various modules is shown in Figure 1.6.

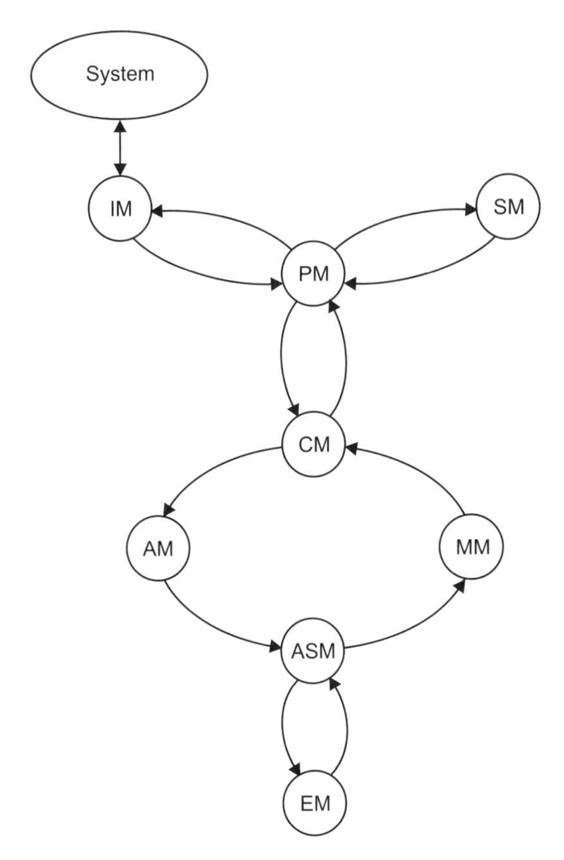

IM = Interface Module
PM = Processor Module
CM = Communication Module
AM = Actuation Module
ASM = Assembly Module
EM = Environment Module
MM = Measurement Module
SM = Software Module

FIGURE 1.6 Block diagram of modules.

1.7.1 Functional Description of Each Module

ENVIRONMENT MODULE

This module is concerned with parameters such as force, temperature, speed, and their effect on the boundaries of the system. This module also deals with the dynamics and existence of the system and the functions. Codes and standardization fall in this module. This module interacts with the assembly module for information exchange.

ASSEMBLY MODULE

Manufacturing, mechanical, and structural realization, and part and system integration are the activities in this module. Input information is received from the actuation module and output is given to the measurement module.

MEASUREMENT MODULE

Sensors and microdevices and transducers are the components of this module which supply information output to the communication module. This module gathers information about system status.

ACTUATION MODULE

Hydraulic, pneumatic, and electric actuators, peizo-electric devices, and microcontrollers are the systems identified in this module. This module receives information from the communication module for execution.

COMMUNICATION MODULE

This module is concerned with the transmission of information between modules within the system. The input and output information reveals the nature of the signal and the distance over which it has to be transmitted, and the operating environment. This module mainly interacts with the processor module.

PROCESSOR MODULE

This module is formed by microprocessors and embedded electronic circuits. This module extracts information from the communication module about measurement parameters, demand settings, and system parameters to be processed. This module interacts with the interface module and the software module for information processing.

SOFTWARE MODULE

This module contains instructions for operating, defined algorithms, and operation control programs of the processor module. The nature and forms of instruction are linked to associate and interact with the processor module.

INTERFACE MODULE

Various levels in the system are interfaced for transfer of information with interaction with the processor module and the system representing the world. This provides man-machine interface for user information. The information is classified by i/p × o/p.

1.8 MECHATRONICS TECHNOLOGY

In mechatronics the key areas of technology integrated are noted and described as follows:

- Engineering Design
- Drives and Actuators
- Sensors and Instrumentation
- Embedded Microprocessor Systems
- Automation and Computerization

Mechatronic technology integrates some or all of the above mentioned technological components in its development, identified by the mode of deployment.

However, few are the mechanical parts used in mechatronic systems. They are designed and optimized for size and shape by static and dynamic analysis, considering environmental conditions. The design process in a mechatronic approach is involved in developing a system in terms of its form and nature of the question and problem. The solid modeling and FEM techniques on computers make engineering design simpler, faster, and aesthetic.

Mechanical drives such as hydraulic and pneumatic actuators, electric drives such as AC/DC motors and stepper motors, and electronic smart drives such as piezo-electric devices incorporated into the mechatronic system according to the application content provide local dedicated execution power enabling independent action of motion control.

Input and output management in terms of processing and accuracy controls and readouts has been provided by utilization of respective sensors and instrumentation that enhances the flexibility of mechatronic systems. Selection and application of proper sensors and instruments in the suitable range of deployment has been the technological interest in mechatronics.

Most mechatronic systems, being different and unique in function, work on custom built and designed electronic circuits which are converted into embedded systems. The embedded microprocessors used in the design and development of a large-scale, distributed mechatronic system act as the main part of the controller. Microprocessor-based embedded mechatronic systems add compactness to the overall system.

The automated design and information transfer are made modular in structure and approach by computerization which may not be a common feature in all mechatronic systems. The top-down modular approach with the techniques of object-oriented programming introduced for software development reduces the processing time and space. Computer communication and software engineering are significant in the industrial automation of the operation and the processes. The decentralized output and centralized input control is the characteristic and advantage of computerization and distributed communication of the mechatronic system in industry.

In mechatronic technology the processes require the mechatronic products and the mechatronic products require mechatronic processes to satisfy the accuracy and quality needs which keep on narrowing with the advent in technology. Hence, mechatronic technology provides propulsive fuel to the improvement of product specifications.

1.9 MECHATRONICS AND ENGINEERING SKILLS

The skillsets for designing mechatronic systems are diversified and vast. The basic skills for a mechatronic professional in an industry or an academic system are broadly identified as:

- Modeling (Mathematical)
- Mechanical Design
- Design of Mechanical Circuits
- Design of Electrical Circuits
- Design of Electronic Circuits
- Development of Control Algorithms
- Design of Control Systems
- Selection of Standard Items
- Instrumentation
- Computer Programming Skills
- Familiarity with Microprocessors
- Signal Processing

- Testing and Installation
- Integration and Communication

The preceeding set of skills have a finer classification as stated in Table 1.1.

TABLE 1.1. Mechatronics and Engineering Skills

Mathematical modeling	Dynamic system representation, numerical integration, linear algebra
Mechanical design	Static analysis, fatigue analysis, creative design, system design, drafting
Mechanical circuits	Design of electrohydraulic circuits, electro-pneumatic circuits
Design of electrical circuits	Electrical network analysis and design harnessing
Electronic circuits	Design of analog circuit design of digital circuits, embedded circuits
Development of control algorithms	Modeling of physical systems, transfer functions, and control laws
Selection of standard items	Selection of actuators, sensors, display units, input panels, indicators, application development
Instrumentation	Test equipment, signal generators, converters, meters
Computer programming skills	Programming languages, software development and testing, customization
Microprocessors	Application and programming microprocessors, interfacing with input and output devices
Signal processing	Analog and digital signal processing, image storage and processing
Testing and installation	Calibration, testing, and installation of all components
Integration and communication	LAN, WAN, VLSI, MAP

1.10 OVERVIEW OF MECHATRONICS

It is the overall study, understanding, and presentation of mechatronics in brief that would give a bird's eye view to readers. As mechatronics has interdisciplinary content, it attracts wide sections of readership and criticism. The broad picture of mechatronics is given through the answers to the following questions.

1.10.1 What Is Mechatronics?

The development of a mechatronic product or a total solution provider is an integrated approach which deals with issues from different disciplines. Mechanical design is intended to possess the minimum possible components achieved through replacement by electronic or electrical alternatives. Input management is accomplished by sensors selected according to the parameters to be measured. Input data is measured to have correction control over the actual output of the system. The desired output is arrived at by feedback and comparison. The execution of action is the assignment of the actuators that produce linear or rotary motion. Hydraulic actuators are selected for higher output. The lower outputs are managed by electric (motors and solenoids), pneumatic, or electronic actuators. The readouts are made attractive through instrumentation with LED and monitor displays. Multiple inputs are processed by a compact microprocessor to give multiple outputs through installed programs. The embedded circuits customized for the given applications automate the solution process. Computerization adds to the automation through data integration, storage, and processing. Data communication for decentralized output is done by computer networks.

1.10.2 Why Mechatronics?

The enhanced quality requirement in products and parts produced, accuracy of results in the systems, repeatability of the processes, the speed in data processing and transfer, acceptable deviations in the desired outputs, flexibility and reconfiguration in operations, and improvement in process capability are some of the technical reasons mechatronics attracts worldwide attention.

The market conditions of changing customer interests, increasing competition, increasing product ranges to satisfy diversified customers, and small product cycles and production runs resulting from obsoletion of features are some of the management reasons that call for the adoption of mechatronics in industry.

Entrepreneurs often think about mechatronics with the intention of import substitution, reputation enhancement of the brand, reverse engineering for faster lead time, and providing a total manufacturing solution (changing from standalone machines).

1.10.3 How Do Mechatronic Systems Work?

The use of FEM packages, CFD tools, solid modeling software in static and dynamic analysis for mechanical design, and optimization of mechanical elements in mechatronic systems makes design efforts easy. The use of computers in design for aesthetic changes of cosmetic features also makes improvement in performance features faster.

To measure parameters such as force, torque, touch, and distance, sensors are used that have more of an electrical and electronic component. The use of timers and microcontrollers not only make the system compact but also eliminate mechanical limitations due to inertia, friction, and bouncing by impact.

Processing of information is done by the microprocessor, the embedded microcontroller, and the PLC. The execution of functions is done by drivers such as motors, solenoids, stepper motors, and display units.

The flexibility to manufacturing and total solution providers is developed by incorporating FMS, CNCS, Robots, LAN, MAP, and TOP mechatronic systems.

1.11 SYSTEMS AND MECHATRONICS

It is clear from the previous sections that mechatronic systems are made up of several systems such as the measurement system, drive and actuation system, control system, microprocessor system, and computer system. A brief definition of each system and its characteristics is given in the following paragraphs.

1.11.1 System

Any mechanical, electrical, or electronic element or set of elements that can give out certain useful outputs under the understandable inputs can be called a system. The system can be purely mechanical, electrical, or electronic requiring compatible inputs. But the mechatronic system is a combination of these systems. The schematic of a system is given in Figure 1.7.

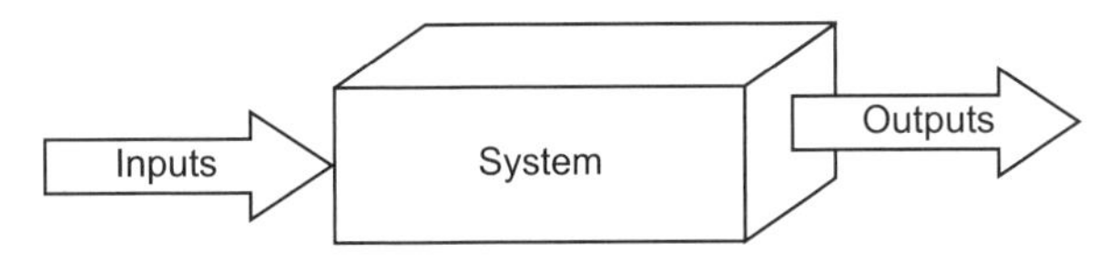

FIGURE 1.7 System schematic.

1.11.2 Measurement System

Any system that measures parameters such as temperature, pressure, force, voltage, current, etc., can be considered a measurement system. A block diagram is shown in Figure 1.8.

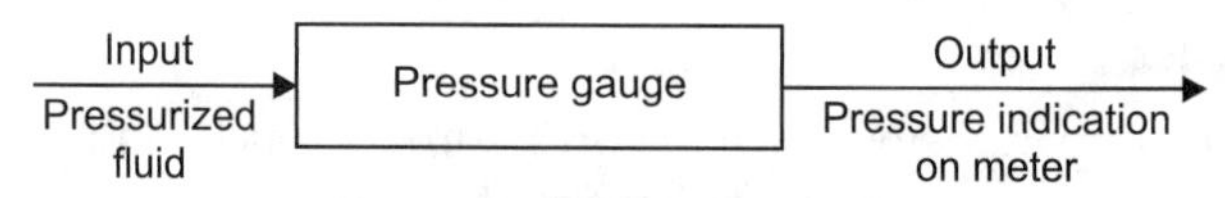

FIGURE 1.8 Measurement system.

A pressure gauge receives pressurized fluid through a pipe and the deformation of a flexible copper tube is converted to dial indication (analog) by the rotation of a pinion. The measurement system and its elements with functions is detailed in Section 1.12.

1.11.3 Actuation System

A drive system that produces linear or rotary motion is an actuator, which may be mechanical such as hydraulic, pneumatic, or electrical such as motors, solenoids, etc. The electric and hydraulic actuators are discussed in detail in later chapters. A simple illustrative block diagram is shown in Figure 1.9.

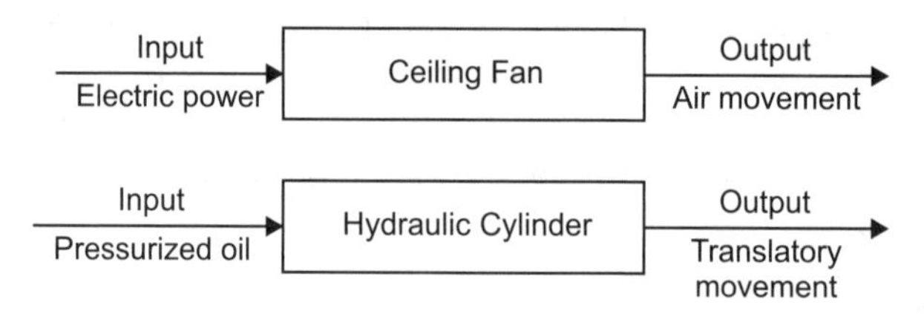

FIGURE 1.9 Actuation system.

The electric power input to an electric motor of a ceiling fan with blades produces rotation that results in the output of air circulation. A ceiling fan is an example of a rotary drive. The pressurized hydraulic oil flowing to the ends of the piston of a hydraulic cylinder makes the piston rod produce reciprocatory motion.

1.11.4 Control System

The system in the control system is characterized by a transfer function. The output produced has no bearing on the input given; such a control system is called an open loop control system. If the error signal between the desired output and the actual output is fedback by a feedback element, such a system is a closed loop control system. An open loop system is shown in Figure 1.10(*a*), and

Figure 1.10(*b*) shows a closed-loop control system. The accuracy, sensitivity, bandwidth, and non-linearity are improved by adopting a closed-loop control system over an open-loop control system. Open-loop systems have the advantage of stability, low cost, and ease of maintenance with simplicity in design and development. For an elaborate study on control systems the reader is encouraged to review books on automatic control engineering as a detailed study is beyond the scope of this book.

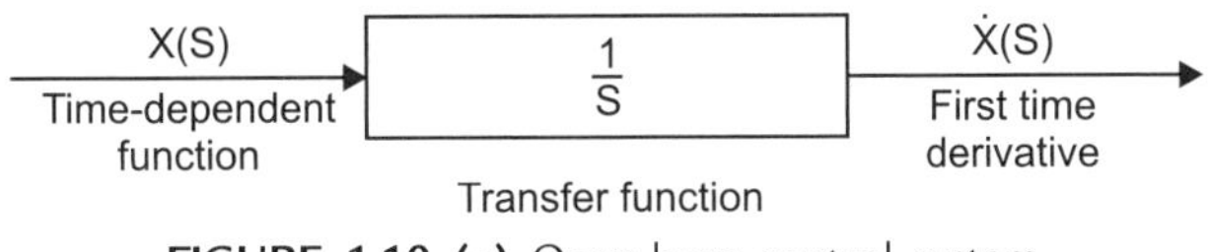

FIGURE 1.10 (*a*) Open-loop control system.

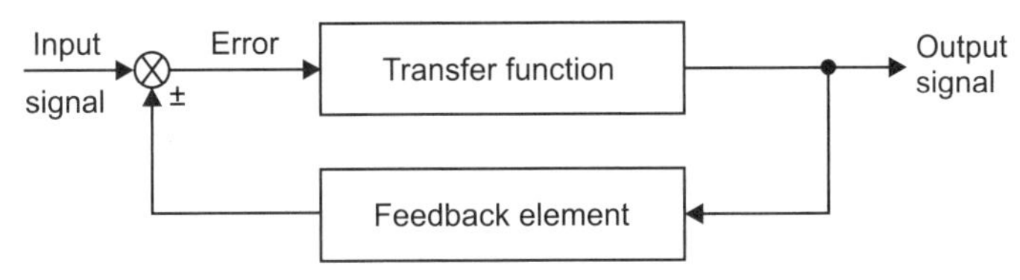

FIGURE 1.10 (*b*) Feedback control system.

1.11.5 Microprocessor System

This system is an electronic system with an integrated circuit compactly installed to form a chip. The circuits are activated by control programs. The conditioned signals are the inputs that can be multiple in number. The outputs are also multiple which through a decoder are given to the drivers and the display. The block diagram in Figure 1.11 shows a microprocessor system.

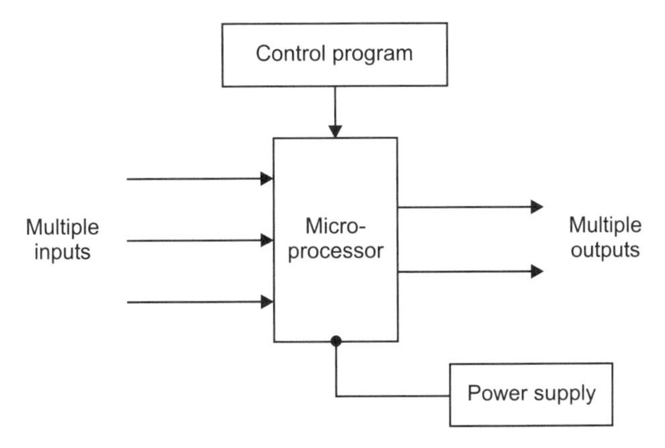

FIGURE 1.11 Microprocessor system.

1.12 MEASUREMENT SYSTEM

The elements of a measurement system, as shown in Figure 1.12, are the sensor, signal conditioner, and display unit.

FIGURE 1.12 Measurement system elements.

- **Sensor:** The input quantities such as force, temperature, etc., supplied to the sensor are processed to give a response in the form of a raw output signal relating to the quantity to be measured.
- **Signal Conditioner:** The raw (weak) signal output from the sensor is manipulated or amplified by a signal conditioner such as an amplifier to give a conditioned output.
- **Display:** The analog or digital indicator is the display unit that gives out the reading of the measured quantity taking the signal from the signal conditioner.

An example of a strain indicator is shown in Figure 1.13.

FIGURE 1.13 Strain indicator.

The resistance strain gauges mounted on a loaded beam in the form of a Wheatstone bridge give a change in voltage by a change in resistance owing to a change in length by deformation. The differential amplifier amplifies the voltage signal that is given to the analog or digital strain indicator to give out the strain reading as the output.

1.13 MICROPROCESSOR-BASED CONTROLLERS

Mechanical mechanisms such as speed governers, cam-actuated valves and switches, and rack-and-pinion driven analog indicators are being replaced by microprocessor-based controllers. The main features of microprocessor controllers are:

- A variety of programs can process the multiple inputs to give multiple outputs.

- The programs can be altered/reprogrammed to change the output specifications.
- There is a programmable memory to store instructions and carry out control functions.
- The processors are integrated chips and are compact in size and embedded in any circuit.

The block diagram in Figure 1.14 shows a microprocessor-based controller.

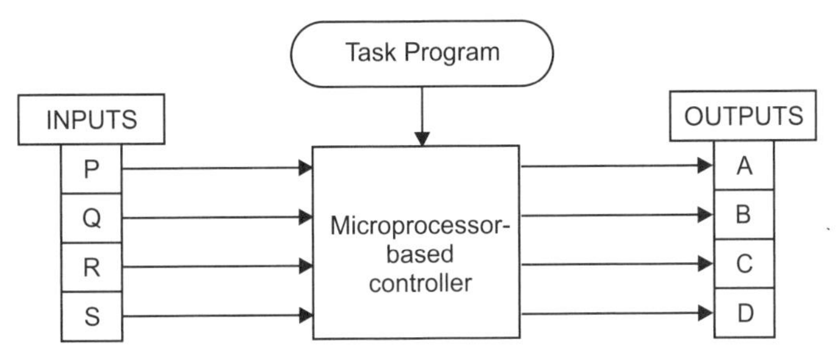

FIGURE 1.14 Microprocessor-based controller.

- Inputs can be from:
  (1) Sensors
  (2) Switches
  (3) Amplifiers
  (4) Encoders
  (5) Batteries
  (6) Computers
- Outputs can be to:
  (1) Actuators or drivers
  (2) Displays
  (3) Decoders
  (4) Solenoids
  (5) Stepper motors
- The task programs include:
  (1) Logic Program
  (2) Sequence Program
  (3) Timing Count Program
  (4) Event Control Arithmetic Functions

1.14 ENGINE-MANAGEMENT SYSTEM

Figure 1.15 illustrates the application of mechatronics to manage an internal combustion engine to improve performance and fuel efficiency. The use of sensors, drives (actuators), and the microprocessor to control and process data is the mechatronic approach to engine management.

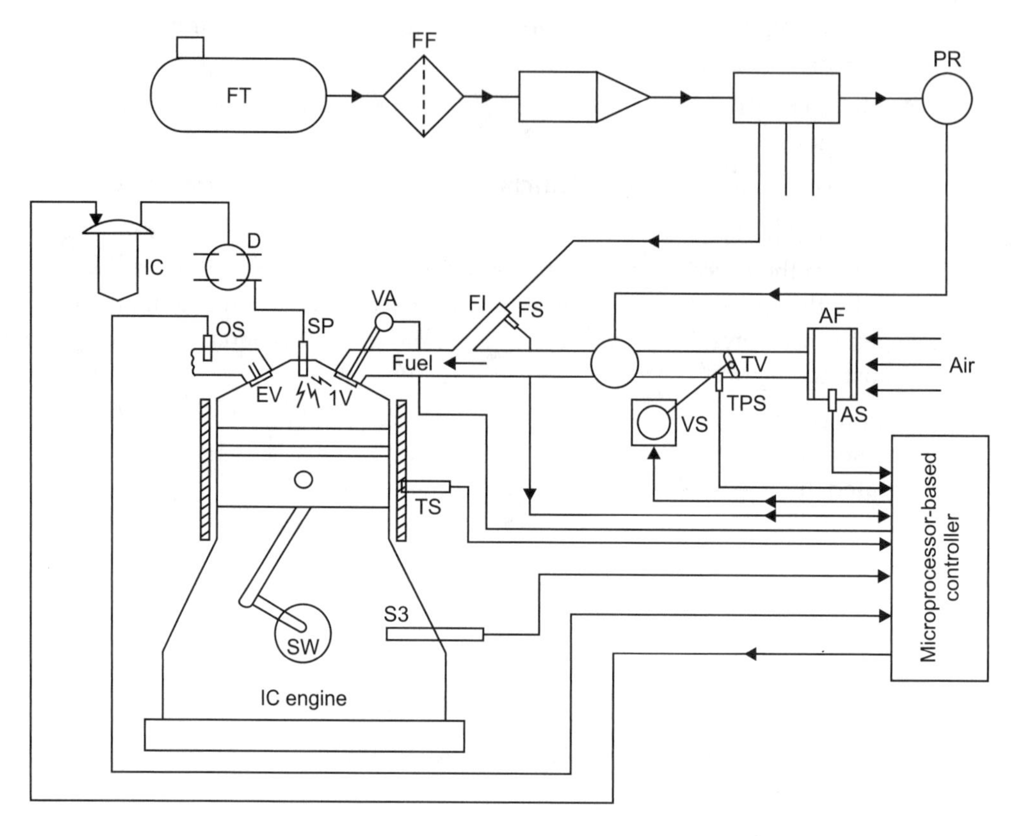

FT = Fuel Tank
FF = Fuel Filter
PR = Pressure Regulator
FI = Fuel Injector
TV = Throttle Valve
VS = Valve Switch
D = Distributor
IC = Ignition Coil
SW = Sensor Wheel
SP = Spark Plug

EV = Exhaust Valve
IV = Inlet Valve
AF = Air Filter
SS = Speed Sensor
TS = Temperature Sensor
AS = Airflow Sensor
TPS = Throttle Position Sensor
FS = Fuel Sensor
OS = Oxygen Sensor
VA = Valve Actuator

FIGURE 1.15 Engine-management system.

The technological advent in mechatronics is to improve fuel efficiency of IC engines effected by speed regulation, proper charge ratio, correct valve operation, timely spark strike, and prevention of fuel loss in the exhaust. The input data from the speed sensor, temperature sensor, throttle valve position, and the airflow rate sensor is processed by the microprocessor to actuate the fuel injector and throttle valve switch that regulates the air fuel ratio to control the speed. The opening and closing sequence of the valve is controlled by the sequence program installed in the microprocessor. The timing program gives the output to strike the spark by the spark plug at the right time by the actuation of the ignition coil. The oxygen sensor present in the exhaust manifold gives the signal to recirculate the exhaust gas if excess oxygen ions are present so that fuel loss is prevented.

In a conventional engine, speed is regulated by the throttle valve connected to a mechanical governor mechanism driven by the crankshaft. The valves are operated by the rocket arm mechanism driven by a cam connected to the crankshaft. The spark timing is related to the end of the compression stroke. The exhaust is used to preheat the air sucked in as is the case of turbo engines which are rare. A block diagram of the microprocessor control of an engine is given in Figure 1.16.

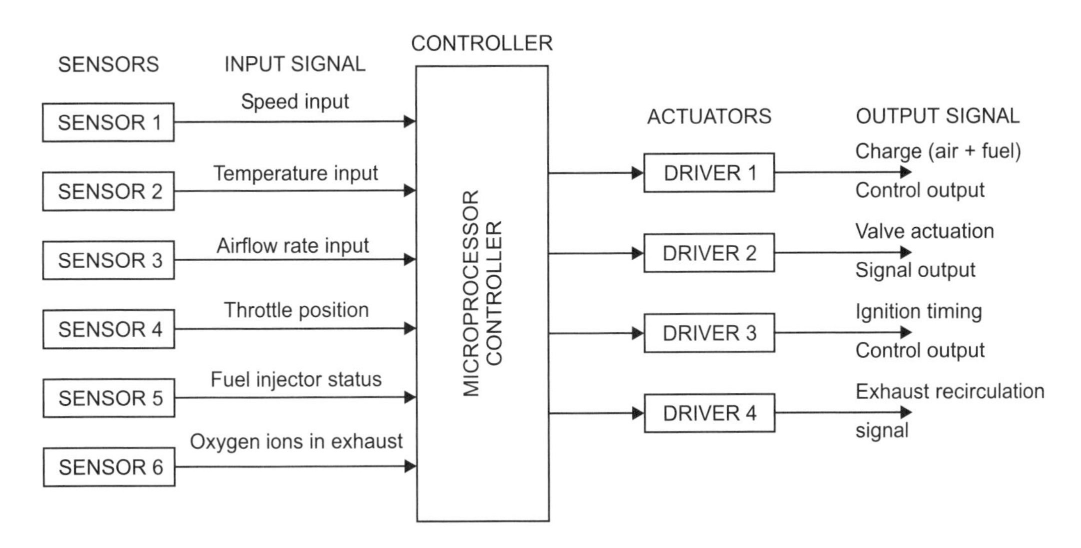

FIGURE 1.16 Block diagram of an engine-management controller.

TABLE 1.2 Comparison of Mechatronic and Conventional Approaches in Engine Management

Mechatronic Approach	Conventional Approach
Speed Regulation: Based on the input signals from the speed sensor, temperature sensor, and hotwire sensor of airflow rate, the speed is either increased or decreased by the opening and closing of the throttle valve controlled by the throttle valve switch receiving signal from the microprocessor-based controller.	**Speed Control:** The governor actuated by the speed of the crankshaft regulates the opening and closing of the throttle valve through a lever attached to it. The speed control has no bearing on the temperature of the engine and the airflow rate.
Valve Actuation: Opening and closing of the inlet and exhaust valve at the appropriate time is effected by the signal from the controller to the valve actuator. The timing sequence is installed by a control program in the microprocessor-based controller.	**Valve Actuation:** The cam-operated rocket arm mechanism controls the opening and closing of the valves. The cam rotates in respect to the rotation of the crankshaft.
Spark Timing: The spark plugs are made to produce sparks at the appropriate time by the supply of current from an ignition coil that receives output signal from the microprocessor that controls through a timing sequence program.	**Spark Timing:** The ignition coils are activated to produce sparks in the spark plug at the end of the compression stroke. Sparks are produced at constant intervals set in the system.
Exhaust: The oxygen sensor identifies the presence of oxygen ions in the exhaust and the microprocessor decides on recirculation for better efficiency.	**Exhaust:** The exhausts are utilized to preheat the air by turbo action.

1.15 AUTOMATIC CAMERA

A modern automatic camera with auto focusing and exposure is shown in Figure 1.17 as a line diagram with elements such as lenses, mirror, viewfinder, battery, flash, aperture, shutter, film, etc.

1.15.1 Auto Focusing

The system actuation switch sets the camera to be ready to take the snap of the object at which the camera is pointed. Input from the range sensor is processed by the microprocessor-based controller to give the output signal to the lens position drive to achieve focusing. Based on the feedback signal from the lens position encoder it is (lens position) modified to the desired position of correct focus.

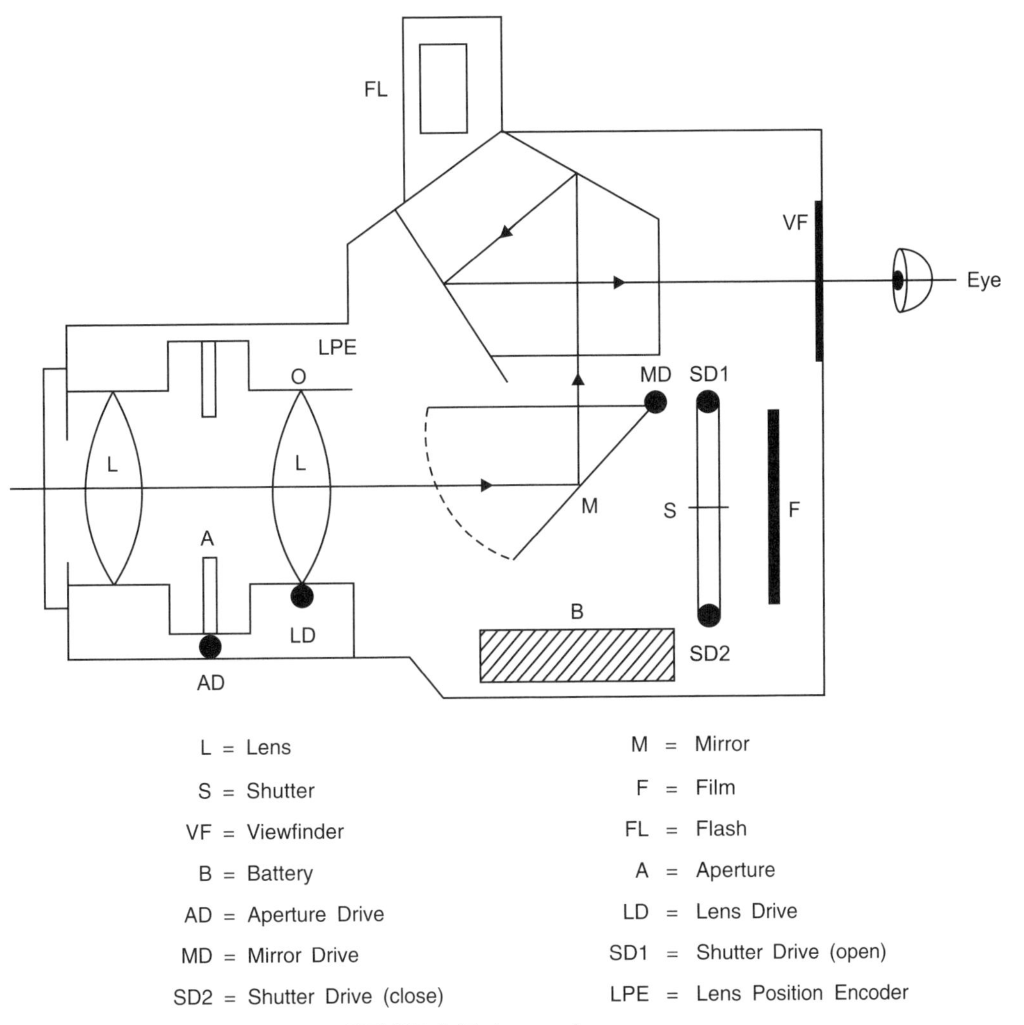

FIGURE 1.17 Automatic camera.

The sequence of program steps for auto focusing are:

1. Send the start command to the lens.
2. Receive input from the range sensor.
3. Determine the initial lens position from the input from the lens position encoder.
4. Calculate the lens movement required.

5. Send lens movement data to the lens position drive.
6. Verify the lens movement based on the feedback signal.

1.15.2 Aperture Control

When the shutter switch is pressed to the first position, the main microcontroller calculates the shutter speed and aperture settings based on the input from the light sensors (also called metering sensors) and gives out the signal to the viewfinder and LCD display. Pressing the shutter switch to the fully depressed second position makes the main microcontroller issue a signal to operate the mirror drive to lift the mirror up and to change the opening of the aperture to the required extent. The shutter is kept open for the required amount of time. When the shutter is closed the film is advanced for the next snap.

The battery test and flash control activities are also carried out by the microprocessor controller. The block diagram in Figure 1.18 shows the microprocessor-based controller operations of an automatic camera.

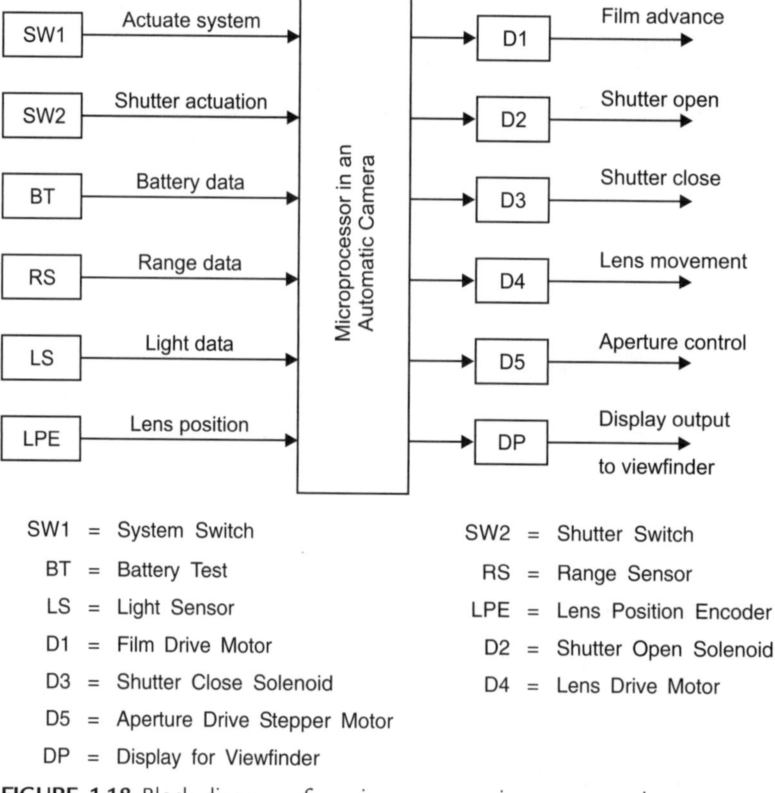

FIGURE 1.18 Block diagram of a microprocessor in an automatic camera.

The sequence of program steps for aperture control are:

1. Read shutter button position first.
2. Take input from metering sensor.
3. Translate into appropriate shutter speed and aperture value setting.

   (If the shutter is preselected the aperture setting is supplied or if the aperture setting is preselected the shutter speed is supplied.)
4. Read second position of the shutter button.
5. Issue signal to the mirror drive.
6. Send signal to shutter drive to open for a calculated time.
7. Supply output signal to close the shutter.
8. Send film advance signal to the film drive motor.

1.16 AUTOMATIC WASHING MACHINE

The washing operation in a washing machine used to wash clothes has a set of sequences which include: 1. Pre-wash cycle; 2. Main wash cycle; 3. Rinse cycle; 4. Spin cycle. To effectively and automatically carry out these cycles the machine is equiped with a microprocessor-based controller, sensors, drivers, and heater.

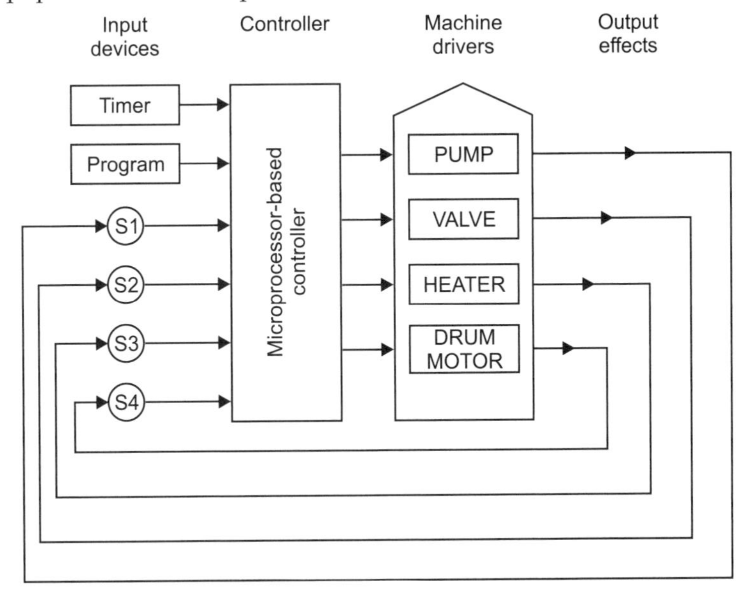

FIGURE 1.19 Automatic washing machine.

The installed timer determines the time for which the cycles are activated. The temperature sensor, water-level sensor, speed sensor, and position (valve) sensor provide inputs to the controller. The pump, heater, valve actuator, and drum motor are the output devices in the machine. The sequence of operations of each cycle is programmed and supplied to the microprocessor which controls the operations of the washing machine.

1.16.1 Pre-wash Cycle

An electrically operated valve opens to allow cold water into the drum for a period of time controlled by the timing program installed in the microprocessor. Clothes in the drum are given a starting wash with the cold water. The water level is sensed and by the signal from the controller the valve is closed due to switching off the current supply.

The program steps are:

1. Opening of the valve.
2. Closing of the valve once the drum is filled.
3. Running the motor (drum) for a specified time period and at a slow speed.
4. Operating the pump to empty the drum.

1.16.2 Main Wash Cycle

When the pre-wash cycle is completed, the microprocessor-activated switch supplying current to the heater heats the water to a temperature of preset value. The set temperature is sensed by the temperature sensor that gives input to the microprocessor which signals the heater switch to put off the heater. The drum motor is operated to a predetermined time with a slow speed. The drum motor is switched off and the discharge pump expels water from the drum.

The program steps are:

1. Opening and closing of water valve to supply water, to a set level.
2. Switching the heater on.
3. Sensing the preset temperature.
4. Switching off of heater.
5. Drum rotation with set speed, for set time.
6. Discharge of water from drum by the pump.

1.16.3 Rinse Cycle

The cold water is supplied to the drum by the opening and closing of the valve controlled by the microprocessor based on the input from the water level sensor. The drum motor is operated to rotate the drum. The water is emptied by the pump. The sequence is repeated several times.

The program steps are:

1. Admit cold water by opening valve.
2. Close the valve when set level is reached.
3. Rotate the drum for a predetermined time.
4. Empty water by the pump.
5. Repeat the above sequence the number of times set.

1.16.4 Spin Cycle

The microprocessor switches on the drum motor which is signaled to rotate at a higher speed than the rinse speed to remove water from the clothes by centrifuge action.

The program steps are:

1. Rotate drum motor with higher speed.
2. Stop motor when water content in clothes reaches a desired extent.
3. Give signal completion.

1.17 AUTOMATIC BATHROOM SCALE

For an automatic bathroom scale, we need to design a mechatronic system for the measurement of the weight of the person to the following objectives:

- To allow a person to stand on a platform.
- To measure the weight of the person.
- To display the weight on an LED screen.
- The readout should be reasonably fast and accurate.
- The measurement should be independent of where the person stands on the platform.

The schematic block diagram of a bathroom scale is given in Figure 1.20.

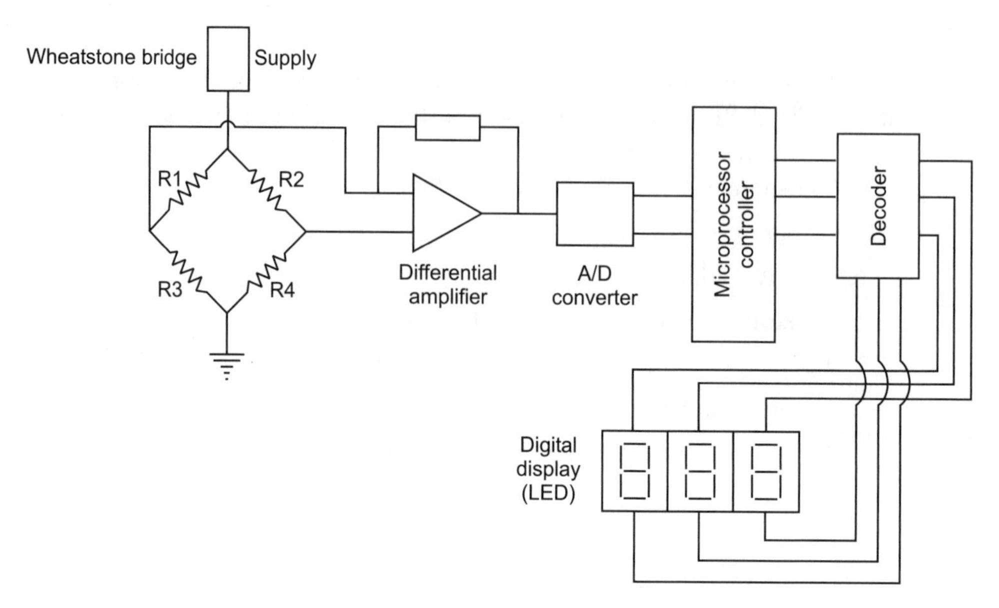

R1, R2: Strain gauges on tension side.

R3, R4: Strain gauges on compression side.

FIGURE 1.20 Automatic bathroom scale (block diagram).

1.17.1 Working Principle

The weighing platform is connected to a load cell with strain gauges mounted. The deflection of the leaf spring results in a change in length of the resistances (strain gauges) leading to the change in voltage output of the Wheatstone bridge. This voltage signal, proportional to the weight, is amplified by a differential amplifier. The analog signal from the amplifier is converted to a digital signal by an analog-to-digital converter before giving it to the microprocessor controller. The output from the controller is given to the decoder, which is connected to the LED display that gives the readout of the weight. A time delay is introduced to have the display remain for a set time.

1.18 ADVANTAGES/DISADVANTAGES OF MECHATRONICS

Although principles of mechatronics have been used effectively in industry, it is important to identify both the advantages and disadvantages as shown in Table 1.3.

TABLE 1.3 Advantages and Disadvantages of Mechatronics

S. No.	Advantages of Mechatronic Systems	Disadvantages of Mechatronic Systems
1.	Effectively serves high-dimensional accuracy requirements.	Improper application and under-utilization can result in losses.
2.	Provides increased productivity on the shop floor.	Maintenance and repair may prove costly.
3.	Reconfiguration feature by presupplied programs facilitate low volume production.	The initial cost is high.
4.	Provides higher level of flexibility required for small product cycles.	Techo-economic estimation has to be done carefully in the selection of mechatronic systems.
5.	Manufacturing lead time is reduced resulting in lowering of unit costs especially in mass production.	Calls for training and re-orientation of the workforce, in design, and in manufacturing.
6.	Results in automation in production, assembly, and quality control.	Technicians and engineers have to be given basic knowledge of two domain disciplines, *i.e.*, precision mechanics and electronics.
7.	Plays major role in total manufacturing solutions rather than standalone machines.	The concurrent integration of mechatronic design requires multidisciplinary knowledge.
8.	Production of parts and products of international standards gives better reputation and return.	
9.	Longer life is expected by proper maintenance and timely diagnosis of faults.	

EXERCISES

1. What are the objectives of mechatronics?
2. Explain the areas of application of mechatronics. What are the advantages and disadvantages of mechatronic systems?
3. Explain the skills required for mechatronics in manufacturing.
4. Justify the interdisciplinary nature of mechatronics.

5. Define microprocessor-based controllers.
6. What are the important criteria in the design of a mechatronic system? Explain a mechatronic-based engine-management system with a block diagram.
7. Define mechatronics and state the major differences between conventional and mechatronic product design approaches.
8. Define a sequential controller and explain with a block diagram the workings of a domestic washing machine.
9. List the advantages of the mechatronics approach in engineering applications.
10. Narrate the evolution of and the historical developments in mechatronics.
11. Explain two types of mechatronic integration.
12. Justify that mechatronics is a formulation of a multi-disciplinary scenario.
13. What are the factors that emphasize the need for mechatronics in industry? Explain.
14. Enumerate the objectives used in mechatronic developments.
15. Exemplify the design aspects of mechatronic systems.
16. With a block diagram explain the various modules that form a mechatronic system.
17. Explain the key areas that are integrated in mechatronic technology.
18. What are the skillsets needed to become a successful mechatronic engineer?
19. Describe the what, why, and how of mechatronics.
20. What is a mechatronic system? What are the different mechatronic systems? Explain with an example.
21. Explain the measurement systems with examples.
22. How does an actuation system function? Explain with an example.
23. What is a control system? What are the types of control systems?
24. Explain the features of a microprocessor system with the help of a simple block diagram.
25. What are the features and elements of a microprocessor-based controller?
26. Compare a conventional engine-management system with a mechatronics system.
27. Give a block diagram depicting microprocessor-based engine management.
28. Explain the functioning of an automatic camera with a block diagram.
29. Explain the operating specifics and program sequences in an automatic camera.
30. Explain the various cycles of operation of an automatic washing machine with the aid of a block diagram.
31. Give the program steps involved in the functioning of an automatic washing machine.
32. Give the mechatronic design of an automatic bathroom scale. What are the elements needed?
33. List the advantages of mechatronics.
34. What are the disadvantages of mechatronics?

CHAPTER 2

SENSORS AND TRANSDUCERS

Mechatronic systems have enhanced flexibility and versatility from the use of sensors and transducers. Sensors and transducers aid in generating information given to the output devices for control action. Sensors also act as a link to the source (input) and the measurement system. They participate in conversion of one domain of energy into another domain with or without the utilization of the modulating domain.

This chapter gives an introduction to sensors and transducers and includes:

- Definition of sensors.
- Description of energy domains.
- Classification of sensors with description of performance parameters.
- Pressure sensors with selection detail.
- Flow sensors with application detail.
- Definition of transducers.
- Classification of transducers.
- Hall effect transducers—principle, application, and advantages.
- Light sensors—principle, types, and application.
- Proximity sensors—types and principle with application.
- Desirable features for sensors and transducers.

2.1 DEFINITION OF SENSORS

A sensor is a part of the measurement system that provides a response to a particular measurable physical parameter, which can be an input energy domain transformed into another form of energy domain with or without the aid of the modulating energy domain.

The simplest example known to a mechanical engineer is the strain gauge. The force on the load cell in the mechanical energy domain is the input. The change in voltage, proportional to the strain, is the output. The electrical voltage supplied to the Wheatstone bridge formed by the resistance of the strain gauge is the modulating energy domain. The domain graph showing the three-dimensional representation of sensor operation is shown in Figure 2.1.

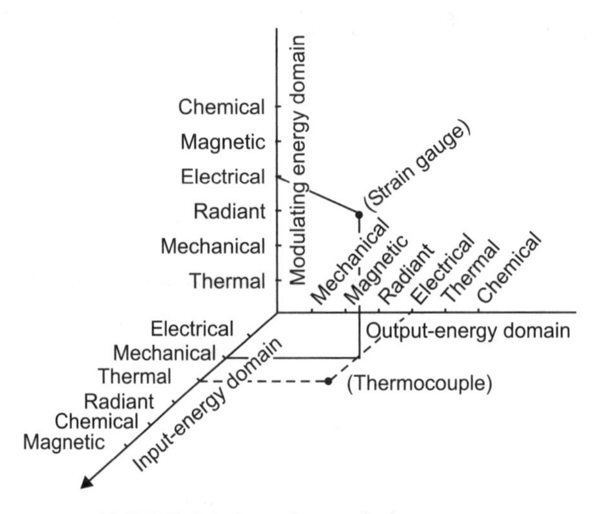

FIGURE 2.1 Domain graph for sensors.

The six energy domains identified to accept, transfer, and modulate by a sensor are:

- **Mechanical.** Distance, velocity, force, acceleration or size, etc., are covered in this domain.
- **Electrical.** Current, resistance, voltage, inductance, and capacitance form the basis of this domain.
- **Magnetic.** Field strength and flux density can be considered in this domain.
- **Thermal.** The effects of temperature such as heat capacity, latent heat, phase changes, sensible heat, and superheating can be identified in this domain.

- **Radiant.** The frequency, phase, intensity, and polarity of electromagnetic radiation fall in this domain.
- **Chemical.** The concentration of chemical substances, crystal structures, and the aggregation of the state concerning the behaviors of the matter exemplify this domain.

For the sensor to respond, the sensor needs the pre-processed inputs. The output from the sensor is post-processed before being measured as a readable parameter. The block diagram shown in Figure 2.2 highlights the association of the sensor with the pre-processing and post-processing stages. The speed of the crankshaft in an engine-management system has an optical light sensor. The slotted disc on the crankshaft provides pre-pocessed light pulses. The pulses sensed are post-processed by the counter and the timer to give the speed measurement. The schematic of the speed measurement is shown in Figure 2.3.

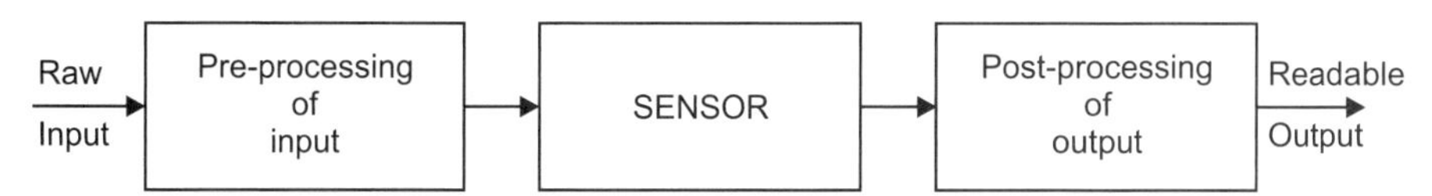

FIGURE 2.2 Sensor association diagram.

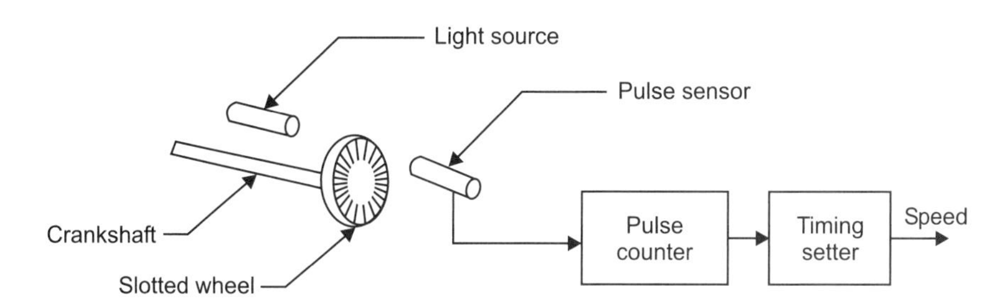

FIGURE 2.3 Speed-measurement system.

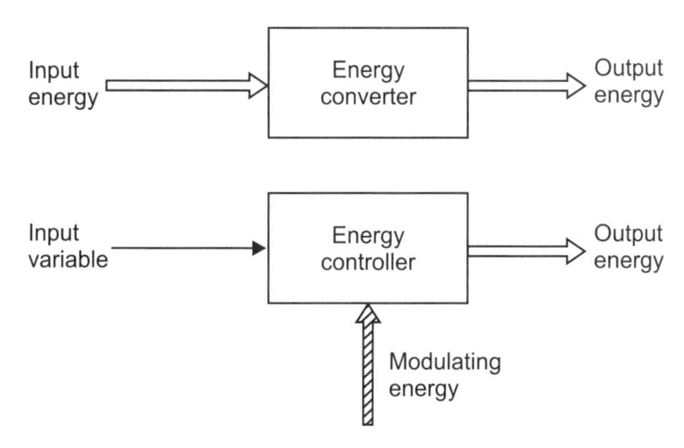

FIGURE 2.4 Sensor energy transformation.

Figure 2.4 depicts the sensor as an energy converter that converts one form of input energy to another form of output energy. It can also be an energy controller aided by modulating energy.

2.2 CLASSIFICATION OF SENSORS

Sensors are classified based on: 1. Functions; 2. Performance; and 3. Output. The parameters of the functional basis are generally mechanical in nature such as displacement, velocity, acceleration, force, torque, dimensions, mass, etc. For a mechanical function the sensor can be a device of mechanical type, electrical type, or an electronic type decided by the element and the principle governing the functioning of the sensor. The input energy, output energy, and the modulating energy domains participating in the work execution process decide the type of sensor.

TABLE 2.1 Classification by Functions

	Domains	Sensing Functions
Sensing parameters	• Displacement	Linear or angular position
	• Velocity	Linear and angular speed or flow rate
	• Acceleration	Vibrational aspect (shock)
	• Dimensional position	Size, thickness, area, volume, deformation, roughness
	• Force	Static, dynamic, absolute forces, differential pressure, etc.
	• Mass	Density, mass, body force (weight)
	• Torque	Power, strength
	• Miscellaneous	Hardness, viscosity

The static and dynamic performance characteristics play an important role in the selection of the type of sensors needed for applications suitable for the mechatronic system. The range, error, accuracy, sensitivity, repeatability, stability, and resolution are some of the static performance parameters to be looked into before selecting the right sensor. Response time, rise time, settling time, and time constant are the main dynamic characteristics usually considered in the mechatronic application of sensors.

Basing the classification of sensors on the type of output form needed to be displayed for the record and processing, the output can be analog data, digital data, coded data, or the frequency waveform. The mechatronic engineer decides the type of sensor useful for the system based on the features.

TABLE 2.2 Classification by Performance

Performance Domain	Performance Parameters
Static characteristics	• Range and span • Error • Accuracy • Repeatability • Resolution • Sensitivity • Stability • Impedance
Dynamic characteristics	• Response time • Settling time • Rise time • Time constants • Peak over shoot • Type of transient response

A detailed study of static and dynamic characteristics is given in books on control theory.

2.3 DEFINITION OF PERFORMANCE PARAMETERS

Performance parameters are classified as static characteristics and dynamic characteristics. These characteristics are considered while selecting the sensors for a particular application. Different applications call for different sensors of different specifications. The specifications of sensors have a bearing on the cost component of the application.

- **Range and Span.** By definition, the range is the limits between which the input or output can vary. The difference between the maximum value and the minimum value is the range.
- **Error.** The difference in the true value and the measured value of the quantity being measured is the error.
- **Accuracy.** The deviation extent of the value indicated by a measurement system is the accuracy.
- **Repeatability.** The ability to give the same output reading when the same input value is applied repeatedly.

- **Resolution.** The incremental step by which the output signal is indicated for a continuous variation of the input signal. It can be considered to be the least count of measurement.
- **Sensitivity.** The ratio of the output value to the input value, whatever the type of output and input unit.
- **Stability.** The ability of the sensor to indicate the same output over a period of time for a constant input.
- **Impedance.** The sensor connected in an electronic circuit presents some form of disturbance known as impedance which is to be known for circuit design.
- **Response Time.** By definition, it is the time elapsed to give the output signal from the point of supply of the input signal.
- **Settling Time.** The time required to come to an output value within the specified error level.
- **Rise Time.** The time taken for the output value to reach a specified percentage of steady-state output.
- **Time Constant.** A fraction of response time (0.632) which is the measure of the inertia of the sensor limiting the speed at which the sensor reacts to the changes in input.
- **Peak Over Shoot.** The maximum value of the output signal before it settles down to the steady state value.
- **Transient Response.** The response of a system depending on its R-L-C characteristics can be critical, over-damped, or under-damped in giving the output signal. This is known as the transient response of a sensor.

TABLE 2.3 Classification by Output

Output Type	Output Characteristic
• Analog output	✓ Continuity in signal ✓ Direct relation to value measured
• Digital output	✓ Serial or parallel form of representation ✓ Information given at regular intervals of time or when needed
• Frequency output	✓ Can be a continuous waveform ✓ Pulsed waveform ✓ Easy convertibility to digital output
• Coded output	✓ Modulation of amplitude ✓ Modulation of frequency ✓ Modulation of pulse width or position

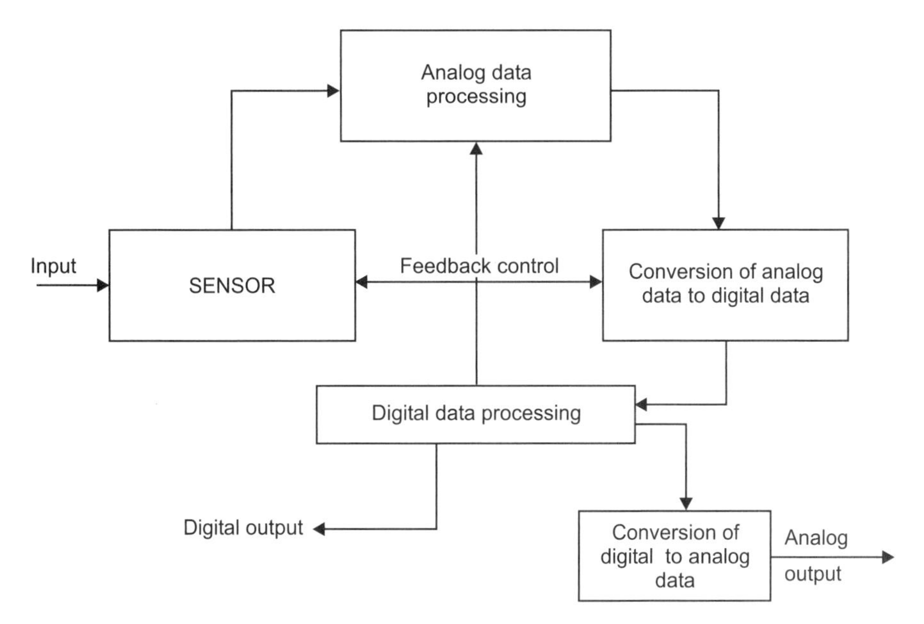

FIGURE 2.5 Sensor function diagram.

2.4 PRESSURE SENSORS

The mechanical pressure gauge with a collapsible tube and analog indication was the first tool used in pressure measurement which still finds application in mechanical industries.

In fluid power systems with dynamic fluctuations electrical pressure sensors are preferred. In these the elastic deformations, depending on the materials' physical properties, are transformed into electrical signal. Strain gauges are frequently used in pressure measurements. An integrated amplifier can be used to amplify the microstrain signal. A voltmeter attached to the sensor unit can suffice for the visual display.

Piezo-electric sensors are most suitable for dynamic pressure measurements and are small in size. For both static and dynamic pressure signals the best solution is piezo-resistive sensors. Vapor-deposited, thick film resistors on silicon wafers constitute the measuring element. The signal gets amplified by the integrated differential amplifier. This has the disadvantage of being sensitive to temperature changes. Table 2.4 highlights the schematic and relative selection and application features of mechanical and electrical/electronic pressure sensors.

TABLE 2.4 Pressure Sensor Features

Function Feature	Mechanical	Strain Gauge	Inductive Sensor	Piezo-electric	Piezo-resistive
Schematic	FIGURE 2.6 (*a*)	FIGURE 2.6 (*b*)	FIGURE 2.6 (*c*)	FIGURE 2.6 (*d*)	FIGURE 2.6 (*e*)
Frequency Range (Hz)	0 – 5	$0 - 10^5$	$0 - 10^4$	$10 - 15 \times 10^4$	$0 - 15 \times 10^4$
Pressure Range (bar)	$0 - 4 \times 10^3$	$0 - 3 \times 10^3$	$0 - 10^3$	$0 - 7 \times 10^3$	$0 - 10^3$
Error %	1	0.25	0.5	1	0.25
Remarkable Characteristic	• Possible to convert into electrical output • Low priced	• The microsignal can be amplified • High priced	• Relatively slow response • Medium priced	• Useful in only dynamic measurements • Highest priced	• Sensitive to the loads beyond range • Medium priced

2.5 FLOW SENSORS

Flow measurement is influenced by parameters such as pressure pulsation, temperature, viscosity of the flow medium, and the degree of contamination by particles other than the medium. The principle of flow measurement is fraught with the effects of flow captured via physical properties and then transformed into a signal proportional to volume flow. Knowing the cross-section of the medium at a point where the velocity of the stream is measured is needed to obtain the flow rate.

A gear pump can be used to measure flow with high accuracy. But they have the drawbacks of higher pressure loss and are prone to seizing due to contamination. By measuring the speed of rotation of the gear pump and the cross-section of the inlet to the pump it is possible to obtain the flow rate of the medium running through the gear pump flow sensor.

A measuring turbine is another mechanical application of flow measurement in which the flow drives the turbine wheel that induces an electrical impulse for each blade. Counting the frequency yields an output signal proportional to the volume discharge rate.

A tapered rotometer that acts against the resistance of a spring actuated by the flow gives a measurable stroke. Knowing the geometry of the rotometer element allows for the measurement of flow through the system installed with this type of sensor.

The inductive flow sensors are effective when the medium is water mixed with additives. The permeated electric-conductive medium with a magnetic field induces voltage proportional to flow velocity. The range of measurement with this type of sensor is wide and does not need the feedback effect of the flowing medium.

Flow measurement with the ultrasound technique works similar to the Laser-Doplet effect. Ultrasound signals emitted into the medium partially reflect the rays due to barrier layers. The phase changes are recorded and evaluated. Table 2.5 shows the schematic brief of these sensors.

TABLE 2.5 Flow Sensors

Characteristics	Schematic
Measuring Range 0-1000 l/min Error—0.5% Dynamics—low Cost—high	Gear Pump 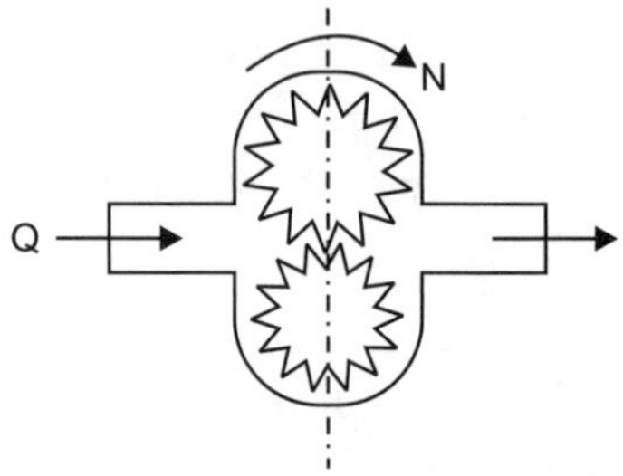 **FIGURE 2.7 (*a*)**
Measuring Range (0.5-20000 l/min) Error—1.5% Dynamics—high Cost—low	Turbine Sensor 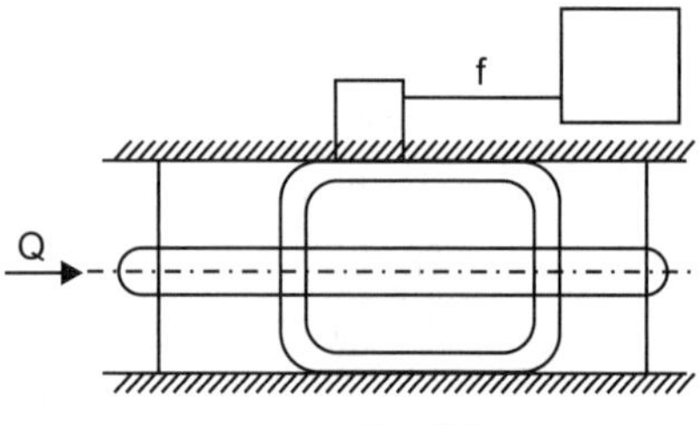 Q = f(v) **FIGURE 2.7 (*b*)**
Measuring Range (1-600 l/min) Error—2% Dynamics—high Cost—medium	Floating Body 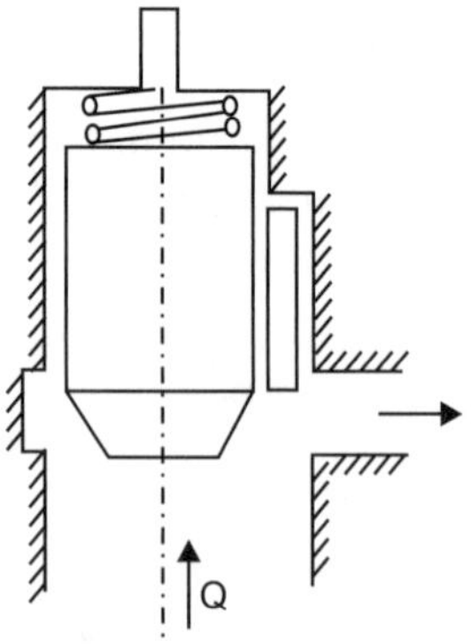 **FIGURE 2.7 (*c*)**

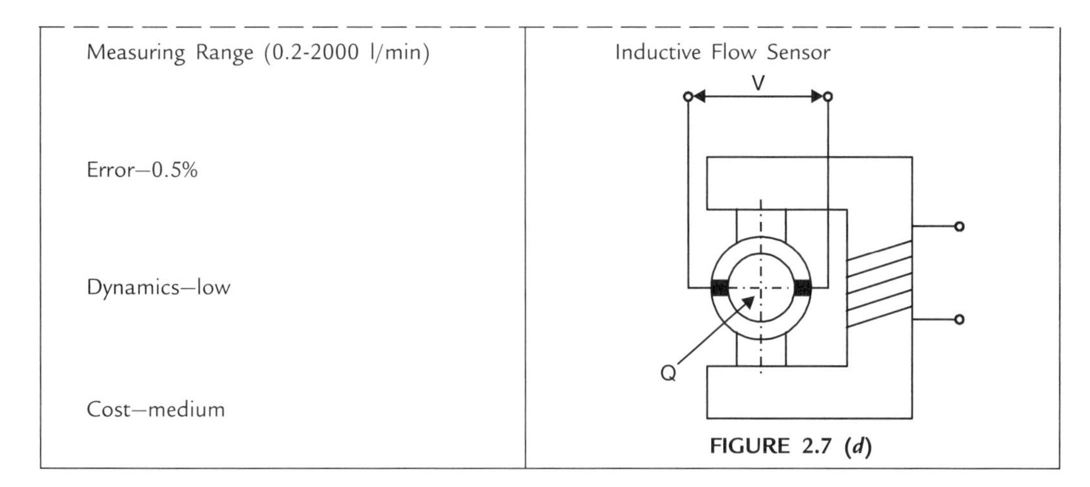

FIGURE 2.7 (*d*)

2.6 TRANSDUCERS

2.6.1 Definition

A transducer is a pedantic word formed from two Latin words. In Latin 'Trans' means 'across' and 'duce' means 'leads.' Hence, transducer leads a physical quantity in one form across to another form. By definition, the transducer is that part of a system that transfers information or data in the form of energy from one part of the system to another with or without changing the form of energy containing the information.

The use of the word transducer is limited to only automatic instruments that do not require human intervention.

Examples Not Considered Transducers	Examples of Transducers
• A desk rule showing length as a number on a scale. • A mercury thermometer showing temperature as a number on a scale. • A voltmeter indicating voltage on an analog scale. • A step down transformer proportionately reducing to the output current.	• A thermocouple that transduces the temperature to electric voltage. • The photo-electric cell that transduces solar energy to electricity. • A pneumatic cylinder that transduces air pressure to mechanical displacement.

2.6.2 Parameters Sensed

Some of the parameters transduced by electro-mechanical devices are as follows:

- Force
- Pressure
- Temperature
- Displacement: Angular and linear
- Proximity: Presence of an object
- Viscosity
- Flow
- Frequency
- Time
- Vibration
- Chemical composition

2.7 CLASSIFICATION OF TRANSDUCERS

The following diagram classifies transducers and their uses.

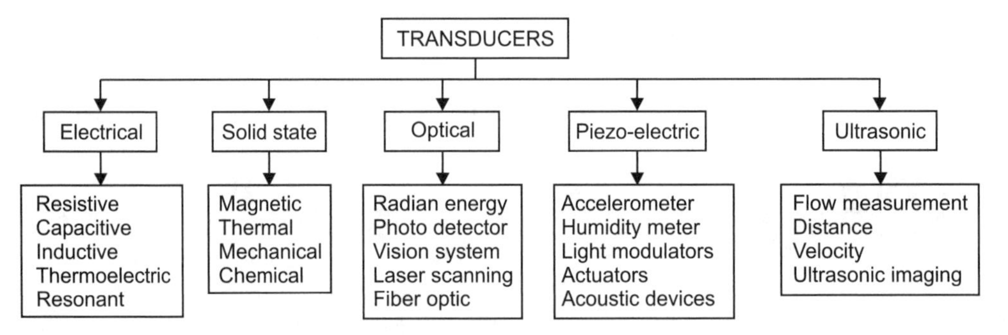

2.8 HALL EFFECT SENSORS

It is important to know about the Hall effect before explaining the functioning of the sensor device based on this effect. The principle of the Hall effect and the functioning and application of Hall effect sensors are explained in the following paragraphs.

2.8.1 Principle of Hall Effect

In 1879 E.R. Hall discovered the Hall effect. "A beam of charged particles passing through a magnetic field experience a force that deflects the beam from the straight line path." This is known as the Hall effect.

Electrons (negatively charged particles) are made to pass through a plate of rectangular cross-section, and a magnetic field is applied at a right angle to the plane of the plate as shown in Figure 2.8. The electrons are deflected towards one side of the plate, making that side negatively charged and the other side positively charged. The force due to the applied magnetic field is called Lorentz force. The mechanism of deflection is governed by the balance in Lorentz force and the force on the charged particle is due to electric field.

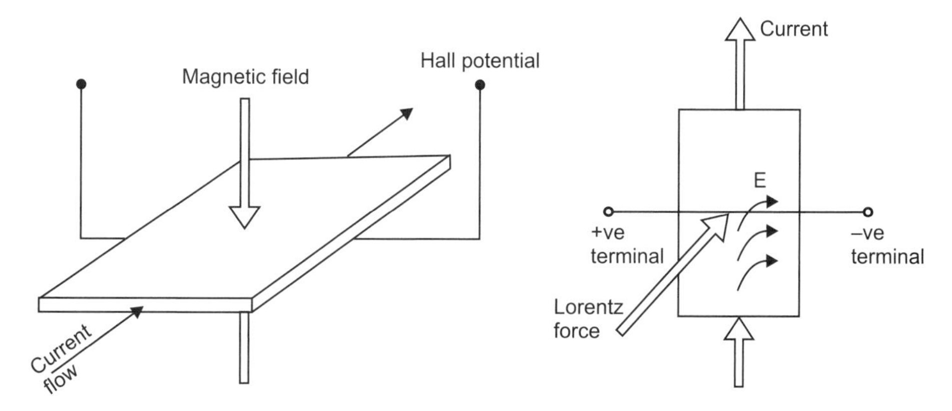

FIGURE 2.8 Principle of Hall effect.

The potential difference, V, created in between the transverse face of the plate, is given by

$$V = H_C \frac{F_L T}{t}, \tag{2.1}$$

where H_C = Hall's coefficient

F_L = Magnetic flux density due to Lorentz force

I = Current flowing through the plate

t = Thickness of the plate

V = Hall voltage

2.8.2 Hall Effect MOSFET (Metal Oxide Semiconductor Field Effect Transistor)

Figure 2.9 shows the MOSFET structure that is adapted to exhibit the Hall effect. Here, the area of the gate channel (plate) is increased and the additional Hall contact is introduced. This leads to Hall voltage proportional to the device current and the magnetic flux density. The Hall plate can be of *p*-type or *n*-type semiconducting material. The small output from the silicon Hall plate can be amplified by integrated circuitry in the chip.

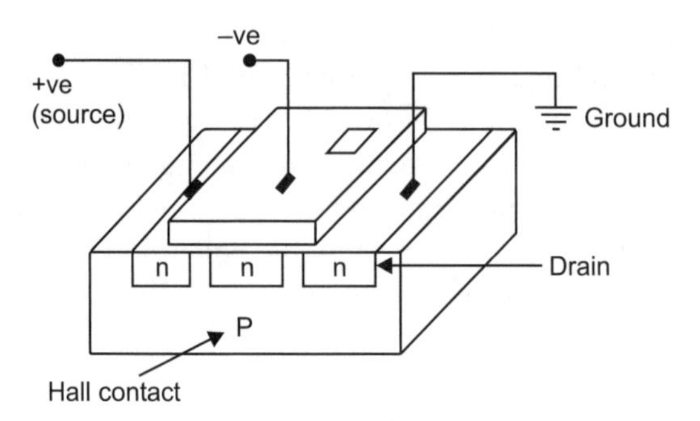

FIGURE 2.9 Hall effect MOSFET.

2.8.3 Types of Hall Effect Sensors

There are two types of Hall effect sensors based on the variation observed in the output voltage:

1. **Linear Hall effect sensor.** In this type the voltage of the output signal varies in a reasonably linear manner with the magnetic flux density. The variation can be as shown by the graph in Figure 2.10.

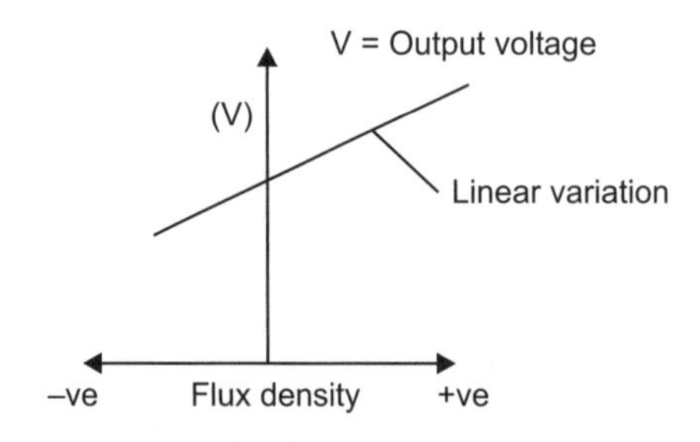

FIGURE 2.10 Linear type.

2. **Threshold Hall effect sensor.** In this type the output voltage shows a sharp drop in the magnetic flux density of a particular value. The graph shown in Figure 2.11 at flux density F_0 shows the output is almost zero.

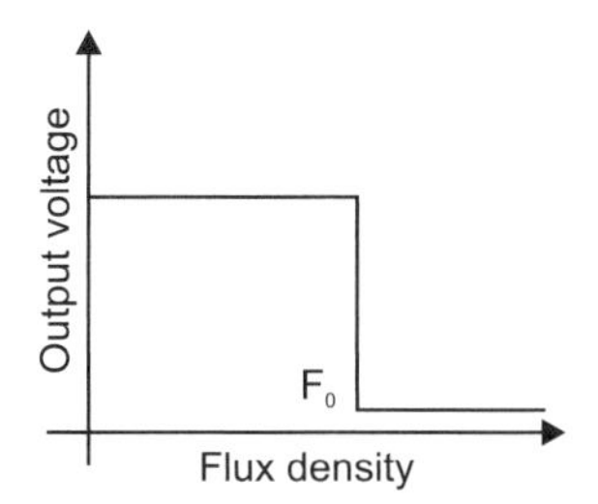

FIGURE 2.11 Threshold type.

2.8.4 Advantages of Hall Effect Sensors

- They can operate as switches of high frequency.
- They cost less than electromechanical switches.
- They are free from the contact bounce problem, as they provide single clear contact.
- They can be used under severe environmental service conditions as they are immune to environmental contamination due to humidity, etc.
- They can be used as proximity, position, and displacement sensors.

2.8.5 Disadvantages of Hall Effect Sensors

Hall effect devices become vulnerable during the offset effect caused by misalignment of Hall contacts and piezo-resistive effects. This drawback can be handled by incorporating offset compensation in the measurement.

2.8.6 Applications of Hall Effect Sensors

1. **Flow meter.** The purpose of a flow meter is to measure the rate of flow of liquid in a pipe. Two electrodes are introduced in the pipe and a magnetic field is applied across the pipe axis. The particles of the liquid flowing across the magnetic field induce a voltage in the electrodes which is proportional to the flow speed. A piezo-electrical crystal generates voltage when subjected to mechanical force.
2. **Fuel level indicator.** A magnet is fixed to float in a fuel tank. The change in the level of the fuel tank changes the distance of the magnet from the Hall effect sensor. This results in Hall voltage that is different for different distances. The Hall voltage is converted to indicate the level of fuel in the fuel tank. Figure 2.12 illustrates the application of the Hall effect to detect the level of fuel in a fuel tank.

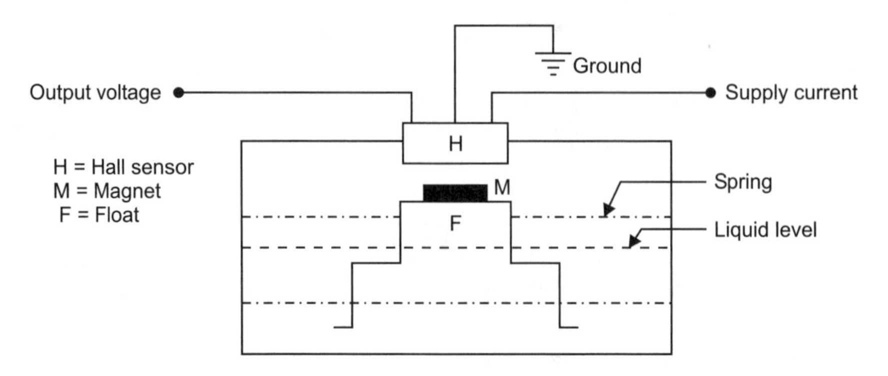

FIGURE 2.12 Fuel level indicator with a Hall sensor.

2.8.7 Application of Hall Effect

A semiconductor transducer is placed in between the legs of a horseshoe magnet. The magnetic field passes through the transducer with leads connected to the connector terminals. The proximity of the object of ferrous material decreases the field strength of the magnetic lines passing the sensor resulting in a change in voltage. The sensor gives binary output to the control devices for further action and processing. The use of a silicon semiconductor enhances the ability to prevent electrical interferences. This type of sensor finds suitable applications in robotics. Such a sensor can be seen in Figure 2.13.

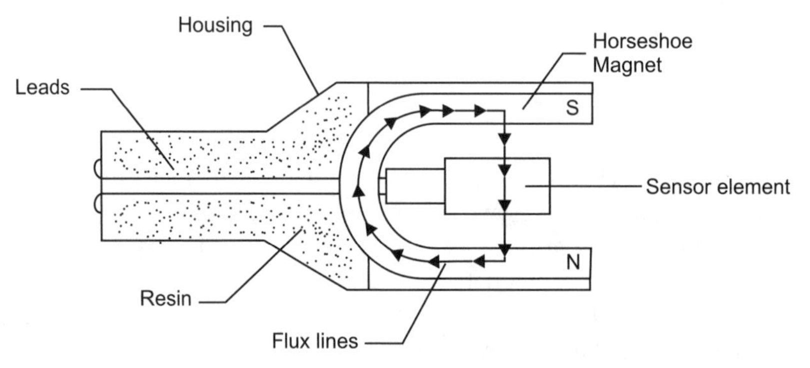

FIGURE 2.13 Hall effect sensor construction.

2.9 LIGHT SENSORS

2.9.1 Principle

Any radiation of appropriate wavelength which falls on the depletion layer of the *p-n* junction develops a potential difference between the junction. The voltage across the layers is proportional to the illumination of the incident radiation. Figure 2.14 shows the incidence on light rays on the *n*-layer of the *p-n* junction. Figure 2.15 is the characteristic curve showing the variation of voltage with the wavelength of radiation.

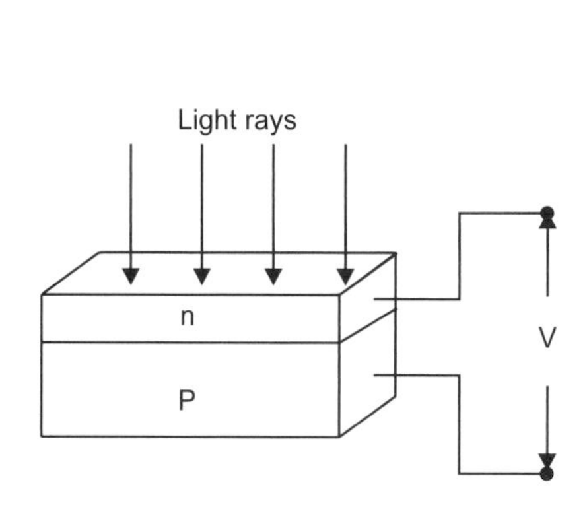

FIGURE 2.14 Photo cell.

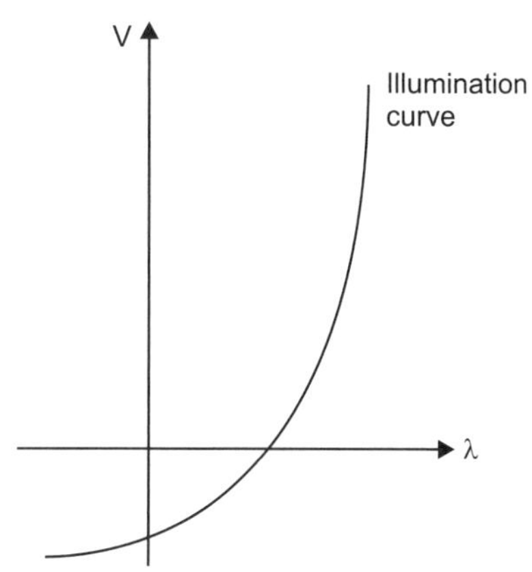

FIGURE 2.15 Characteristics.

2.9.2 Types of Light Sensors

Photo diodes. The configuration of a photo diode is shown in Figure 2.16. The reverse bias is applied against the *p-n* junction that results in a very high resistance. The light ray is made to fall on the *n*-layer (depletion layer). This results in a decrease in resistance of the diode, developing a reverse current due to the sweep up of the electron-hole pair. The reverse current is the measure of intensity of the incident radiation. Photo diodes with different responses can be applied to detect from infrared through ultraviolet wavelength ranges. The response of photo diodes is quick, and they can be used as variable resistance devices.

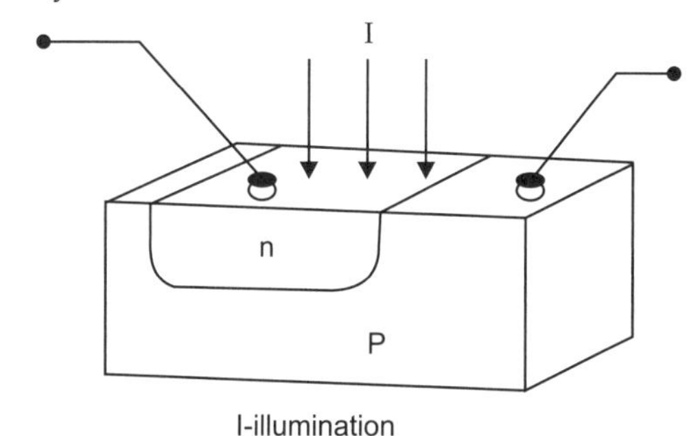

FIGURE 2.16 Photo diode.

Photo transistors. A base collector arranged in parallel to the photo diode of a bipolar transistor is formed as shown in Figure 2.17(*a*). In this *p-n* junction the collector base is photo sensitive. The incident radiation on the base results in the reverse current of the photo diode which is the base current of the transistor. The base current is proportional to the intensity of light. The internal signal processing amplifies the base current to give higher sensitivity to the photo transistors. The application is demonstrated in Figure 2.17(*b*).

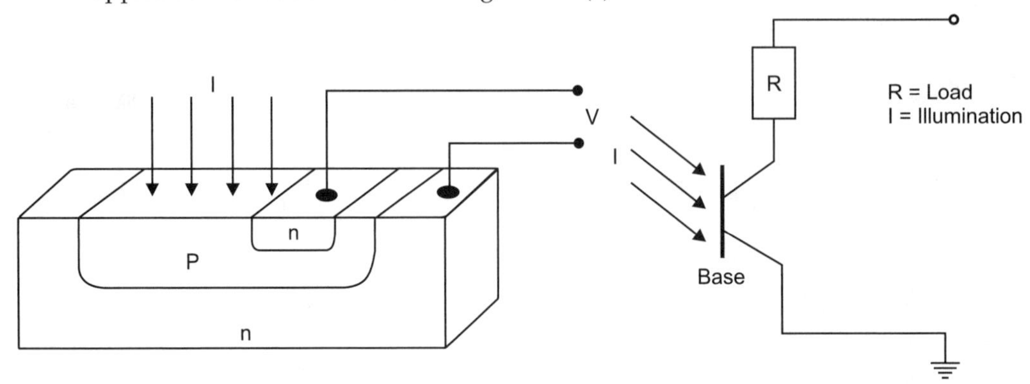

FIGURE 2.17 (*a*) Photo-transistor structure. **FIGURE 2.17 (*b*)** The circuitry.

Photo conductors. By coating a layer of indium antimonide (InSb) or cadmium sulphide (CdS) on a layer of silicon dioxide (SiO_2) photo resistors are produced. The base is the *p*-type material (SiO_2) and the *n*-type materials (In or CdS) are diffused on it. The incident light ray illuminating the *n*-type layer results in a change of conductivity. The bridge circuit arrangement detects this change by the change in output voltage proportional to the intensity of incident light. The photo resistor design and the required circuitry are shown in Figure 2.18.

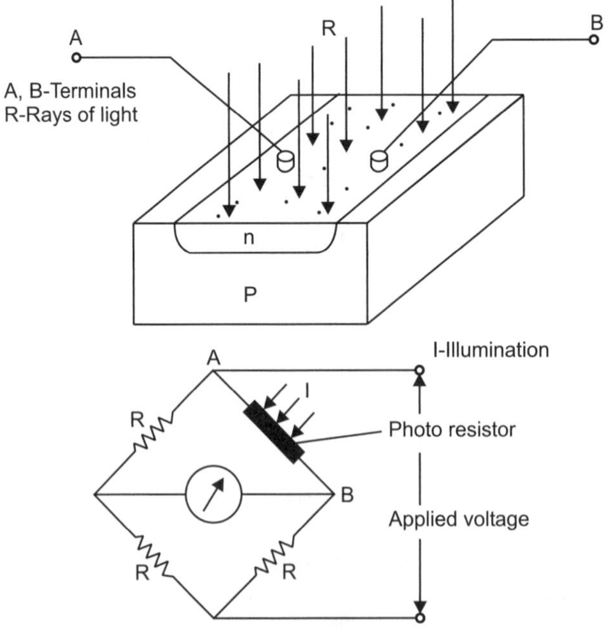

FIGURE 2.18 Photo conductor, structure, and bridge.

2.10 PROXIMITY SENSORS

The presence of an object within the vicinity of job operation is recognized by the output of the proximity sensor. These sensors are used to generate information in object grasping and obstacle avoidance. Some examples of proximity sensors commonly used are inductive sensors and ultrasonic proximity sensors. Inductive sensors can be effective only when the objects are made of ferro-magnetic materials. The objects of non-ferrous metals are recognized by ultrasonic sensors.

2.10.1 Inductive Proximity Sensors

Principle

The ferro-magnetic material brought close to this type of sensor results in flux lines of the permanent in the sensor. This leads to a change in the inductance of the coil. The induced current pulse in the coil is proportional to the rate of change of flux lines in the magnet which is the result of a foreign object present in the field.

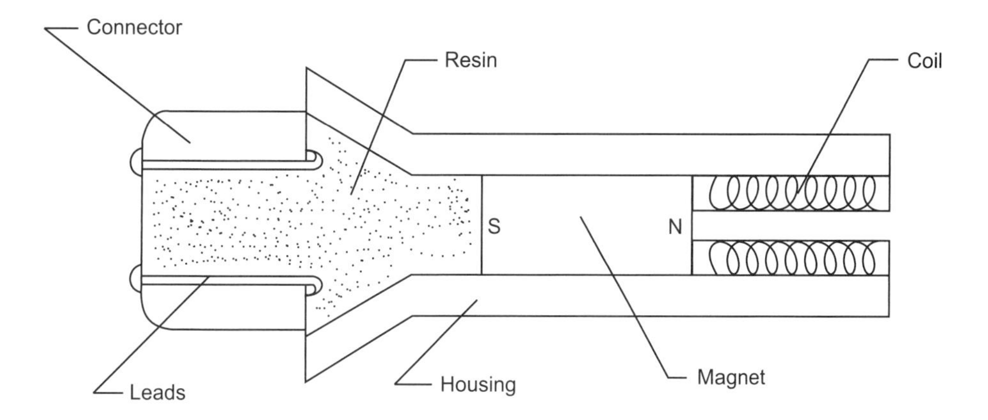

FIGURE 2.19 Inductive proximity sensor.

Construction

The proximity inductive sensors basically consist of a wound coil located in front of a permanent magnet as shown in Figure 2.19. The permanent magnet is encased in rugged housing. The change in current in the coil is output through the leads embedded in the resin. The leads connected to the display through a connector give a signal for the presence of an object in the vicinity.

2.10.2 Ultrasonic Sensors

Principle

The accoustic waves emitted by the transmitter in the sensor hit the object and the reflected waves are sensed by the receiver to generate information about the presence of an object of any material. This is the ecomode of operation. In the opposed mode if the waves are not reflected the receiver does not get the signal due to blocking of the transmitted waves.

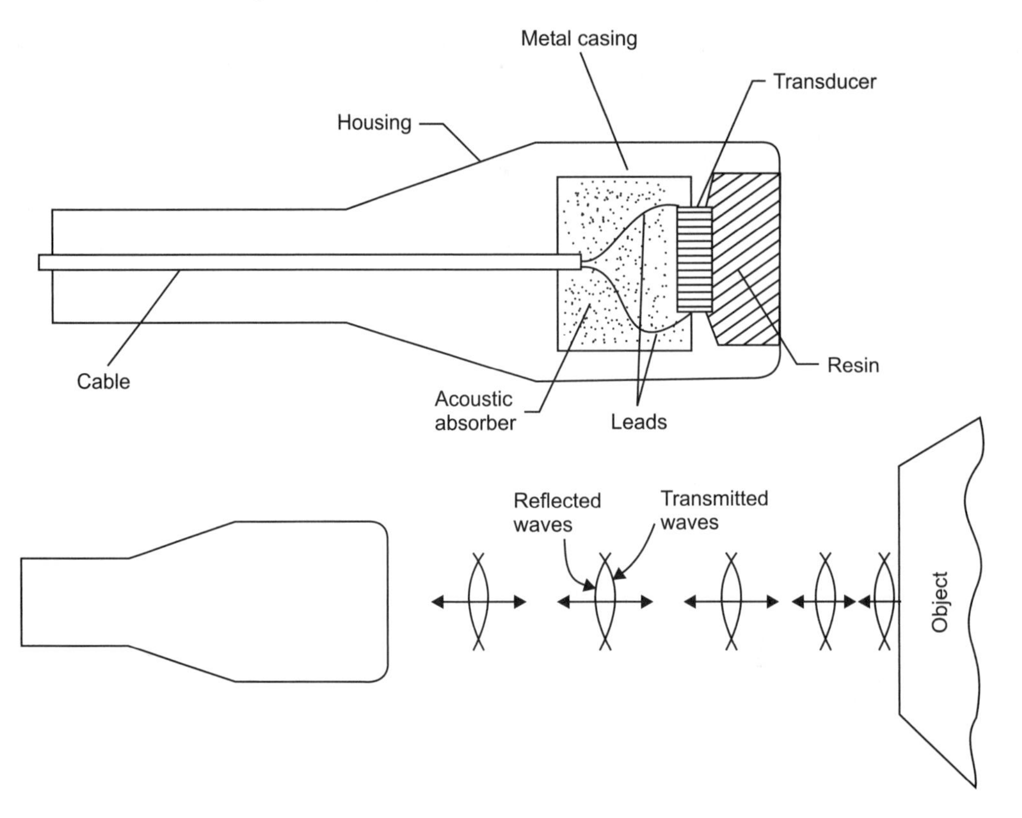

FIGURE 2.20 Ultrasonic proximity sensor.

Construction

The main part in this type of sensor is the transducer element which can act both as the transmitter and the receiver. The transducer is covered by the resin block which protects it from dust and humidity. The absorber material behind the transducer performs the function of acoustic damping. The general protection is provided by the metallic housing. Information about the presence of the object is carried from the receiver through the leads and cable. Figure 2.20 shows the construction of an ultrasonic proximity sensor.

2.11 DESIRABLE FEATURES FOR SENSORS AND TRANSDUCERS

The following table details some desirable features for sensors and transducers.

Features	Functions
• Precision	✓ Should be as high as possible.
	✓ Deviation in measurement reading should be minimum.
• Accuracy	✓ Should be very high.
	✓ Error between sensed and actual value should approach zero.
• Speed of response	✓ Time taken to respond should be minimum.
	✓ Response should be instantaneous.
• Operating range	✓ Wide operating range.
	✓ Good accuracy over the range.
• Reliability	✓ Life should be long.
	✓ Frequent failures are not acceptable.
• Calibration	✓ Should be easy to calibrate.
	✓ Drift should be minimum.
• Cost and case	✓ The purchase cost should be low.
	✓ The installation and operation should be easy.

2.12 OPTICAL SENSORS

The proximity of the object is detected by the action of the traveling light. The light emitted by the transmitter focuses on the object which reflects to be received by the receiver photo diode. The constructional features of an optical sensor are shown in Figure 2.21. The light from the emitting diode is focused by the transmitter lens, onto the object surface. The reflected light waves travel back and are received by the solid state photo diode through the receiver lens. The object within the range of the sensor can detect the presence. The focal length of the sensor lenses decide the range within which the proximity of the object is detected.

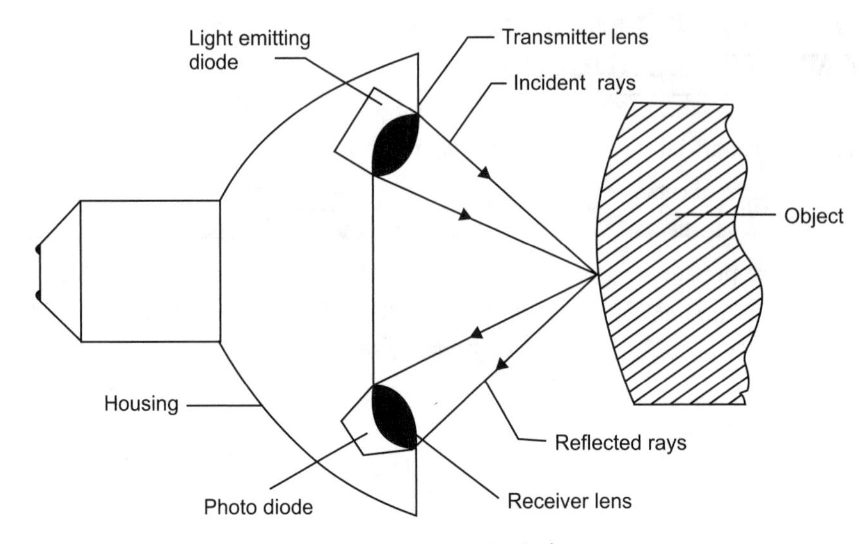

FIGURE 2.21 Optical sensor.

EXERCISES

1. Explain the working principle of a Hall generator.
2. Explain the principle of an eddy current proximity sensor.
3. List the different types of internal and external sensors used in a mechatronic system and briefly explain them.
4. State, in general, the principle of operation of transducers and highlight their differences with sensors.
5. Briefly explain any two types of transducers.
6. Explain the principle of operation of Hall effect transducers.
7. Explain how sensing is achieved by an incremental optical encoder.
8. Explain the following terminology related to transducers:
   (*a*) Accuracy
   (*b*) Repeatability
   (*c*) Stability
   (*d*) Sensitivity
9. Explain the following terms related to transducers: (*a*) Draft, (*b*) Linearity, (*c*) Time constant, (*d*) Settling time.
10. Explain the principle of the workings and application of a photo-transistor.
11. Define a sensor. What are the energy domains input, modulated, and output by a sensor? Explain the domain graph with an example.

12. Explain a speed measurement system to signify the use of a sensor.
13. Classify sensors based on the function performed by them.
14. Give the classification of sensors based on the performance of the sensor.
15. What are the static and dynamic performance parameters in sensors? Briefly explain.
16. Considering the outputs, categorize sensors.
17. With a block diagram explain how sensors function in a mechatronic system.
18. What are pressure sensors? Explain any two with a sketch.
19. List the factors on which pressure sensors are selected.
20. Distinguish between mechanical and electromechanical pressure sensors.
21. What are flow sensors? How do they function?
22. Explain any two flow sensors used as mechatronic elements.
23. Define a transducer. Give examples.
24. Give examples to distinguish between transducer systems and non-transducer systems.
25. What are the parameters that are transformed by transducers?
26. Give the classification of transducers.
27. What is the Hall effect? Explain Hall effect sensors.
28. Explain with a sketch the Hall effect MOSFET.
29. Explain two types of Hall effect sensors.
30. What are the advantages of Hall effect sensors?
31. Explain the use of a Hall effect sensor in a fuel level indicator.
32. Give the applications of Hall effect sensors.
33. Explain the characteristics of a photo cell as a light sensor.
34. What are different types of light sensors? Explain any two.
35. What are proximity sensors? Give examples.
36. Explain the principle of an inductive proximity sensor.
37. Explain the construction of an inductive proximity sensor.
38. Explain ultrasonic proximity sensors.
39. Give the construction and workings of ultrasonic proximity sensors.
40. Specify the requirements of functional features for sensors and transducers.
41. Explain the construction and workings of an optical sensor.

Chapter 3

Hydraulic Systems

The energy transfer for actuation using hydraulic fluid needs the systematically arranged hydraulic components to form the hydraulic system known as the hydraulic power pack. The performance of the hydraulic system is based on pressure control, flow control, and direction control. Prior to the development of a power pack there is a need for the design of a suitable hydraulic circuit. This chapter on hydraulic actuation covers the following topics:

- Definition of actuator and actuator system.
- Classification of actuators.
- Classification of hydraulic cylinders.
- Construction and details of hydraulic cylinders.
- Application of hydraulic cylinders.
- Hydraulic motors classification, construction, workings, and application.
- Valves and their classification.
- Types of valves—pressure, flow, and direction control valves.
- Symbols of hydraulic system components.
- Explanation of a general hydraulic circuit.

- Various hydraulic circuits with explanation of workings including:
  - Relief circuit
  - Regulating circuit
  - Counter-balance circuit
  - Sequence circuit
  - Meter-in and meter-out circuit
  - Direction control circuit
  - Hydrostatic transmission

3.1 INTRODUCTION TO HYDRAULIC SYSTEMS

Useful work is done by a mechanical system with the consumption of energy. The prime intention of a mechanical system is to produce motion. The motion given out as output by a mechanical device can be translatory or rotary. The energy is transferred to the system in the form of electrical energy, hydraulic energy, or pneumatic energy. The pressurized oil produces hydraulic energy. The compressed air gives pneumatic energy. The electrical generator outputs electrical energy. Suitable devices are designed to convert these various forms of energies into translational or rotational motion to produce the required useful work.

The systems that transform energies into motion are the *actuators*. Based on the type of motion developed, they are categorized as linear actuators or rotary actuators. Linear actuators give out translatory motion. Such linear actuators are used in machine tool slides, robotic manipulators, and material handling systems. Rotary actuators produce rotational movement. The application of such actuators can be observed in shaft rotation employed in machine tool spindles, revolute joints of manipulators, screw shafts, and wheel rotation in material handling machinery.

Both linear and rotary actuators are designed for the input of electrical, hydraulic, or pneumatic energies. Actuators (linear) that make use of electrical energy are called *solenoids*. Rotary actuators that use electrical energy are electric motors. Actuators, linear or rotary, driven by hydraulic energy are called hydraulic actuators. Pneumatic actuators can also be linear or rotary with the aid of energy supplied by compressed air. Electric actuators take different forms of electric energy, *i.e.*, Direct Current (D.C.) or Alternating Current (A.C.), and are called D.C. motors and A.C. motors, respectively.

Furthermore, the motion produced should match certain specifications. The work done should conform to the following basic specifications:

- **Force.** The amount of force depends on the maximum load to be moved.

- **Distance.** The length through which the load to be moved decides the distance.
- **Velocity.** Velocity is decided by the time taken to move the load the required distance.
- **Torque.** Torque is the rotational load to be moved.
- **Radius.** The radius of rotation depends on the torsional moment to be developed.
- **Speed.** The angular velocity is given by the number of rotations produced in the given time.

The above specifications, in combination, give the derived specification "power" of the actuator; that is, the rating used in the selection of the actuator. The above specification requirements have lead to different configurations in the design of actuators. These classifications are detailed in the following sections.

3.2 DEFINITION OF ACTUATORS AND ACTUATOR SYSTEMS

Any mechanical, electrical, or electromechanical system that produces linear or rotary motion to drive mechanical events such as shafts, screws, slides, or a manipulators can be called an *actuator*.

Along with the actuator other elements such as connecting parts, fixtures, attachments, and hardware that serve the purpose of generation of motion either linear or rotational can be called the *actuator system*.

A hydraulic cylinder with a connecting yoke or clevis and mounting flange or trunnion can be considered an actuator system. An electrical motor with an output shaft provided with a key or flanged coupling and mounting plate is also an actuator system.

The purpose of an actuator is to convert one form of energy into mechanical work in the form of motion. The energy comes in the form of electrical energy or mechanical energy. Electrical conductors are the transfer elements for electrical energy. Hydraulic oil or air is the medium used to carry the mechanical energy. The actuator that works on the pressure energy supplied by the hydraulic oil is called the hydraulic actuator.

3.3 CLASSIFICATION OF ACTUATORS

The first level of classification is based on the type of motion produced by the actuator, *i.e.,* linear or rotational. In the next level of classification the type of

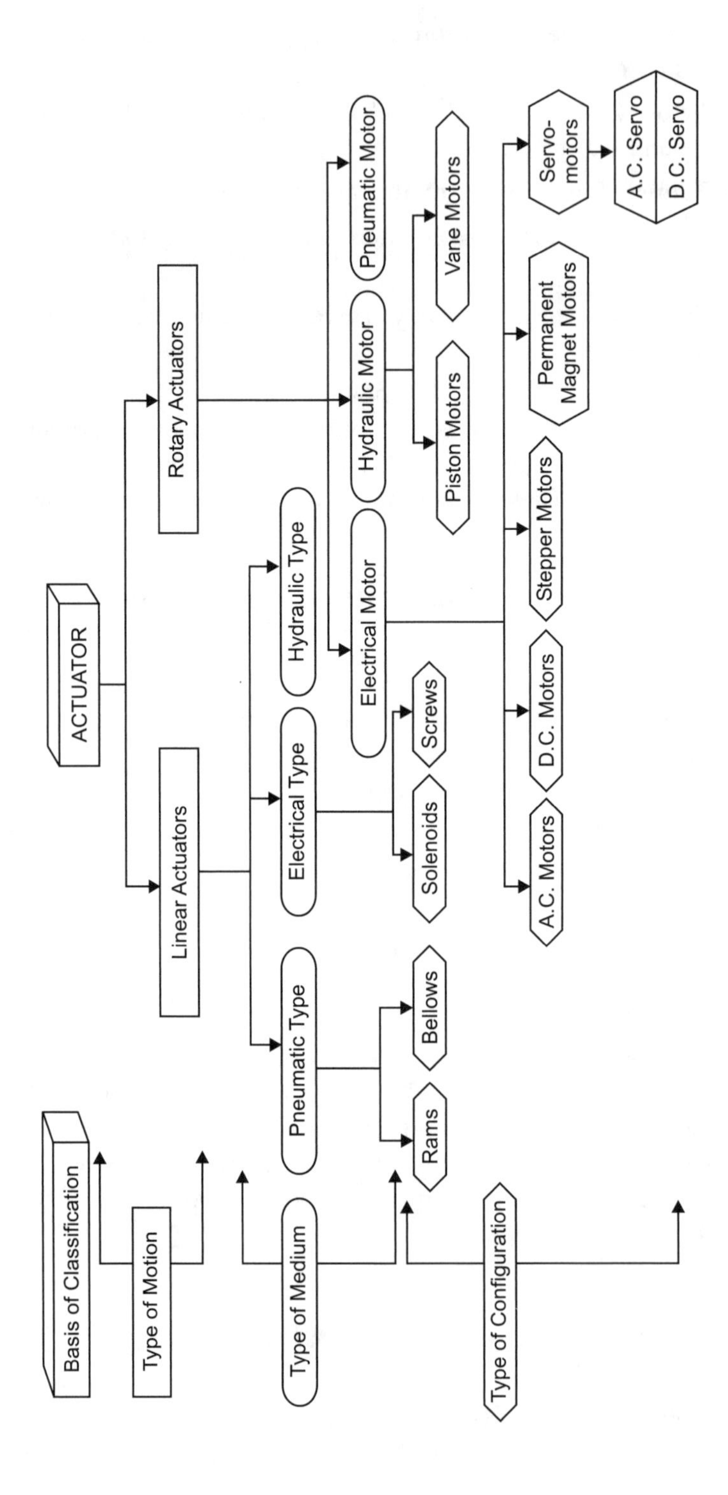

FIGURE 3.1 Tree diagram of classification of actuators.

energy and the energy medium is used for classification. In the third level, the configuration of design forms the basis of classification.

In the first level of classification the type of motion bifurcates the actuators into linear and rotary actuators. Translation is the function of linear actuators, whereas rotation is the function of rotary actuators. Sometimes rotatory actuators with the aid of mechanical elements produce translatory motion in a mechatronic system. In some mechanical systems the reciprocatory motion is converted to rotary motion by cranking action. The output of translation is the displacement and force exertion. The rotary actuation results in the output of angular displacement and torque. Both types of actuations are responsible for some form of power development that can perform work.

The type of medium conveying energy to the actuating system further classifies each actuator in this level. Electrical energy is the most commonly used form of energy that directly and indirectly assists the function of actuation. Pressurized hydraulic oil and compressed air are other kinds of media that help in carrying energy to the hydraulic and pneumatic actuators. It is the electrical energy that is used to create the pressure in hydraulic and pneumatic media.

The third level of classification uses the design configurations as the basis. The functional requirements, the output specifications, and the characteristic performance indicators are taken as the base to configure various designs. This basis of classification is application dependent. The style, manufacturing considerations, cost considerations, reliability considerations, and size considerations form the basis.

3.4 CLASSIFICATION OF HYDRAULIC CYLINDERS

The following table classifies hydraulic cylinders.

Hydraulic Cylinders			
Based on piston and rod style		**Based on cylinder style**	**Based on mounting style**
1. Single-acting	Gravity return	1. Tie rod construction	1. Flange mounting type
	Spring return	2. Threaded construction	• Bottom flange
2. Double-acting	Single-rod end	3. Welded construction	• Top flange
	Double-rod end	4. Threaded and welded	• Lug type flange
3. Ram type		constructions	2. Clevis type
4. Telescopic		5. Bolted cover type	3. Trunnion type
Rod connections	- Threaded type rod		• Bottom trunnion
	- Clevis type rod		• Center trunnion
	- Yoke type rod		• Top trunnion

3.5 HYDRAULIC CYLINDERS

A hydraulic cylinder is a linear actuator that produces linear displacement. The pressurized hydraulic oil provides the medium for energy input. The pressure is converted to force. Based on the length of displacement and the force output different configurations of design are possible in hydraulic cylinders.

3.5.1 Constructional Features

The main parts of a hydraulic cylinder are the piston, piston rod, cylinder tube, and the end covers. The piston rod is connected to the piston and the other end of the rod extends out of the cylinder. The piston divides the cylinder into two chambers; namely, the rod end side and the bottom end side. The piston seals prevent the leakage of oil between these chambers. The passages in the end covers supply pressurized oil to the chambers. The cylindrical tube fitted with end covers form the cylinders. In between the rod end covers are the rod seals used to prevent the leakage of oil. The rod end extending outside the cylinder is provided with the connector element. The end covers may be fixed to the tube by threaded connection, welded connection, or tie rod connection. The piston seals between the piston and tube not only prevent leakage but also avoid metal-to-metal contact. The seals are replaceable. The leakage that may result between the rod and piston can be avoided using a nylock nut for fastening. A tapered spigot at the end of the piston provides cushioning action at the end of the stroke to prevent the possibility of impact with the end cover.

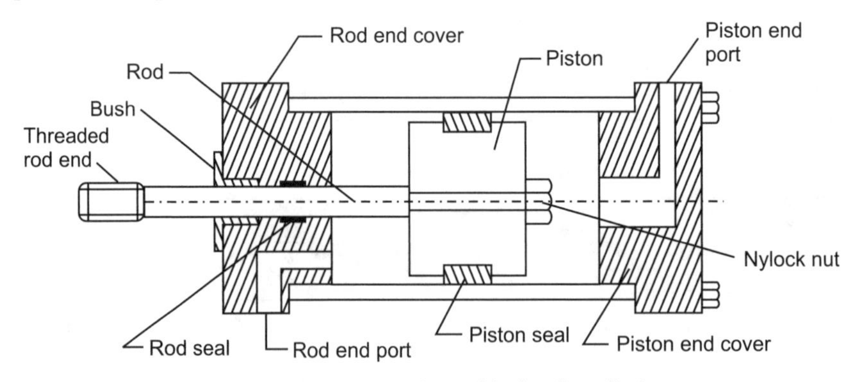

FIGURE 3.2 Construction of hydraulic cylinder.

3.6 CONFIGURATIONS IN HYDRAULIC CYLINDERS BASED ON ROD AND PISTON STYLES

The following table demonstrates various configurations with rods and pistons.

Deliberations	Diagrams
1. **Single-acting gravity return:** The pressurized oil is admitted to the bottom end of the vertical cylinder. When stopping the supply, the piston returns due to the gravity of the load on the platform. As the oil is permitted only on one side of the piston the cylinder is called single acting.	1. 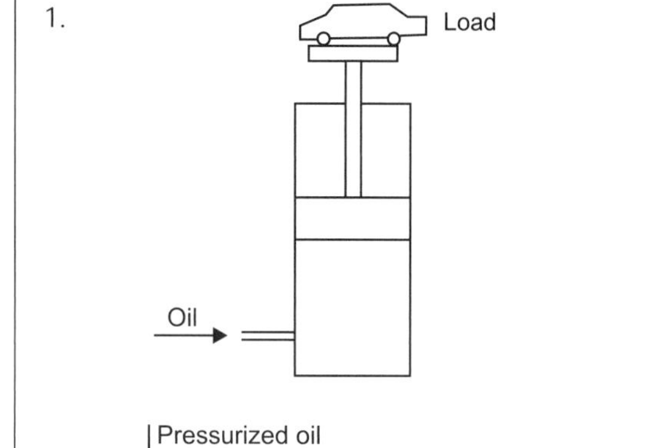
2. **Single-acting spring return:** The oil acting can be either on the piston end or the rod end. One stroke is by the action of the pressurized oil, and the return is by the spring action. In the power stroke the spring is compressed. The return is actuated by the energy stored in the spring.	2. 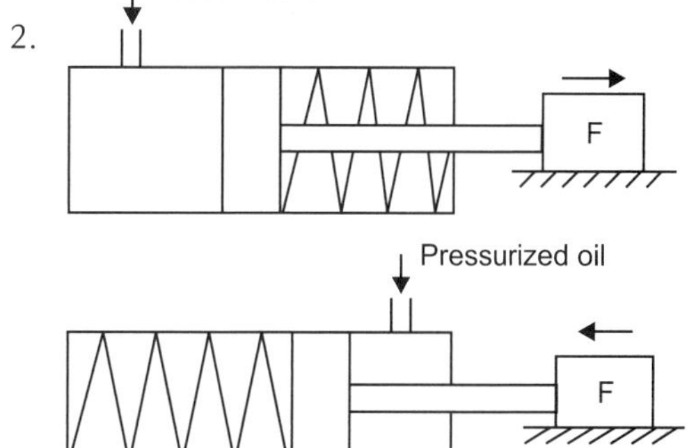
3. **Double-acting—single rod end:** The pressurized oil acts on both sides of the piston. The forward stroke is actuated by permitting oil to the piston end and the return takes place by the action of oil admitted on the rod end. The force exerted on the forward stroke is more than that on the return stroke, but the speed is more on the return stroke for a fixed discharge rate.	3. 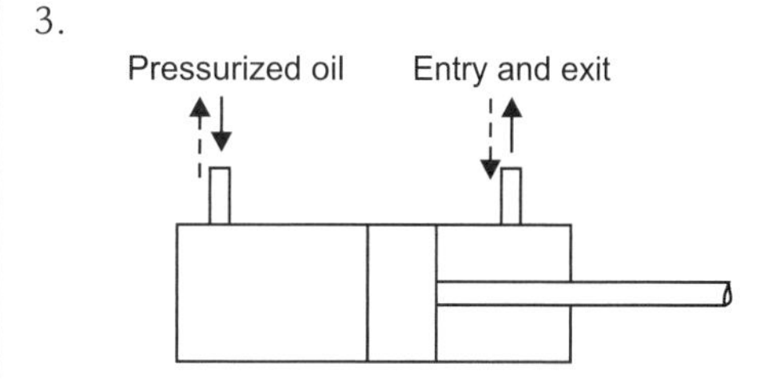
4. **Double-acting—double rod:** The rod is provided on both sides of the piston. Hence, the force and speed are the same in both the forward and return strokes.	4. 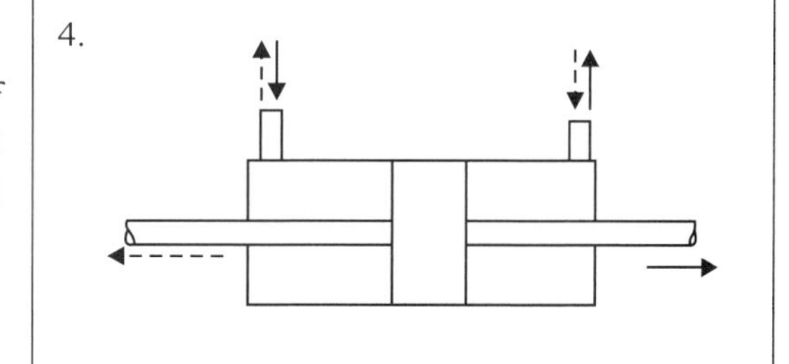

5. **Ram type cylinder:**

   In this cylinder there is no difference between the diameter of the piston and the rod. The piston rod acts as the piston and the oil acts on the bottom side of the rod. The lifting platforms make use of this kind of cylinder. The enhanced diameter gives high strength to withstand the load.

6. **Telescopic cylinder:**

   The concentrically arranged rods extend sequentially by the action of the pressurized oil at the bottom giving a very long stroke length in this type of cylinder.

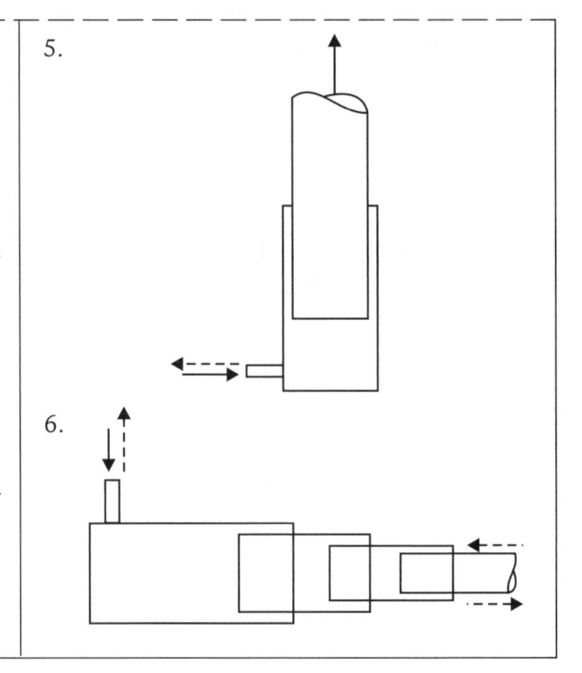

3.7 CONFIGURATIONS BASED ON CYLINDER STYLE

The following table demonstrates various constructions in cylinders.

1. **Tie rod construction:**

   The piston end cover and the rod end cover hold the cylinder tube in between them. The fastening is done with the four tie rods that are nothing but long studs. When the length of the cylinder is small this kind of connection is highly useful.

2. **Threaded construction:**

   The internal threads in the cylinder tube and the external threads in the end covers facilitate the formation of a closed chamber bifurcated by the piston inside. Long cylinders are designed with this feature and leakage is the problem faced with this design.

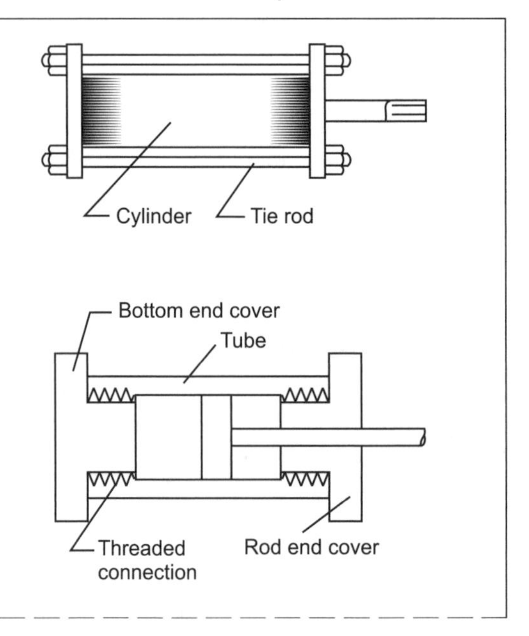

3. **Welded construction:**

The bottom end cover generally welded to the tube and the rod end cover is threaded to the tube in the process of piston assembly. The leak-proof welded joints are required as the high-pressure fluid acts inside the tube. Both side welded construction does not favor the repair and replacement of seals on the piston and rod.

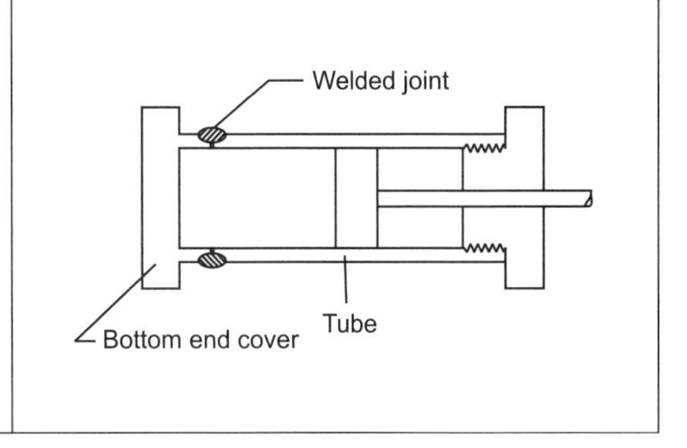

3.8 CONFIGURATIONS BASED ON MOUNTING STYLE

The following table demonstrates various mounting types used with cylinders.

1. **Flange mounting:**

According to the application of the hydraulic cylinders the mounting styles change. A flange with four bolting holes is fixed to the cylinder either to the front cover or to the rear end cover.

2. **Lug mounting:**

A lug is an L-bracket cut from an angle or formed by bending. The lugs have to be provided at both front and rear ends. The bolting holes provided in the lug facilitate the mounting of the cylinder to the machine frame.

3. **Clevis mounting type:**

The clevis, as shown, is fixed to the rear end. The plates of the clevis will have holes. The eye in the mounting structure mates with the clevis and is connected by a pin. The cylinder can oscillate about an arc while producing linear motion.

4. **Trunnion mounting type:**

A sleeve welded to the cylinder carries pin-like spigots called trunnion which can be located at the front or in middle or at the rear end. They also provide oscillation to the cylinder.

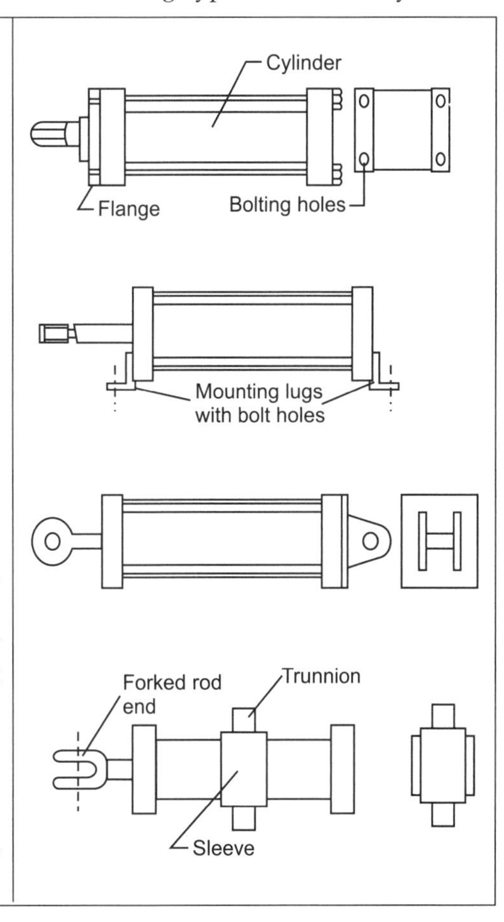

3.9 APPLICATIONS OF HYDRAULIC CYLINDERS

- *Applications in machine tools:*
  - Table and slide drives in grinding machines.
  - Ram drives in broaching and honing machines.
  - Clamping applications in jigs and fixtures.
- *Positional control devices* use linear hydraulic actuators for feed drives.
- *Material handling applications—earth movers:*
  - Arm movements and bucket movements in excavators and loaders.
  - Blade movements in bulldozers and morter graders.
- *Lifting platforms.* Automobile garages use hydraulic lifting platforms in servicing that make use of hydraulic cylinders.
- *Applications in robotics.* The heavy duty load application of continuous oscillations of robotic manipulators calls for the use of hydraulic cylinders.
- *Hydraulic doors and jacks.* The automatic closing of doors and automobile seats are equipped with hydraulic cylinders.
- *Space applications.* Launching pad fixtures are operated with hydraulic cylinders.
- *Antenna mechanisms.* Heavy antenna make use of hydraulic cylinders in their mechanisms to till the positions.
- *Load simulation applications.* The simulators of vibration test rigs have the platforms provided with hydraulic cylinders to simulate the loads.
- *Stearing mechanisms.* Heavy duty stearing and braking applications make use of hydraulic cylinders.

3.10 HYDRAULIC MOTORS

Hydraulic motors are fluid power motors which transform hydraulic energy (pressure energy in the oil) to mechanical rotation. The function of hydrostatic rotatory motors is based on the displacement principle. Using the geometry of displacement volume, these motors are distinguished into piston motors, vane motors, and gear motors. The displacement of the piston by pressurized hydraulic fluid results in rotation, as is the case of piston motors. The impact of high pressure oil on the vanes produces rotation in vane motors. The flowing high pressure oil moves the gears in mesh to give rotary motion to the shaft connected to the gears in gear motors.

3.11 CLASSIFICATION OF HYDRAULIC MOTORS

If the type of displacement forms the basis for the first level of classification, the design configuration of the mechanical displacement elements further classifies hydraulic motors into different types. Some contemporary designs are discussed in the following sections.

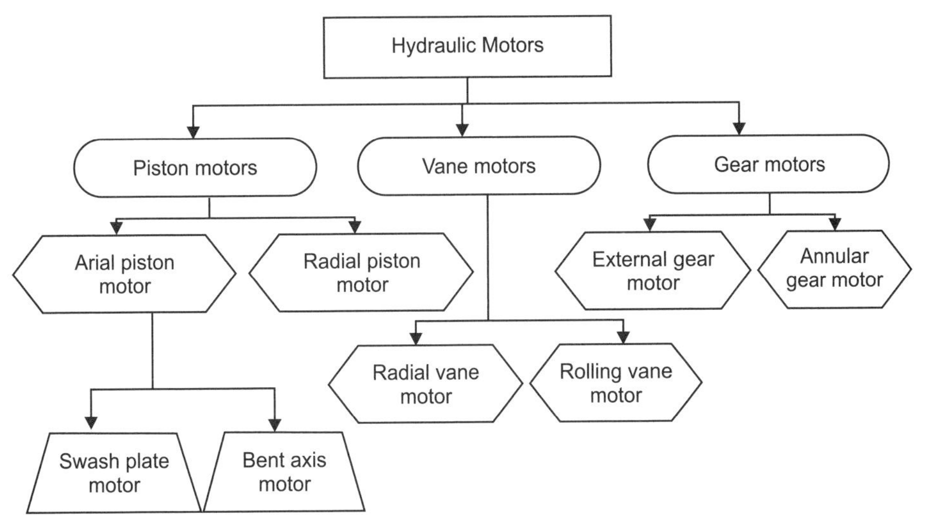

FIGURE 3.3 Classification of hydraulic motors.

3.12 SWASH PLATE MOTORS

3.12.1 Construction and Workings

In swash plate motors, the pistons that are hardened and ground are mounted on the wedge-shaped circular swash plates by a ball joint forming the bearing. The pistons are located in the holes of the piston housing. The holes are internally connected to the inlet and outlets of the motor. The whole assembly is covered by the casing. The details are shown in Figure 3.4.

The high pressure oil pushes the piston back and forth resulting in the rotation of the swash plate. The shaft connected to the swash plate rotates to give mechanical work.

3.12.2 Features

- Universally applicable.
- Highly efficient in a wide range of operations.

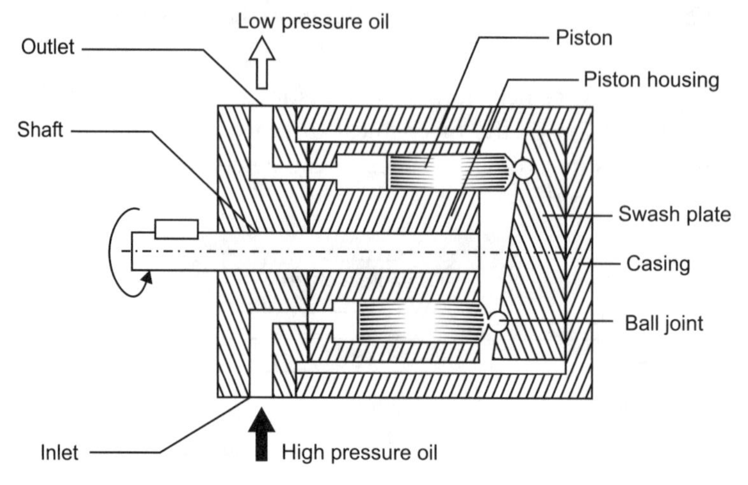

FIGURE 3.4 Swash plate motor.

- Do not depend on pressure, speed, and torque.
- Highly suitable for high-performance applications.
- Typically high speed motors.

3.12.3 Application and Selection Parameters

- Constant and variable speed operations.
- Volume of displacement: 25-800 cm^3.
- Speed range: 800-8000 rpm.
- Working pressure up to: 400 bar.

3.13 BENT AXIS MOTORS

3.13.1 Construction and Workings

A bent axis motor is also a piston motor. The axis of the piston is inclined to the axis of the output shaft. The pistons are located in the housing that is interconnected, and they also have connection with the inlet and outlet parts. The pistons with ball joint bearings are mounted on the plate (flange) on the output shaft.

The pressurized oil received from the pump results in displacement of the pistons in a phased manner leading to the rotation of the flange. The flange in rigid connection with the shaft gives rotational output. The oil losing its pressure is delivered through the outlet to the reservoir.

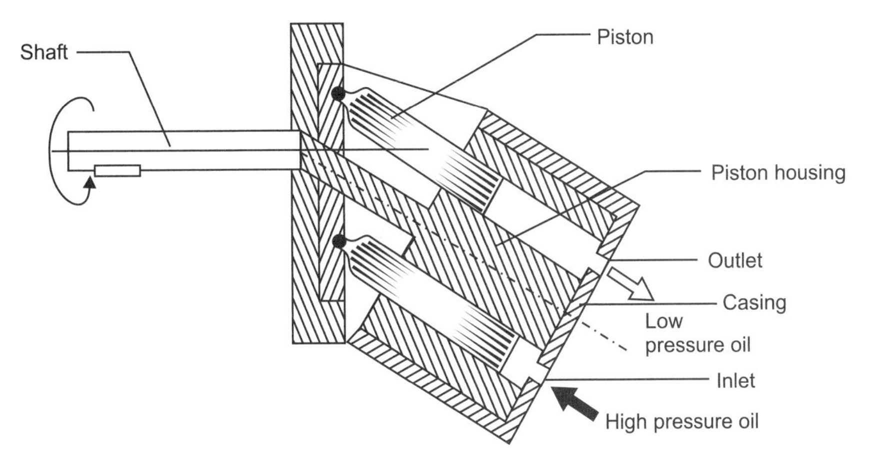

FIGURE 3.5 Bent axis motor.

3.13.2 Features

- Highly efficient.
- Wide operational range.
- Specialized for low speed and high torque applications.
- Highly performing motor.
- Constant as well as variable speed options.

3.13.3 Application and Selection Parameters

- Volume displacement range: (25-800) cm^3.
- Rotational speed range: (0-8000) rpm.
- Operational pressure range: (0-400 bar).

3.14 RADIAL PISTON MOTORS

3.14.1 Construction and Workings

In a radial piston motor, the construction is characterized by the stator and rotor constructions with their axis set by the eccentricity. The pistons arranged radially on the rotor are located in the holes in the stator. The inlet is provided for the pressurized oil directed from the pump. The delivery of oil is followed through the motor outlet provided in the stator housing.

The high pressure oil displaces the radial pistons in sequence because of eccentricity, and this is responsible for the movement of the rotor. In this process the pressure of the oil is converted to displacement and then to rotation needed to actuate the work.

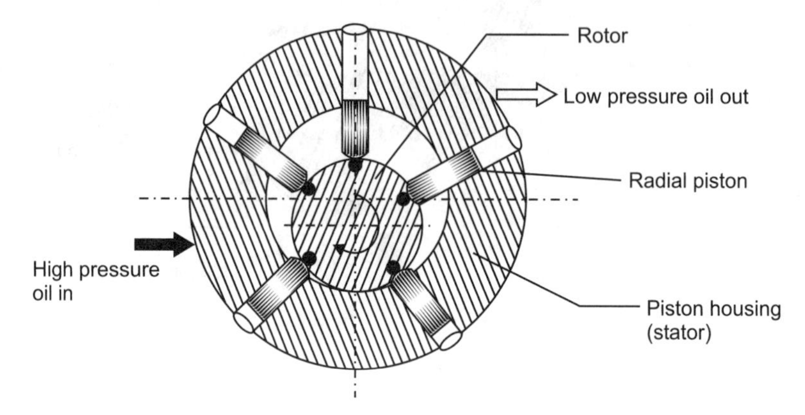

FIGURE 3.6 Radial piston motor.

3.14.2 Features

- Universally applicable motor.
- Very good efficiency is achievable.
- Can be suitably used in low speed and high torque requirements.
- Characteristics do not diminish even at low speeds.

3.14.3 Application and Selection Parameters

- Applicable both as a constant and a variable speed motor.
- Volume displacement range: (5-7000 cc).
- Speed of rotation: (500-2000 rpm).
- Working pressure: (200-350 bar).

3.15 VANE MOTORS

3.15.1 Construction and Workings

In vane motors, a permanent casing with an inlet and outlet for the oil is fitted with a replaceable lining. The rotor with vane housing stays eccentric with the casing axis. The radial vanes that are located in the slots in the vane housing can move radially inward and outward by the contact with the casing.

The pressurized oil supplied through inlet impinges on the vanes. The peripheral displacement of the vanes transfer the movement to the rotor producing rotational work. The transformation sequence here is also pressure force to displacement of vanes to rotation of shaft.

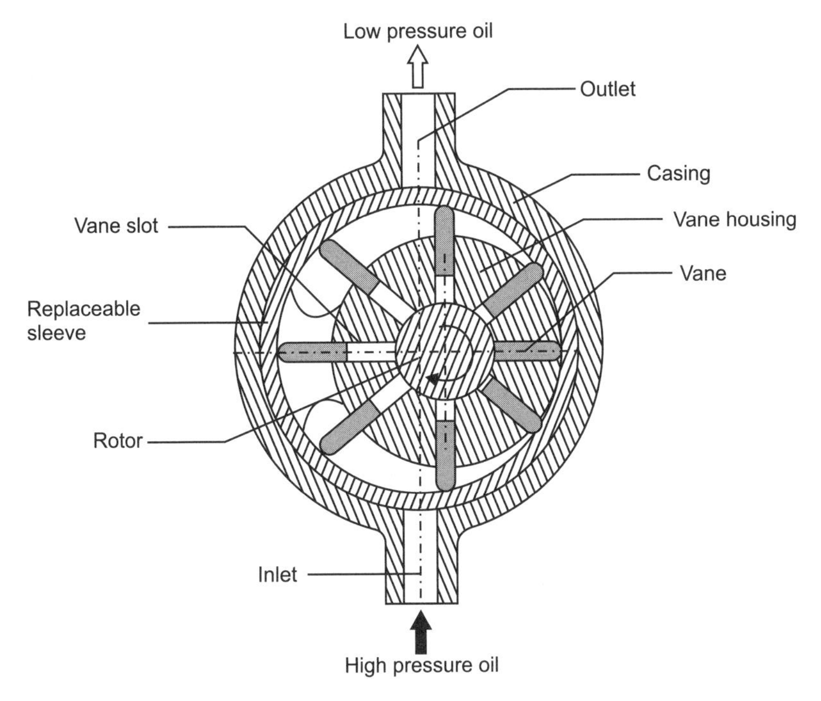

FIGURE 3.7 Vane motor.

3.15.2 Features

- Medium power output capacity.
- Quiet in operation.
- Low degree of irregularity.
- Specially suitable for servodrives.

3.15.3 Application and Selection Parameters

- Used in both constant and variable speed applications.
- Volume displacement range: (5-2000 cc).
- Rotational speed: (0-3000 rpm).
- Working pressure range: (10-200 bar).

3.16 GEAR MOTORS

3.16.1 Construction and Workings

In gear motors, two meshing spur gears with shafts are mounted on the bearing located in the closed housing. The housing is provided with an inlet and outlet port. One of the gears acts idle and the other is mounted on the output shaft projecting out of the casing.

The high pressure oil led from the pump forces the gear in the mesh to move in opposite directions. The rotational displacement of the gear is transferred to the output shaft. The low pressure oil is delivered out through the outlet port.

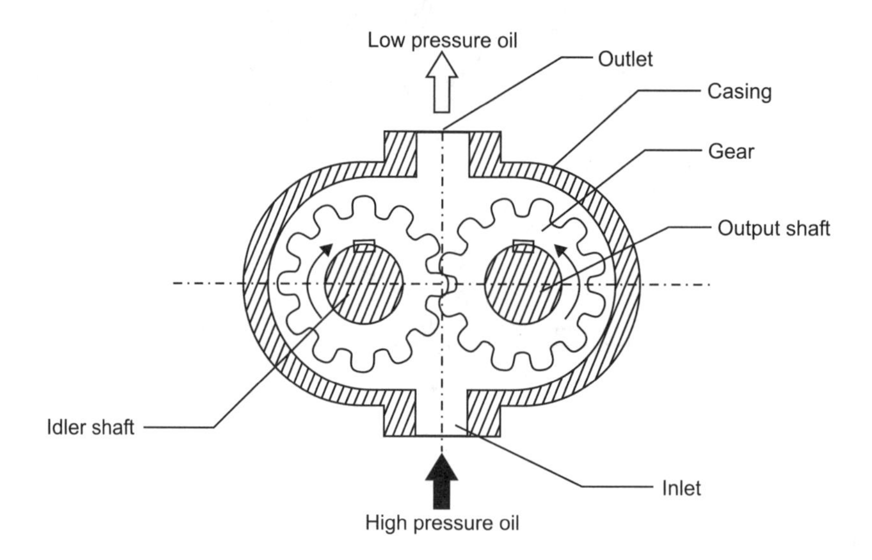

FIGURE 3.8 Gear motor.

3.16.2 Features

- Medium operational power motor.
- Simple in design.
- Not highly efficient.
- Pressure, speed, and torque do not have much effect on efficiency of operation.
- Suitable for medium speed and low torque applications.

3.16.3 Application and Selection Parameters

- Constant speed motor.
- Low displacement range: (5-300 cc).
- Speed of rotation: (200-300 rpm).
- Working pressure: (10-300 bar).

3.17 ANNULAR GEAR MOTORS

3.17.1 Construction and Workings

In annular gear motors, the outer casing, which is stator, acts as an internal annular gear. The external annular gear is mounted on the rotor. The number of teeth on the internal annular gear is less than the number of teeth in the internal annular gear, in order to create for the oil. The pressure force results in the rotational displacement of the internal annular gear. The displacement is transferred to the shaft to give output actuation.

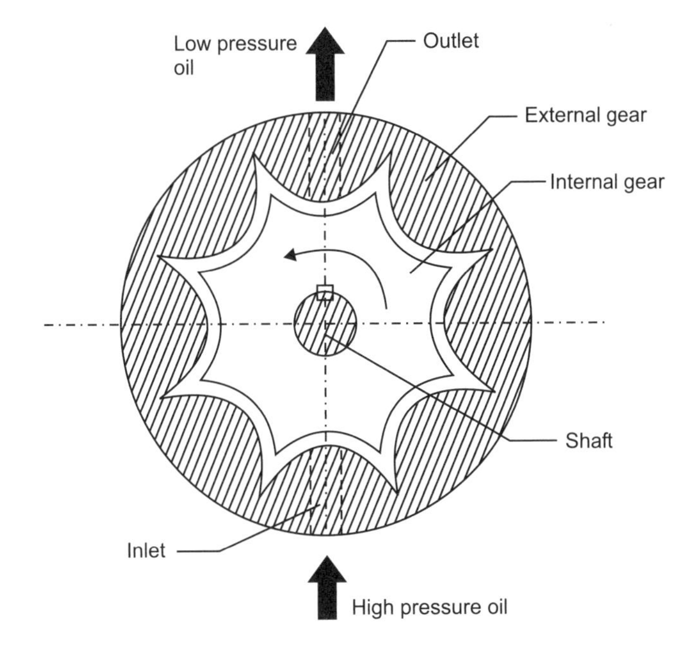

FIGURE 3.9 Annular gear motor.

3.17.2 Features

- Medium power range.
- Quiet in operation.
- Suitable for low speed and high torque applications.
- Constant speed motor.

3.17.3 Application and Selection Parameters

- Volumetric displacement: (50-900 cc).
- Rotary speed: (10-1000 rpm).
- Working pressure: (10-250 bar).

3.18 VALVES

To effect various functions in a hydraulic power pack and to control the operations of the actuators *hydraulic valves* are used. The following subjective aspects need to be controlled in the hydraulic circuits according to application requirements:

- Pressure control
- Flow control
- Direction control

The system pressure should not exceed a certain limit for the safety of the pump. The different branches in the circuit need different pressures in the hydraulic oil supplied to the actuators. The pressure should be retained in a certain actuator function without exhibiting drift in the position. The actuators are sequenced by pressure control to operate in a required order. The relief, regulation, balancing, and sequencing are the pressure control aspects carried out by pressure control valves.

The amount of oil specified by the volumetric flow rate reaching the actuator is regulated by the flow control valves. For an actuator of a given geometrical configuration the speed of actuation has a direct bearing on the rate of flow of oil. The position of the flow control valve in different segments of the circuit executes the different flow control actions needed for the application.

The actuators (linear or rotatory) are reversed in operation or stopped at a certain stage for idling. The reversal in direction is effected by the direction control valves. Different aspects and duties and different configurations in the directional valve exist.

3.19 CLASSIFICATION OF VALVES

The following diagram classifies valves.

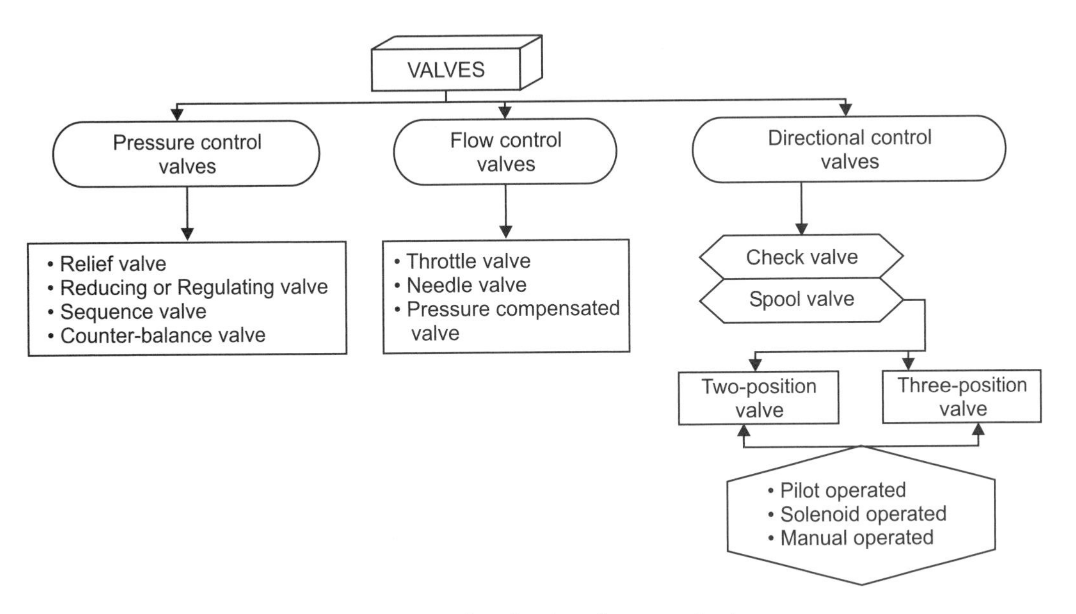

FIGURE 3.10 Classification diagram of valves.

3.20 PRESSURE CONTROL VALVES

In the following sections, we discuss pressure control valve characteristics.

3.20.1 Construction

Pressure control valves have a three-piece construction. The split is to ease manufacturing processes. The top block has a bore to accommodate the force adjusting spring and the adjustment screw. The middle block accommodates the spool in the bore machined in it. Two ports in the middle block act as an inlet and outlet to the relief valve. The bottom block has holes for the pilot line and also acts as a cover to the middle block from the bottom. The spool is located in the pilot hole at the bottom and is pressed against a spring at the top. Figure 3.11 shows the construction of a relief valve.

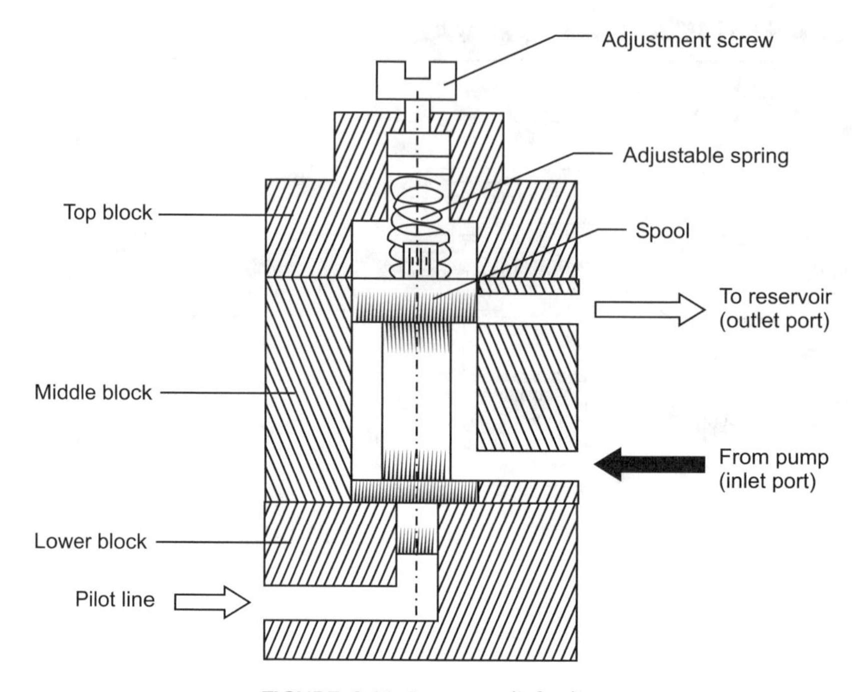

FIGURE 3.11 Pressure relief valve.

3.20.2 Workings

When the system pressure exceeds the set pressure, the spool acted on by the pilot pressure increase lifts to open up port A to relieve the oil to the reservoir. The movement of the spool is controlled by the spring force which can be adjusted by the screw provided to set the required system pressure beyond which the valve would crack to relieve the excessive pressure. The relief valve is generally located and connected in parallel to the pump. The relief valve regulates the system pressure in the hydraulic circuit.

3.21 REDUCING/REGULATING VALVES

3.21.1 Construction

These valves have a top block housing the adjustable spring with an adjustment set screw or knob. The lower block has the stepped bore to accommodate the spool. The lower block also has two ports. One port connects the valve to the pressure line up the stream and the other port acts as an outlet supplying the oil to the actuator at reduced or regulated pressure. A range of pressure can be set to the valve using the knob with the screw actuating the spring connected to the spool. A reducing or regulating valve is shown in Figure 3.12.

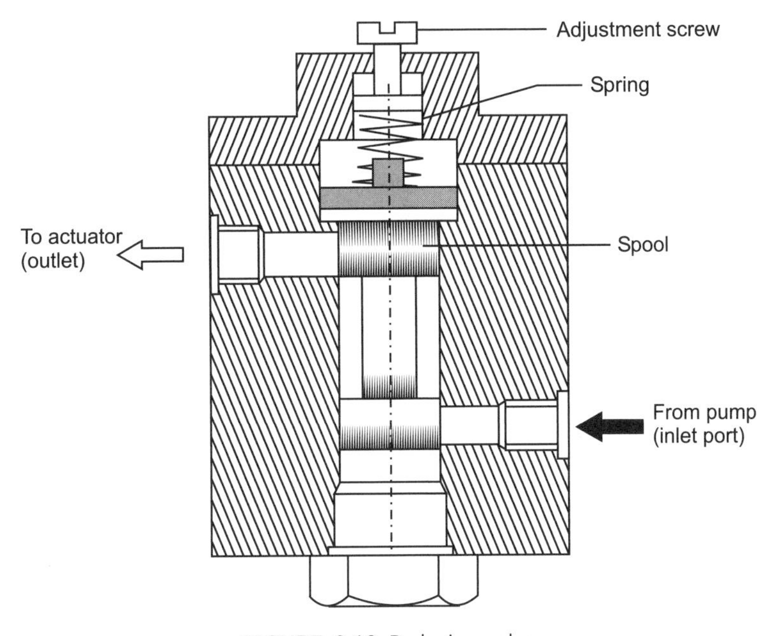

FIGURE 3.12 Reducing valve.

3.21.2 Workings

The oil with higher pressure from the pump enters port A (inlet port). So when the pressure is less than the set pressure the spool does not move. Once the set pressure is exceeded the spool moves against the spring allowing extra flow. Since the power, which is the product of pressure and flow rate, has to be constant, the increase in flow rate reduces the pressure. Hence, the required pressure is supplied down the stream. Thus, the regulation or reduction in pressure is effected by the valve situated in that branch of the hydraulic circuit.

3.22 SEQUENCE VALVES

The construction of the sequence valve is basically in line with the relief valve in all respects as depicted in Figure 3.13. This type of valve is used to sequence the operation. To establish sequence and time delay between primary and secondary lines the sequence valve is used. It maintains a preset pressure in the primary line upstream in the circuit. The valve conducts a function in the secondary line by sequencing the downsteam pressure. On reaching the set pressure in the primary performing one function, the spool is lifted to divert the flow to the secondary circuit executing the next function in the sequence of operational process actuation.

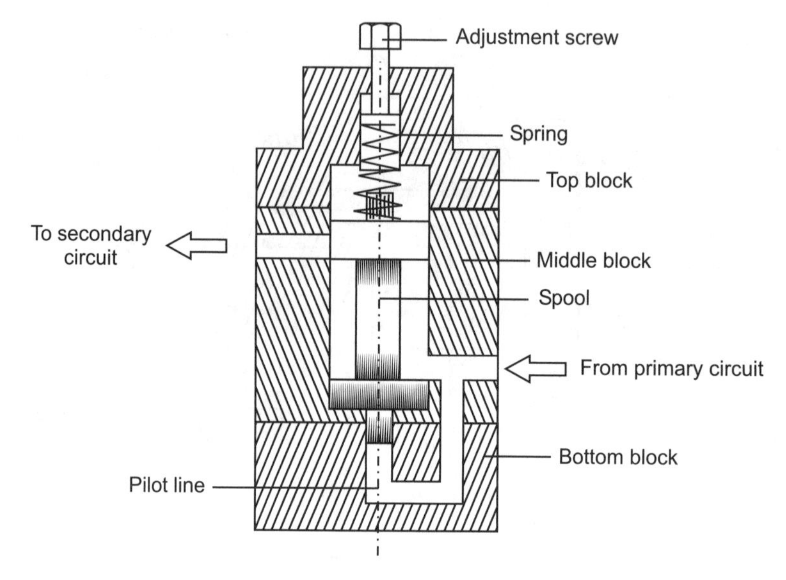

FIGURE 3.13 Sequence valve.

3.23 FLOW CONTROL VALVES

Fixed flow valves, variable flow valves, and pressure compensated flow control valves are the options available for flow control. The discussion on throttle valves (needle valves) is within the scope of this book. The flow control in the circuit is necessitated by the requirement of speed control of the systems using linear or rotary actuators. Figure 3.14 shows a needle valve.

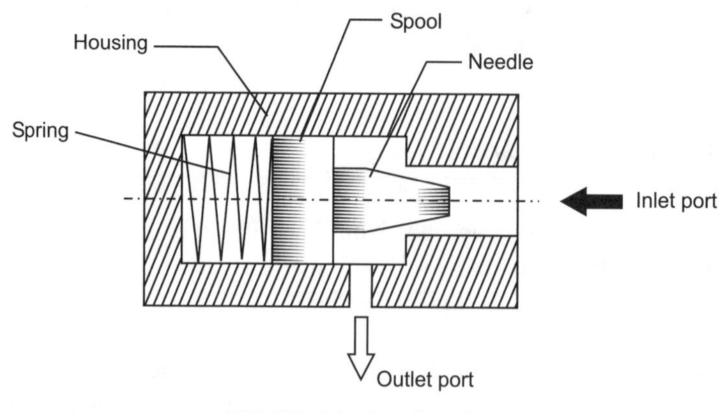

FIGURE 3.14 Needle valve.

3.23.1 Construction

A poppet with a tapering needle moves in the bore of the manifold block. The tapering needle is projected in the inlet passage. The poppet is actuated by the spring fitted at the back. An outlet is taken from the bore chamber in the manifold which serves as the outlet. The valve can be made the variable type by accommodating a screw that can adjust the spring force to set the required flow rate.

3.23.2 Workings

The tapering needle in the inlet passage can move in and out based on the flow rate changing the annular space between the inlet and the needle. The changed flow cross-section varies the flow. The outlet is connected downstream to regulate the speed of actuation.

3.23.3 Applications

- Used in fine metering such as in cushioning of the cylinder at the end of a stroke.
- Used in fine feed control in machine tools.
- Used in meter-in and meter-out circuits.

3.24 DIRECTION CONTROL VALVES

3.24.1 Construction

In direction control valves, the manifold block has ports (two) connected to the pump line and two ports connected to the actuator. One port is connected to the return (tank) line. All the ports are connected internally through a central bore. A hardened ground spool moves inside the bore. One end of the spool is connected to the solenoid and the other end is mounted with a helical spring. The schematic is shown in Figure 3.15.

3.24.2 Workings

In the energized position of the solenoid, port P1 is connected to port A1 which delivers the oil to one end of the actuator. The oil from the outlet of the actuator flows through the valve connecting A2 to the tank port T.

In the de-energized position of the solenoid port P2 is connected to port A2 which in turn supplies oil to the other end of the cylinder for retraction. The returning oil of the actuator goes to the tank through A1 being connected to the tank port T.

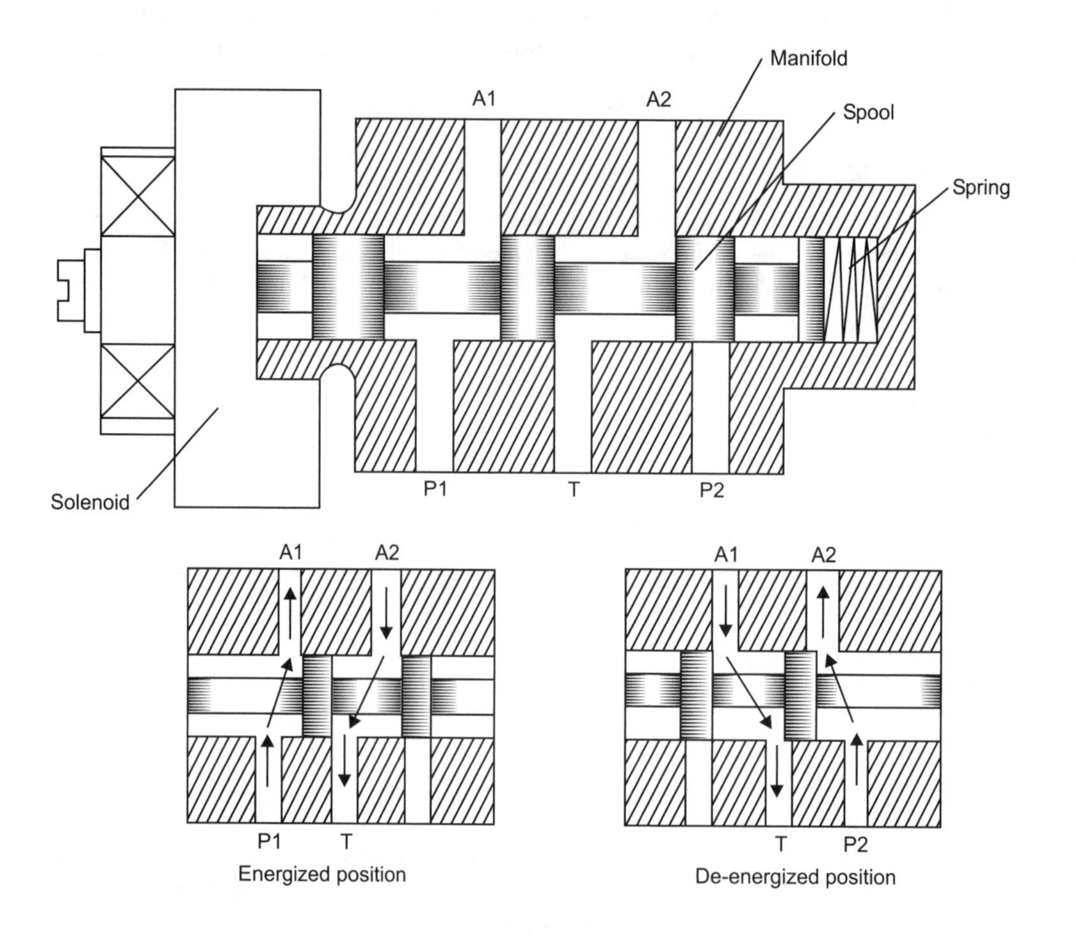

FIGURE 3.15 Direction control valve.

3.24.3 Application Features

- The solenoids are operated either in A.C. or D.C.
- The improper closure of the spool because of spool sticking may result in heavy drawn current. This may lead to the burning of A.C. solenoids.
- A.C. solenoids give better control due to easily derived control voltage.

- Frequency of operations ranges between 1500 to 2000 operations/hr (energization and de-energization cycle).
- Since the stoke and the force that a solenoid operated valve can generate is small, the pilot operation is suggested in heavy duty applications.

3.24.4 Requirements of a Direction Control Valve

1. Minimum interport leakage. This can be accomplished by a perfect fit between the spool and the manifold bore. A quality manufacturing process is of high importance in terms of heat treatment and machining.
2. Low pressure drop. This has due importance to minimize power loss in the circuit.
3. Fast response of operation. The cycle time of the operation of the actuator is reduced because of the fast response of the direction control valves.

3.25 CHECK VALVES

3.25.1 Construction

In a check valve, a ball is fitted in the bore of a housing. A helical spring between the wall of the housing and the ball acts in actuation exerting force on the ball. The hydraulic oil at high pressure acts against the spring force allowing the flow in one direction. The flow in the other direction is checked by the ball closing the inlet. Hence, it is called a check valve.

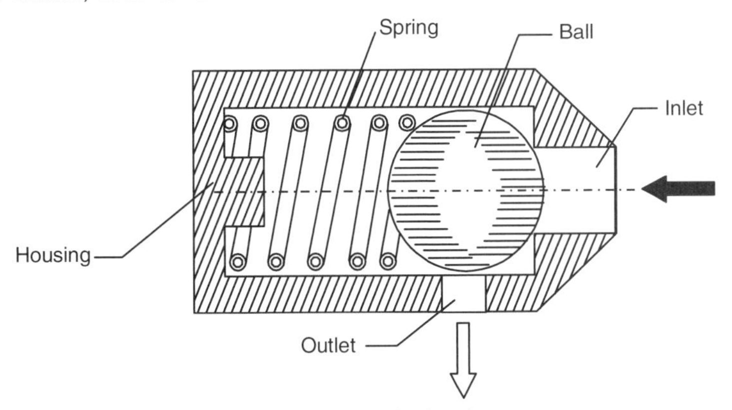

FIGURE 3.16 Check valve.

3.26 SYMBOLS FOR HYDRAULIC SYSTEM COMPONENTS

3.26.1 Pumps

1. Fixed displacement pump.

2. Variable displacement pump.

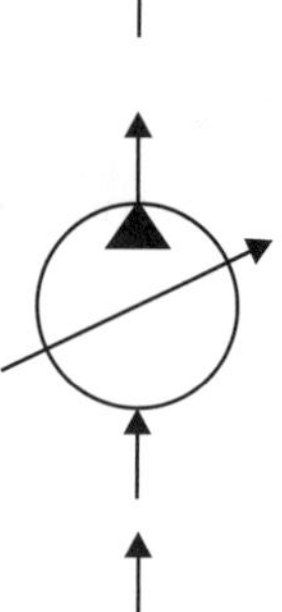

3. Pressure compensated variable displacement pump.

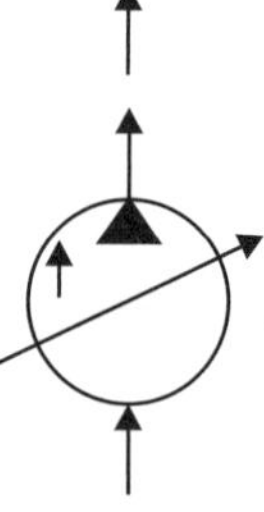

3.26.2 Linear Actuators

1. Single-acting cylinder.

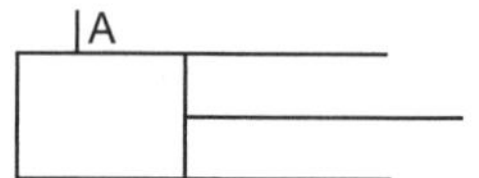

2. Double-acting single rod cylinder.

3. Double-acting double rod cylinder.

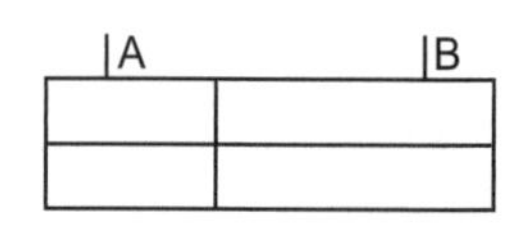

3.26.3 Rotary Actuators

1. Fixed displacement motor.

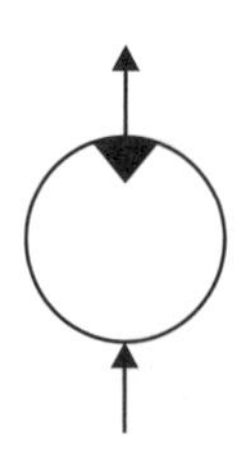

2. Variable displacement motor.

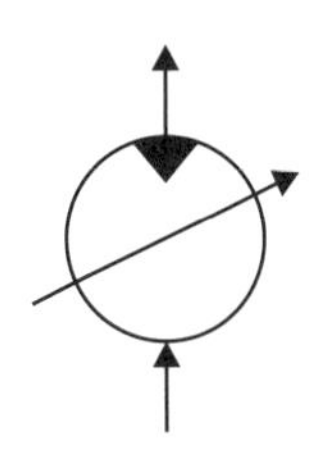

3.26.4 Pressure Control Valves

1. Simple relief valve.

2. Pilot operated relief valve.

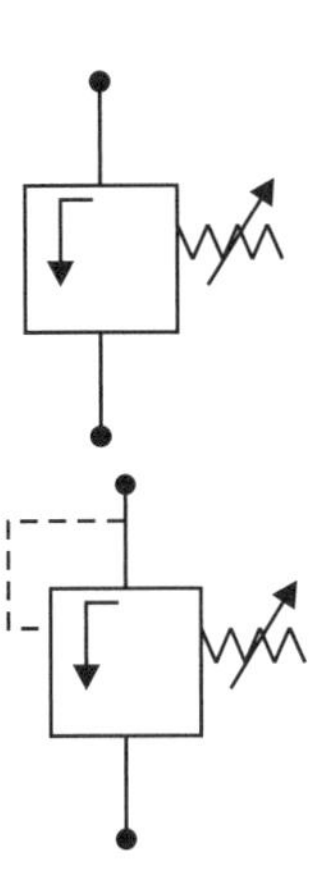

3.26.5 Flow Control Valves

1. Throttle valve (Needle valve).

2. Variable throttle valve.

3. Pressure compensated throttle valve.

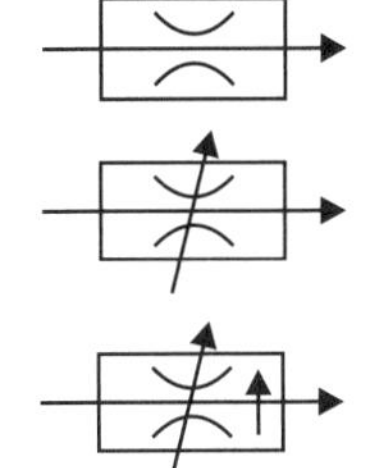

3.26.6 Direction Control Valves

1. Check valve.

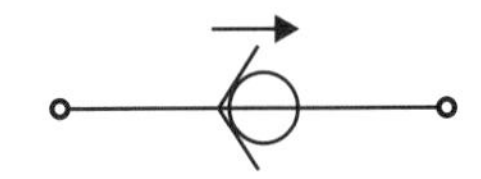

2. Spool valves.

   Two-position valves:

   (*a*) Two-way valve (2 × 2)

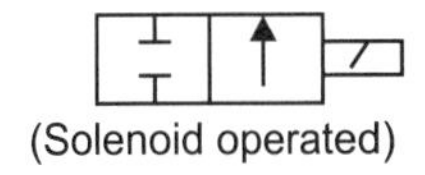

(*b*) Three-way valve (3 × 2)

(Pilot operated)

(*c*) Four-way valve (4 × 2)

(Solenoid operated spring return)

Three-position valves:

(*a*) Closed port closed center

(*b*) Open port closed center

(*c*) Open port open center

(*d*) Closed port center

3.26.7 Miscellaneous Elements

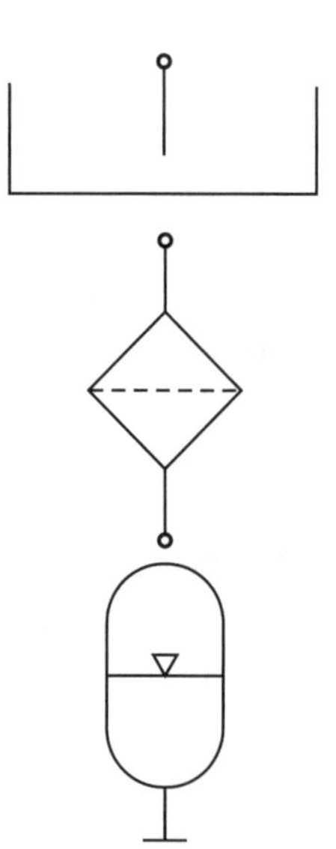

1. Hydraulic reservoir.

2. Filter.

3. Accumulator.

3.26.8 Operations

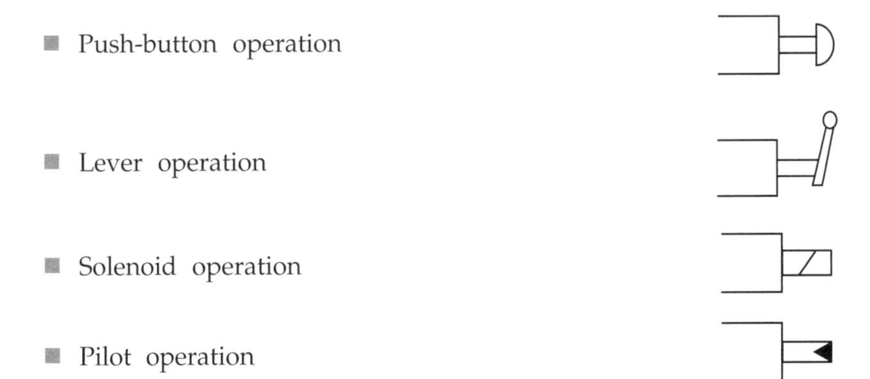

- Push-button operation
- Lever operation
- Solenoid operation
- Pilot operation

3.27 GENERAL HYDRAULIC CIRCUIT

The following diagram shows a general hydraulic circuit.

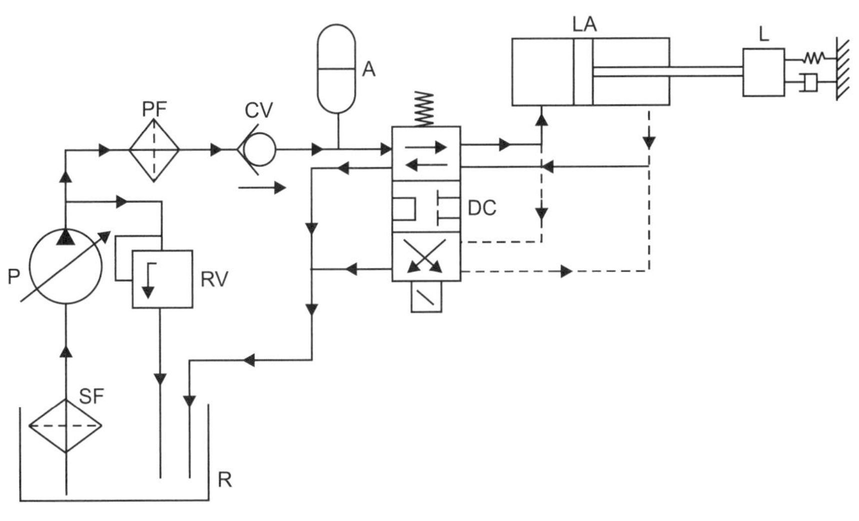

R = Reservoir	CV = Check valve
SF = Suction line filter	A = Accumulator
P = Pump	DC = Direction control valve
RV = Relief valve	LA = Linear actuator
PF = Pressure line filter	L = Load

FIGURE 3.17 General hydraulic circuit.

TABLE 3.1 Components and their Features

Components	Functional Features
1. Reservoir	■ Holds sufficient fluid for all system operations in the hydraulic power pack. ■ Returning hot hydraulic oil is cooled. ■ Prevents turbulence and suppresses the bubbles formed.
2. Filter (suction or pressure line)	■ Removes metallic or non-metallic foreign particles from the fluid. ■ Prevents blocking of system operations. ■ Minimizes system wear and tear. ■ Specified by the allowed pressure and size of particles to be trapped.
3. Hydraulic pump	■ Positive displacement pumps: • Sealed between inlet and outlet. • Prevent back flow of fluid. • Flow rate remains constant irrespective of system construction. • Control flow and pressure with high efficiency. *Example:* piston pumps. ■ Non-positive displacement pumps: • Not sealed, hence, back flow with system blockade. • Losses are more and are less efficient in operation. • Used in low pressure applications. • Flow rate and pressure are not controlled. *Example:* gear pumps.
4. Relief valve Q p_o p_i $\Delta p = (p_o - p_i)$	■ Used to keep system pressure constant, irrespective of changes in flow. ■ Difference in pressure between inlet and outlet is high for small flow rates. ■ For high flow rates Δp is low. ■ Also used for branching, sequencing, and regulating purposes.

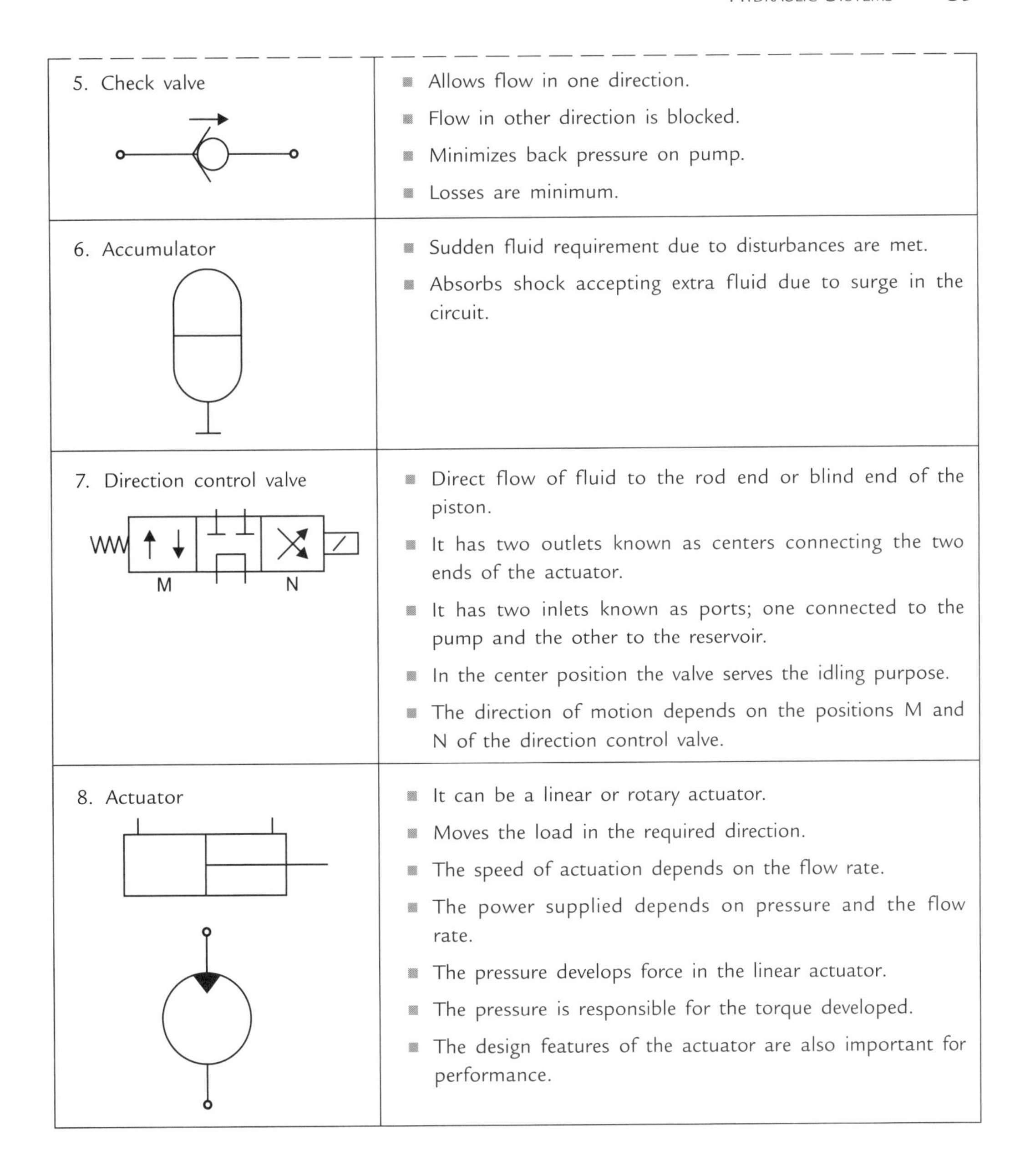

5. Check valve	▪ Allows flow in one direction. ▪ Flow in other direction is blocked. ▪ Minimizes back pressure on pump. ▪ Losses are minimum.
6. Accumulator	▪ Sudden fluid requirement due to disturbances are met. ▪ Absorbs shock accepting extra fluid due to surge in the circuit.
7. Direction control valve	▪ Direct flow of fluid to the rod end or blind end of the piston. ▪ It has two outlets known as centers connecting the two ends of the actuator. ▪ It has two inlets known as ports; one connected to the pump and the other to the reservoir. ▪ In the center position the valve serves the idling purpose. ▪ The direction of motion depends on the positions M and N of the direction control valve.
8. Actuator	▪ It can be a linear or rotary actuator. ▪ Moves the load in the required direction. ▪ The speed of actuation depends on the flow rate. ▪ The power supplied depends on pressure and the flow rate. ▪ The pressure develops force in the linear actuator. ▪ The pressure is responsible for the torque developed. ▪ The design features of the actuator are also important for performance.

3.28 HYDRAULIC CIRCUITS

The hydraulic power pack is produced with the aid of the hydraulic circuit showing the position and connectivity of various components by the respective symbols with connections shown by the line diagram. The design of hydraulic circuits has three functional activities:

- Positioning various elements
- Establishing connectivity between the elements
- Selecting suitable specifications for the elements

The first two activities decided by the task are performed by the hydraulic power pack. The third activity is derived by the performance detail of the function executed by the hydraulic system. The primary functions that are based for positioning and connecting are:

- Pressure control
- Flow control
- Direction control
- Contamination control
- Temperature control

The basis for the selection of specifications for hydraulic components are:

- Force/torque to be developed
- The speed of actuation
- The pressure drop across the element
- The heat generation
- The size of the foreign particles allowed
- The shock and vibration of operation

The suction line filters, pressure line filters, and return line filters provide contamination control. A cooler can be provided for temperature control. The pressure, the flow, and direction control can be studied through the following examples of hydraulic circuits:

Relieving circuit	Pressure control
Reducing/regulating circuit	
Counter-balance circuit	
Sequencing circuit	
Meter-in circuit	Flow control
Meter-out circuit	
Extension and retraction circuit	Direction control
Hydrostatic transmission circuit	

The modeling of hydraulic components, which provides the basis for the selection, is given in Chapter 5. The layouts of the circuits with the functions of the above circuits are given in the following paragraphs.

3.29 RELIEVING CIRCUITS

3.29.1 Functioning of Circuit

The relief valve connected in parallel with the pump is set to the system pressure which is the maximum pressure of the system. If, by some obstruction or overload, the system pressure exceeds the set pressure in the relief valve, the valve cracks and relieves the oil to the hydraulic reservoir. By this, the damage to the pump and other hydraulic elements is prevented. Generally, the relief valve is pilot operated except in certain light duty applications. Refer to Figure 3.18.

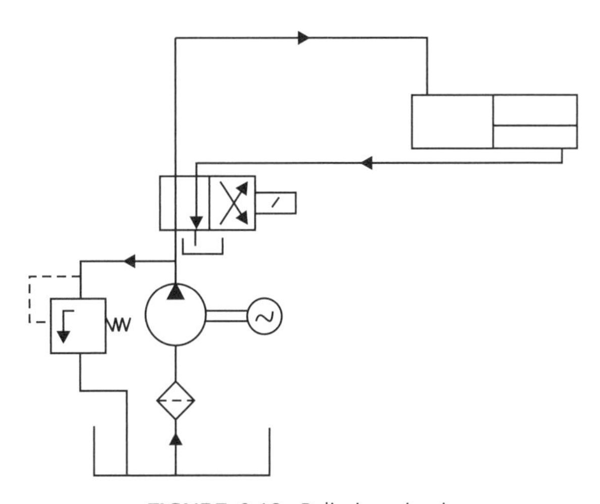

FIGURE 3.18. Relieving circuit.

3.30 REDUCING/REGULATING CIRCUITS

3.30.1 Functioning of Circuit

In this circuit the relief valve is connected in series in between the pump line and the actuator. The flow of oil to the actuator is through the reducing valve and the return of oil is through a bypass line with a check valve as shown in Figure 3.19 as the regulating valve does not allow the reverse flow of oil. The relief valve is set to a pressure less than the system pressure. Hence, the circuit is called a reducing or regulating circuit. Hence, the valve supplies the actuator with the needed reduced pressure.

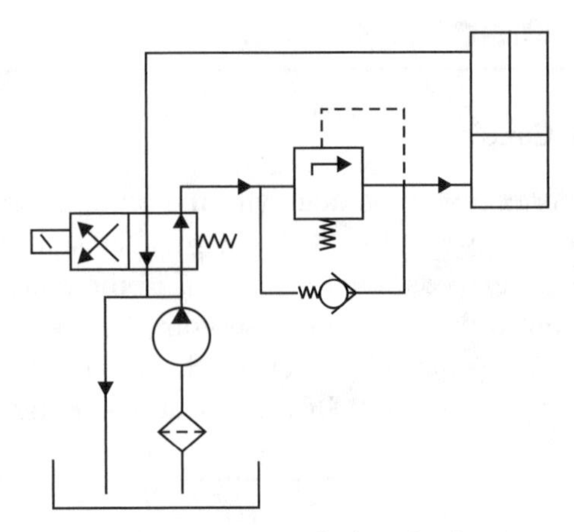

FIGURE 3.19 Reducing circuit.

3.31 COUNTER-BALANCE CIRCUITS

3.31.1 Functioning of Circuit

The relief valve is connected in series in the return line. The cylinder's bottom end is connected to the inlet of the relief valve and the outlet of the relief valve returns the oil to the tank through the direction control value. The flow of the oil to the bottom of the cylinder is through the check valve in the bypass line. The back pressure due to the load does not exceed the set pressure in the relief valve, the flow does not occur, and the cylinder is balanced without drift. On the back, pressure exceeding the valve cracks and allows the flow through to the tank. Hence, it is a called counter-balance circuit. The circuit arrangement is as shown in Figure 3.20.

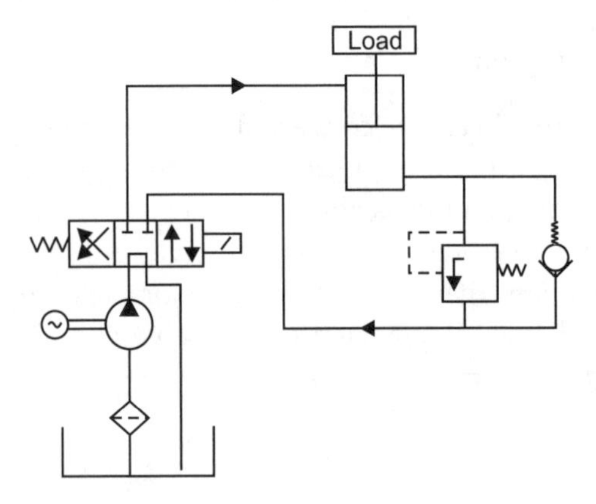

FIGURE 3.20 Counter-balance circuit.

3.32 SEQUENCE CIRCUITS

3.32.1 Functioning of Circuit

The typical application of a relief valve to a sequencing circuit is shown in Figure 3.21. In the circuit a relief valve is connected in series between the pump line and the drill cylinder which allows flow for developing the set pressure. Another relief valve is connected in series between the return line of the drill cylinder and the clamp cylinder. This arrangement ensures the idle movement of the drill bit before clamping of the drill bit, and declamping after the drilling action by the drill bit is over. Check valves are provided in the bypass line of both the relief valves. The circuit arranges the execution of the operations in the following sequence:

- Idle travel of the drill bit
- Clamping of the work piece
- Drilling of the hole
- Declamping of the job

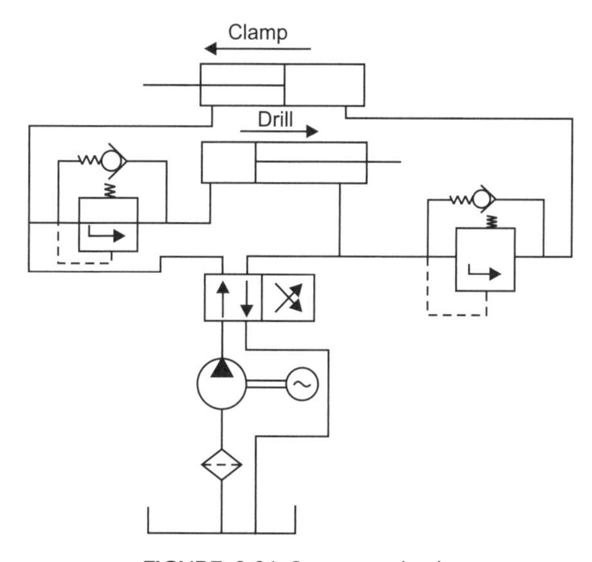

FIGURE 3.21 Sequence circuit.

The circuit is called a sequence circuit. The return flow from both the drill and clamp cylinders is effected by the check valves in parallel with the sequence valves.

3.33 METER-IN CIRCUITS

3.33.1 Functioning of Circuit

The oil from the pump directed by the direction control valve is supplied to the cylinder through the throttle valve that controls the flow rate of the oil flowing into the cylinder. This controls the speed of extension of the cylinder. Hence, it is called the *meter-in control circuit*. The return flow takes place through the check valve connected in parallel with the throttle valve. The check valve prevents the loss of control of the feed system under negative load. The meter-in circuit is shown in Figure 3.22.

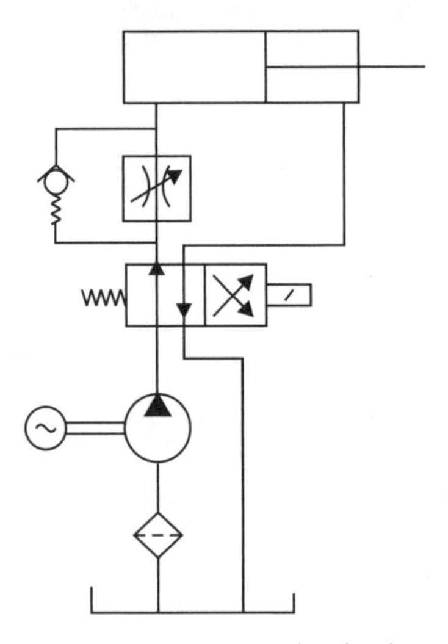

FIGURE 3.22 Meter-in circuit.

3.34 METER-OUT CIRCUITS

3.34.1 Functioning of Circuit

When the flow from the cylinder during the work stroke is metered or controlled the circuit is called a *meter-out control circuit*. The throttle valve is positioned in series between the outlet of the cylinder and the direction control valve connected to the tank. The meter-out circuit provides back pressure to avoid lunging of the load giving the cushioning action. A check valve in parallel with the throttle valve gives a bypass for the flow into the cylinder. The meter-out circuit is shown in Figure 3.23.

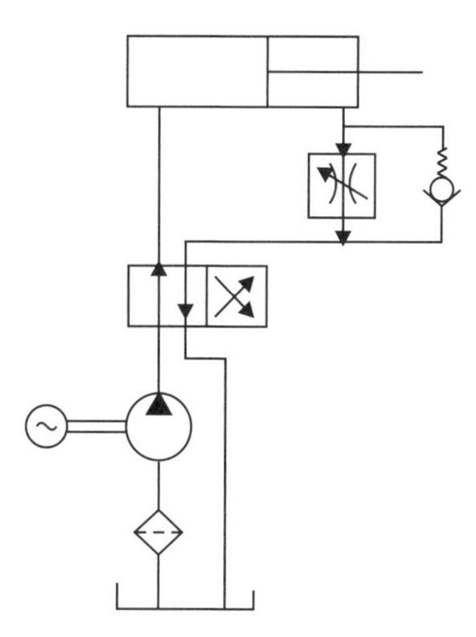

FIGURE 3.23 Meter-out circuit.

3.35 DIRECTION CONTROL CIRCUITS

3.35.1 Functioning of Circuit

A three-position, four-way direction control valve is located between a hydraulic cylinder and the pump delivering the oil at a higher pressure. In position A of the direction control valve the extension stroke of the cylinder takes place and in position B of the direction control valve the retraction of the cylinder takes place. The middle C position of the valve gives idling action sending oil from the pump to the tank. Figure 3.24 shows the application of a direction control valve.

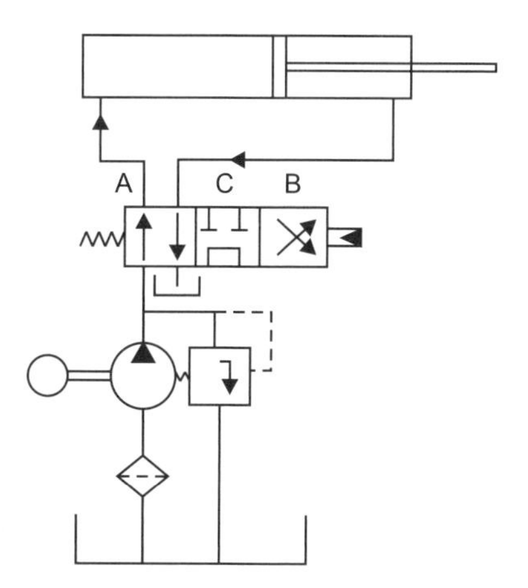

FIGURE 3.24 Direction control circuit.

3.36 HYDROSTATIC TRANSMISSIONS

Figure 3.25 shows the schematic of a hydrostatic transmission. A hydrostatic transmission is basically a pump-motor combination which may be integral or remotely connected using the hydraulic drive and control, and the torque transmission is effected. The pump and motor in this type of circuit can be a fixed or variable displacement type decided by the output characteristics desired for the application. A constant torque system is suitable for a feed drive system. The reversal of the motor is effected by the reversal of pump displacement. The output reversal is also possible by using a direction control valve as shown in Figure 3.26.

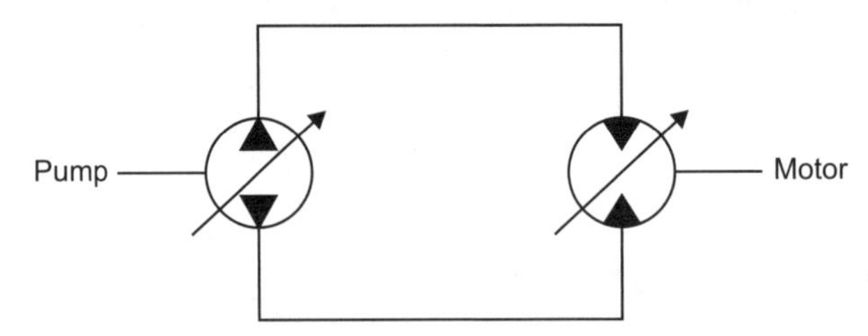

FIGURE 3.25 Hydrostatic transmission.

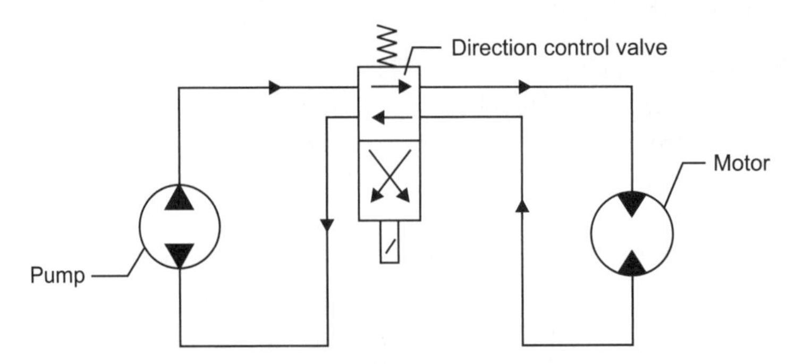

FIGURE 3.26 Hydrostatic transmission with direction control valve.

EXERCISES

1. Sketch and explain the workings of a directional control valve used in a hydraulic system.
2. State and explain the workings of a hydraulic motor.
3. Draw a will-dependent and travel-dependent sequence circuit.
4. What are three types of pressure control valves? Explain the workings of any two of these valves.

5. Explain the workings of a double-acting cylinder.
6. Define rotary actuators.
7. Explain with an example the application of a solenoid operated 2/2 direction control valve with a sketch. (Consider a lift application.)
8. Define an actuator. What is an actuator system? Give an example.
9. Give a detailed classification of the actuators used in a mechatronics system.
10. Classify linear actuators.
11. Explain with a sketch the construction of a hydraulic cylinder (linear actuator) that is double acting.
12. Explain with sketches different configurations of hydraulic cylinders.
13. Give the application of a hydraulic cylinder as a drive system in mechatronics.
14. Give the brief classification of hydraulic rotary actuators (motors).
15. Write the construction and workings of a swash plate hydraulic motor.
16. Explain the application and selection features of a swash plate motor.
17. Explain the workings and construction with a sketch of a bent axis motor.
18. What are the advantageous features of a bent axis motor?
19. List the selection parameters of a bent axis motor.
20. Explain the construction of a radial piston motor.
21. Explain the working features of a radial piston motor.
22. List the features and selection parameters of radial piston motors.
23. Describe with a sketch the constructional features of a vane motor.
24. What are the application features and selection parameters of a vane motor?
25. Distinguish between pistons and vane motors from the point-of-view of application and selection.
26. Justify the importance of hydraulic motors as the drive systems in mechatronics.
27. Detail the construction of gear motors and discuss their application.
28. What are the features of gear motors? List the selection basis.
29. Explain the construction, workings, and application features of annular motors.
30. What are the different types of control actions performed in a hydraulic power pack? How are they accomplished?
31. Briefly classify valves.
32. What are the different types of pressure control valves?
33. Explain the construction and workings of the following valves:

    (*a*) Relief valve (*b*) Regulating valve (*c*) Sequence valve
34. Describe the construction, workings, and application of needle valves.
35. How are solenoid-operated direction control valves constructed?

36. Describe the workings of a direction control valve with the direction of the flow pattern.
37. List the application features of direction control valves.
38. What is the function of a check valve? How is it constructed?
39. Give the graphic symbols for:
    (*a*) Variable displacement pump
    (*b*) Double-acting linear actuator
    (*c*) Variable hydraulic motor
40. Give the graphic symbols for:
    (*a*) Pilot-operated relief valve
    (*b*) Pressure compensated flow control valve
    (*c*) Check valve
41. Give the graphic symbols for:
    (*a*) Solenoid-operated 4 × 2 direction control valve
    (*b*) Pilot-operated 4 × 3 open port, closed center D.C. valve
    (*c*) Suction filter
    (*d*) Accumulator
42. Describe the general hydraulic circuit used to control a linear actuator.
43. What is the basis for selection of hydraulic components in a circuit designed for a power pack?
44. Write and explain the following circuits:
    (*a*) Relieving circuit
    (*b*) Regulating circuit
45. How is counter-balancing done using a relief valve?
46. Give the circuit used to carry out the sequence-clamp, drill, and declamp in a drilling machine.
47. Distinguish between meter-in and meter-out circuits.

CHAPTER 4

ELECTRICAL ACTUATION SYSTEMS

Linear and rotary actuation is also effectively done using an electrical medium. Certain advantages over hydraulic actuators make them useful in mechatronic systems. The switching and control of electrical actuators form part of mechatronic systems. This chapter covers the issues related to electrical actuation through the following topics:

- Mechanical switches—types.
- Bouncing in mechanical switches and their prevention.
- Solid state switches such as diodes, thyristors, triacs, and transistors.
- Linear electrical actuators such as solenoids and relays.
- Rotary electrical actuators:
    - D.C. motors—permanent magnet, and brushless.
    - Synchronous and asynchronous motors.
    - A.C. induction motors—with applications.
    - Stepper motors—applications and characteristics.
    - Servomotors—A.C. and D.C. servomotors with controls.

The comparison of various motors has been tabulated for quick reference on the following pages.

4.1 MECHANICAL SWITCHES

Mechanically opening and closing electrical circuits by breaking or making through one or more contact pairs is done by *mechanical switches*. The opening is transmitted by a '0' signal and closing is transmitted by a '1' signal.

The terminals of the separate electrical circuits that are to be switched are called "poles." The individual contacts for each pole are called "throws." Based on the number of throws and poles in a switching action, switches are classified into the following types:

TABLE 4.1 Mechanical Switches

Type of Switches	Configuration	Contact Detail
1. Single Pole-Single Throw (SPST) contact	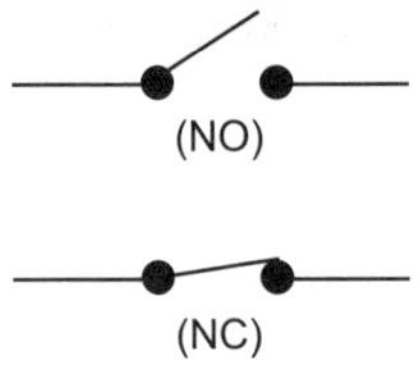 (NO) (NC)	SPST arrangement has one normally open (NO) and one normally closed (NC) contact.
2. Single Pole-Double Throw (SPDT) contact	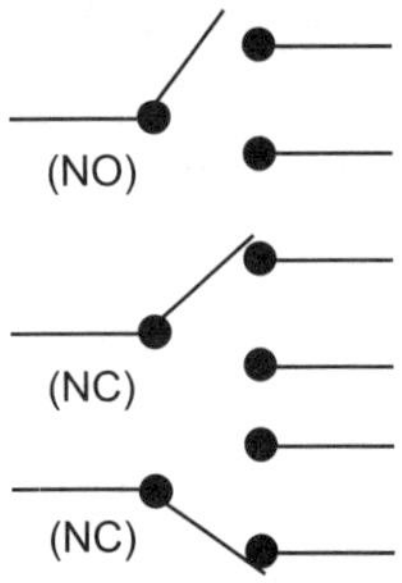 (NO) (NC) (NC)	SPDT contact has one normally open (NO) and two normally closed (NC) arrangements possible.
3. Double Pole-Double Throw (DPDT) contact	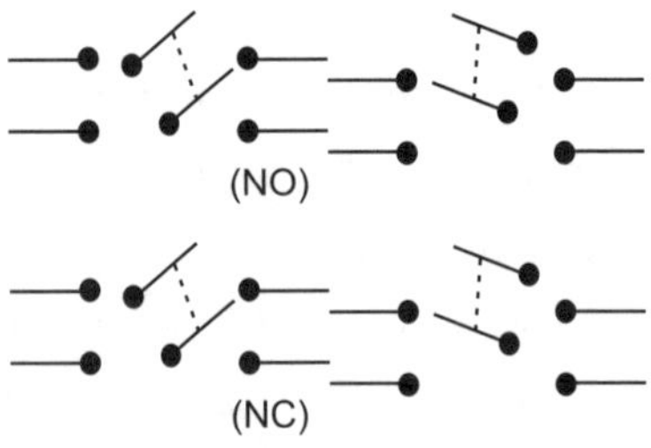 (NO) (NC)	In this arrangement there are two normally open (NO) and two normally closed (NC) possibilities.

4.2 DESIGN VARIETIES OF MECHANICAL SWITCHES

The following table gives some of the design varieties of mechanical switches.

Switch Types	Features
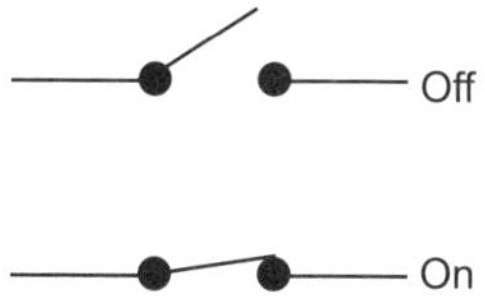 Off On	1. *Toggle Switch.* This comes with a lever actuated by a spring which can be arranged in two or more positions of connections and disconnections. The common light switch used in household wiring is a good example.
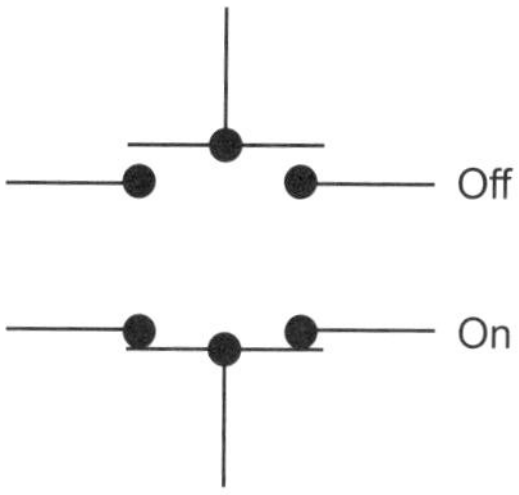 Off On	2. *Push-button Switch.* These are two-position devices actuated by a button that is pressed or released. They have an internal spring mechanism. Some push-button switches will latch alternately on or off with every push of the button.
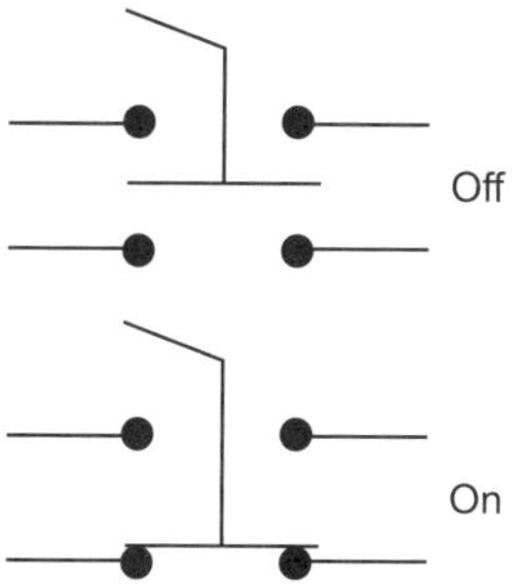 Off On	3. *Selector Switch.* These are actuated with a rotary knob or lever of some sort to select one of two or more positions. They rest in any of their positions or contain a spring return mechanism for momentary action.
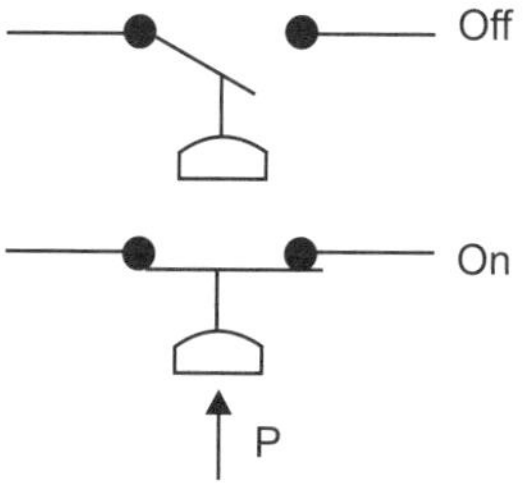 Off On P	4. *Pressure Switch.* Gas or liquid pressure can be used to actuate the switch mechanism by applying the pressure to a piston, diaphragm, or bellows that convert pressure into mechanical force.
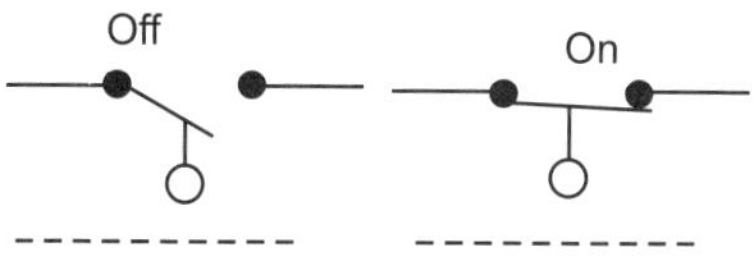 Off On	5. *Liquid-level Switch.* A floating object in the liquid makes contact with the rise in level of the liquid. A low level will break contact.

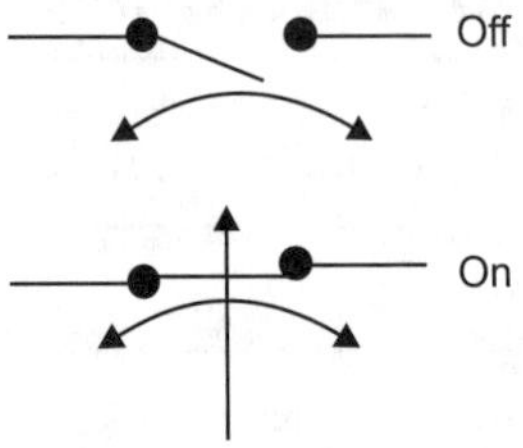 Off On	6. *Speed Switch.* These switches sense the rotary speed of the shaft either by a centrifugal action or a weight mechanism mounted on a shaft. This can be seen in governers used for speed control.
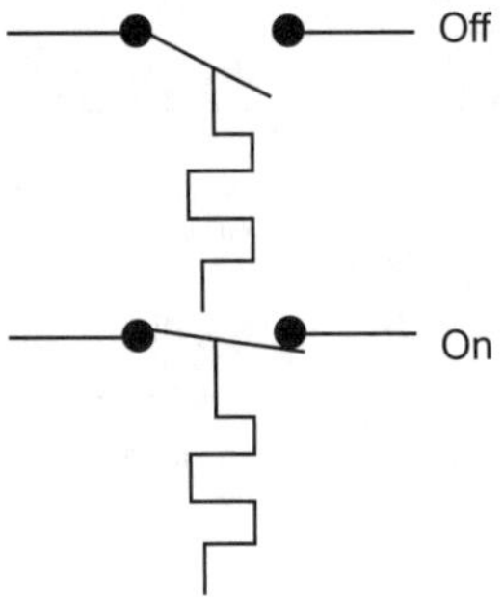 Off On	7. *Temperature Switch.* A bimetallic strip of two metals each having a different coefficient of thermal expansion, joined back to back, act as a switch due to temperature variations. The bending of the strip is used to actuate the switch contact mechanism.
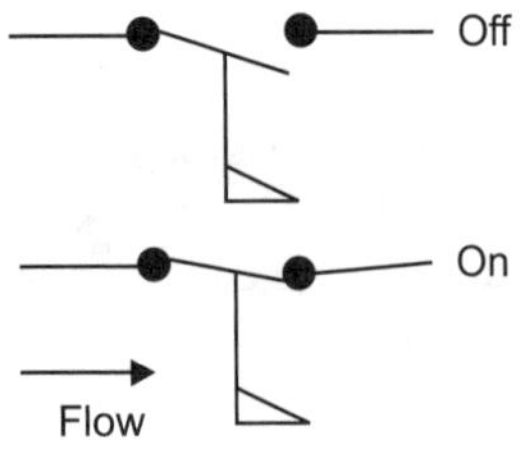 Off On Flow	8. *Liquid-flow Switch.* Inserted into a pipe, a flow switch detects any gas or liquid flow rate in excess of a certain threshold, usually with a small paddle or vane which is pushed by the flow, actuating a lever mechanism.
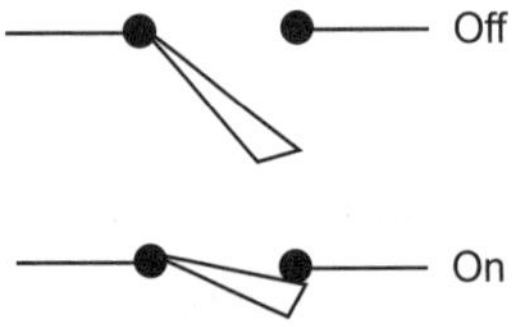 Off On	9. *Proximity Switch.* These sense the approach of a metallic machine part either by a magnetic or high-frequency electromagnetic field. Simple proximity switches use a permanent magnet to detect the presence of ferrous metal by the action of the Hall effect.
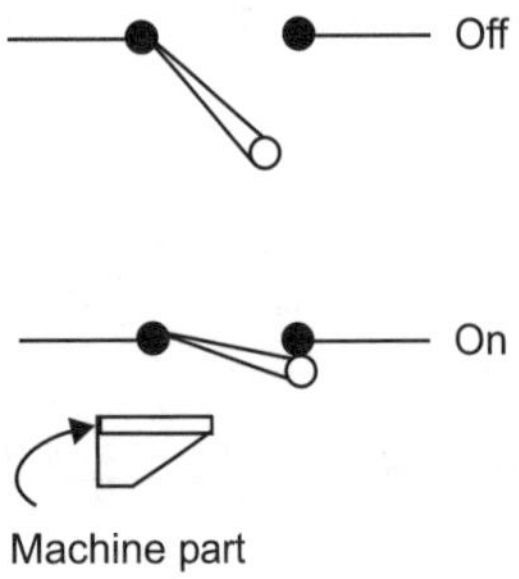 Off On Machine part	10. *Limit Switch.* These resemble the toggle or selector switch where the lever is actuated by contact with the moving machine member. The diagramatic representation can be seen in Figure 4.1.

4.3 LIMIT SWITCHES

Before the external force is applied the actuator will be in a free position. On contact with the target the actuator lever will travel a certain distance without actuation which is known as *pre-travel*. In the operating position the limit switch changes from a normally open (NO) state to a normally closed (NC) state. The actuator can travel a certain distance in the NC position beyond the operating point, safely, which is known as *over travel*. The release position is achieved between the operation state and the normal state during which the actuator makes the *differential travel*. Between the release position and the free position the distance traveled is the *release travel*. The limit switches are used in machine tools to make and break contacts as the machine comes in contact with the actuator of the switch. The limit switches are available in different shapes and designs which are selected to suit the application. The arrangement of a limit switch is shown in Figure 4.1.

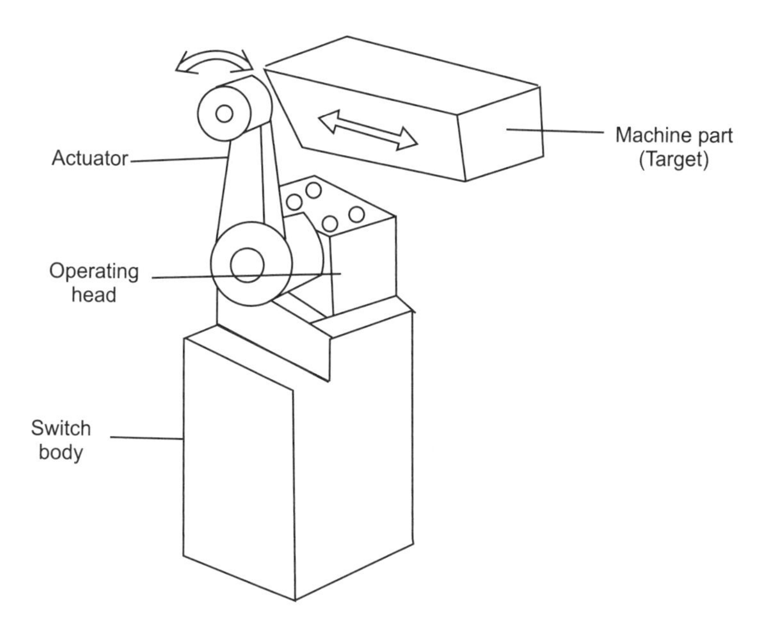

FIGURE 4.1 Limit switch.

4.4 CONTACT BOUNCE

The levers and the mechanism of actuation in a mechanical switch should establish contact under the force of actuation and provide continuity in a single, crisp movement. But unfortunately switches often fail to achieve this goal due to the static and dynamic effects of the contacting material and mechanisms. The flexibility of the mechanical elements tend to deflect the contacting element under contact force. The attempt to increase the stiffness by increasing the dimensions leads to an increase in mass giving an inertial effect. The switches bounce on contact to establish closure which persists for about 20 milliseconds before coming to full rest and providing unbroken contact. In many applications such as switching of incandescent lamps, the switch bounce is of no consequence. However, the bounce in switches used to give signal to an electronic amplifier or electronic circuit (having fast response time) may produce highly noticeable and undesirable effects.

A closer look at the oscilloscope display in Figure 4.2 reveals a random set of makes and breaks when the switch is actuated once.

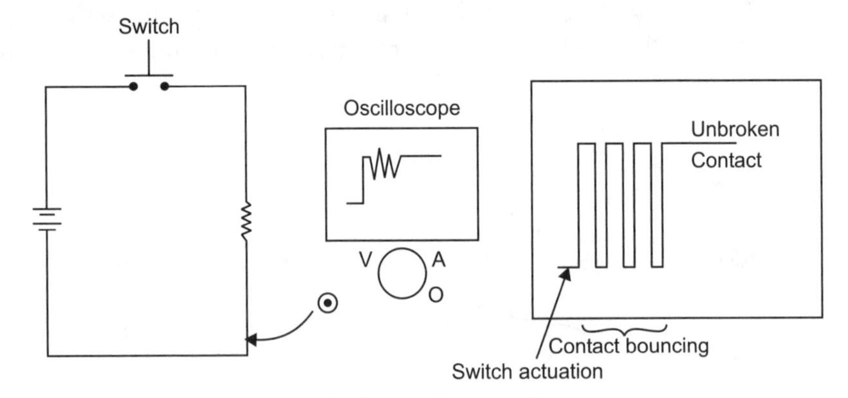

FIGURE 4.2 Contact-bounce phenomena.

Example. If a contact-bouncing switch is used to provide a clock signal to a digital counter circuit, so that each actuation of the switch is supposed to increment the counter by a value of 1, the switch will give signal to count several increments for one actuation. This is undesirable as far as the application requirement is concerned.

The debounce needed as shown in Figure 4.3 can be accommodated in the hardware or software design to be discussed later.

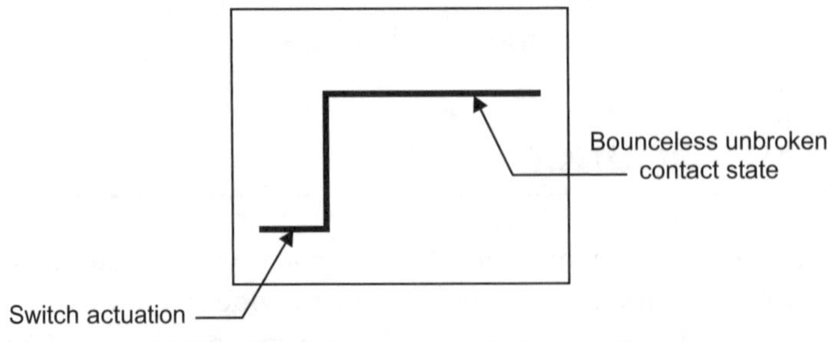

FIGURE 4.3 Bounceless switch operation.

4.5 TIPS FOR MINIMIZING CONTACT BOUNCE

- Design the switch for sliding contact rather than direct impact. Knife switch designs have sliding contact.
- Dampen the switch mechanism by an air or oil shock absorber.
- Use sets of contacts in parallel, different in mass and contact gap, so that one of the sets is in contact.
- Use a buffer spring to absorb the force of impact.
- Wet the contact surface with liquid mercury in the sealed environment. The surface tension of the mercury maintains continuity of contact and prevents contact bounce.
- Reduce the force of impact and the kinetic energy of the moving contact to minimize the bouncing.

4.6 METHODS TO PREVENT BOUNCING (HARDWARE SOLUTION)

The following table shows several methods for preventing contact bounce.

Method	Use of SR-Flip-Flop	Use of D-Flip-Flop	Use of Schmitt Trigger
Arrangement of circuit	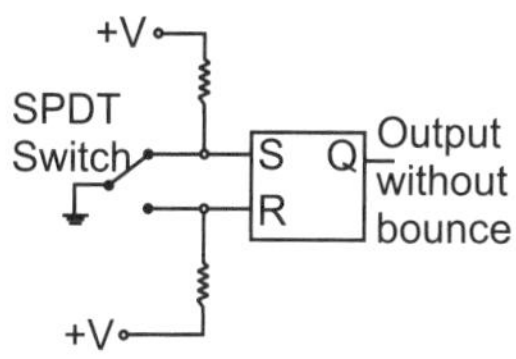 +V, SPDT Switch, S, Q, R, Output without bounce, +V	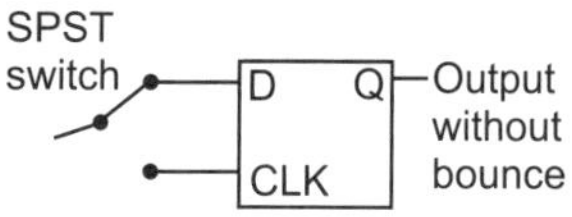 SPST switch, D, Q, CLK, Output without bounce	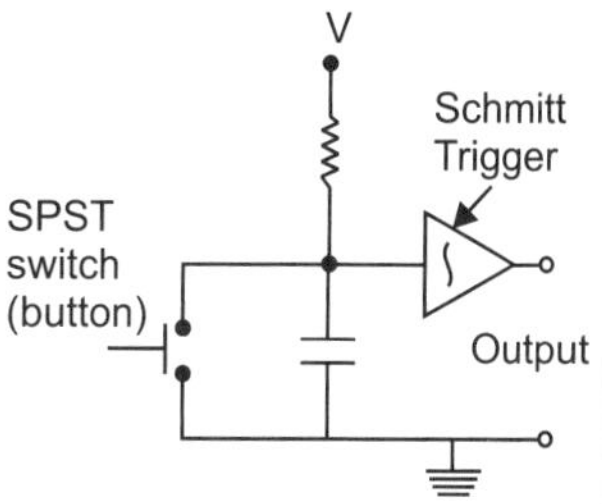 V, Schmitt Trigger, SPST switch (button), Output
Working procedure	In the shown position, S is at '0' and R is at '1' so the output is '0'. When the position is shifted to a lower pole, the S becomes 1 and R to 0 giving an output 1. Bouncing the signal does not change S from 1 to 0 or 0 to 1, keeping the output unaltered. By using two NOR or two NAND gates the SR flip-flop can be devised.	The output from the D-flip-flop is changed by the clock signal which has a clock period greater than the time for which the bounce of the switch persists (say 20 min.). By this attempt the bounce signal is ignored and the output without bounce is received.	In the NO position of the switch the capacitor gets charged and the input voltage to the trigger is high which gives low output. On the bounce of the switch, the NC position discharges the capacitor and no input is given to the trigger as there is no charge during the bounce and the bounce of the switch gets ignored.

4.7 SOLID-STATE SWITCHES

To electronically switch the circuits, the following principles of solid-state power switching devices are used:

1. Diodes
2. Thyristors
3. Triacs
4. Transistors (Bipolar)
5. Power MOSFETs

4.7.1 Diodes

Diodes can be regarded as unidirectional elements that allow the flow of current in one forward direction only. Diodes are characterized by the curve given in Figure 4.4. The diode has an anode and cathode. When the anode is positively charged with respect to the cathode, *i.e.*, forward biased, it allows the flow of current, but breaks down when the bias is reversed.

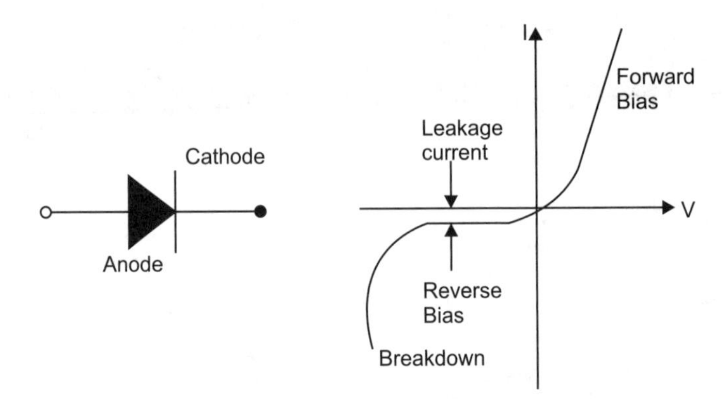

FIGURE 4.4 Diode characteristics.

If an A.C. voltage is applied to the diode, it is switched on only in the forward bias part of the input. It turns off in the reverse bias component of the A.C. signal. Hence, the current through the diode is half rectified which is shown in Figure 4.5.

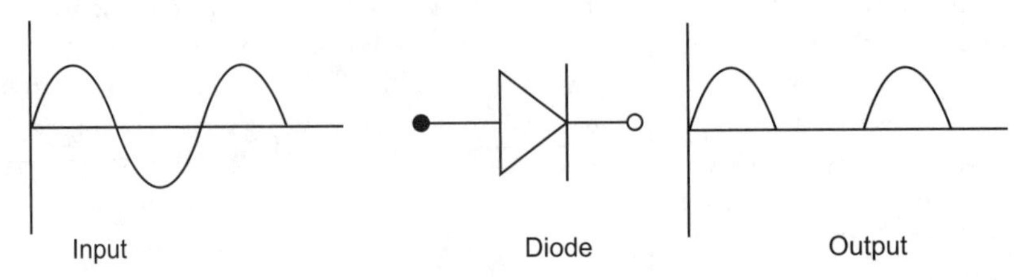

FIGURE 4.5 Diode operation.

4.7.2 Thyristors

The other name for a thyristor is Silicon Controlled Rectifier (SCR), and it can be considered a diode with a gate. With the help of the gate the thyristor can be switched in the reverse bias also. When the gate current is zero the thyristor acts as a diode that is unidirectional. Figure 4.6 shows the characteristics of a thyristor in general.

Even when forward biased the current is negligible with the forward leakage current. At a high forward breakdown current the thyristor switches on, allowing the flow of current, and the voltage drops down to a low value. The higher the gate current the lower the breakdown voltage. Hence, the switching of the thyristor depends on the gate current. Figure 4.7 shows the switching characteristics of a thyristor.

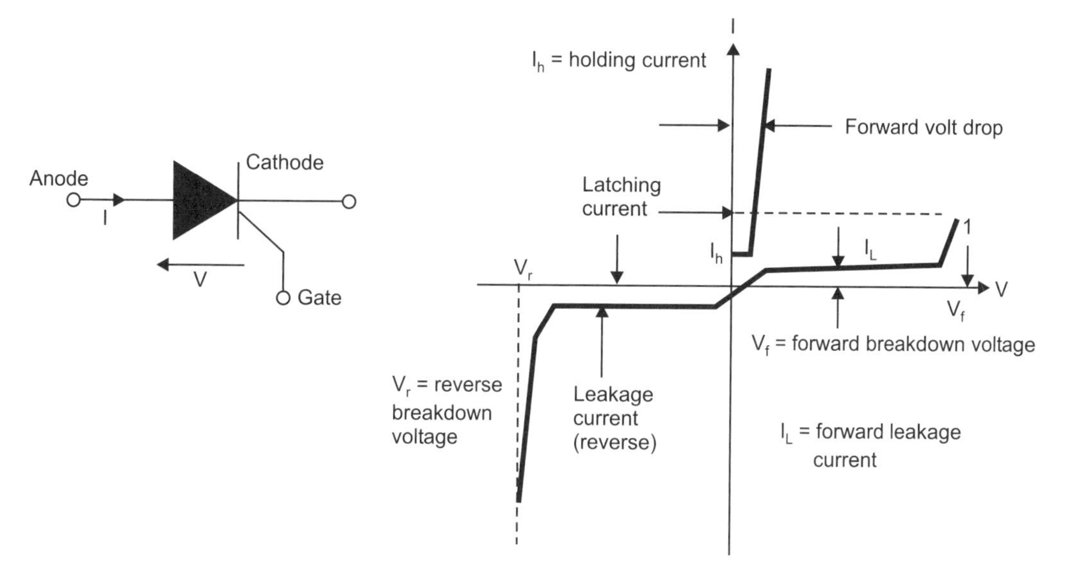

FIGURE 4.6 Thyristor characteristics.

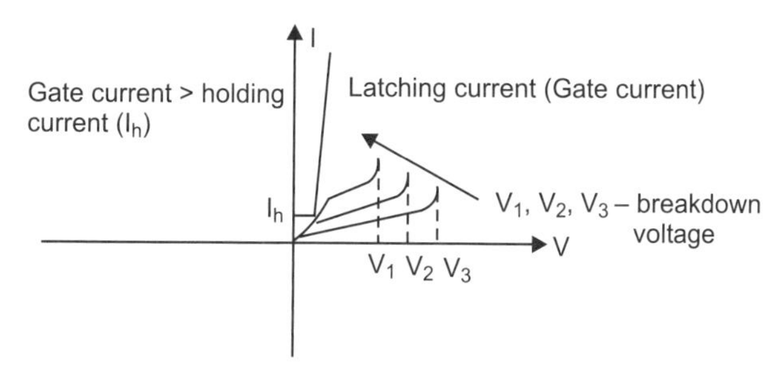

FIGURE 4.7 Switching characteristics of a thyristor.

TABLE 4.2. Some Thyristor Switching Arrangements

Circuit Diagram	Input and Output
(1) High side arrangement 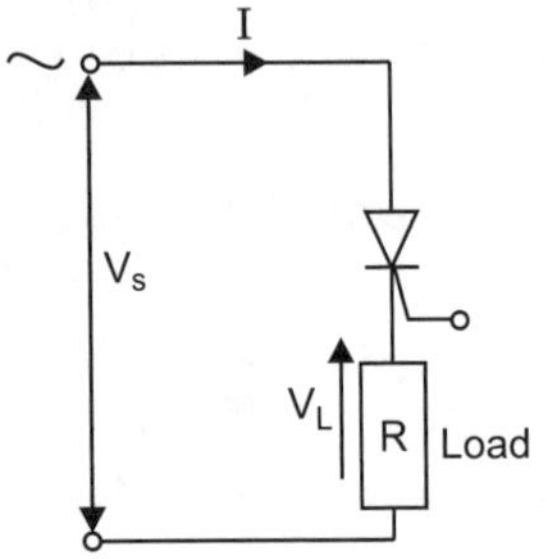	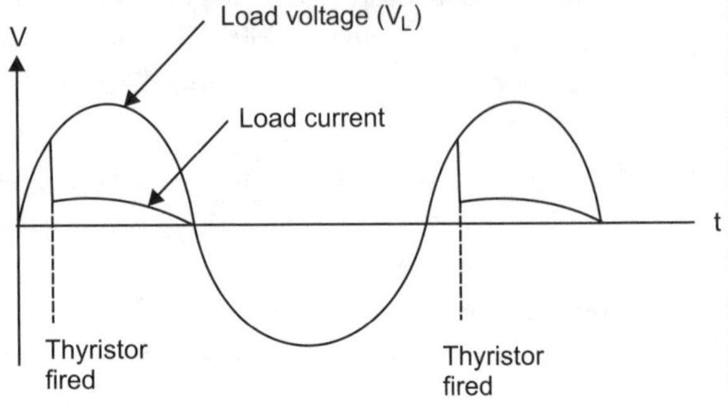 Load is connected to supply when thyristor is conducting which is preferred.
(2) Low side arrangement 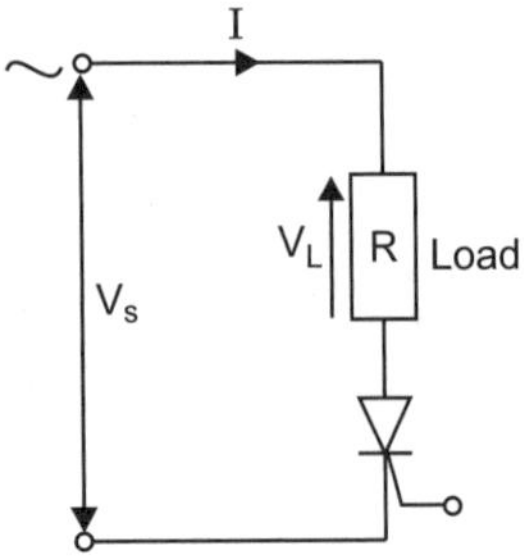	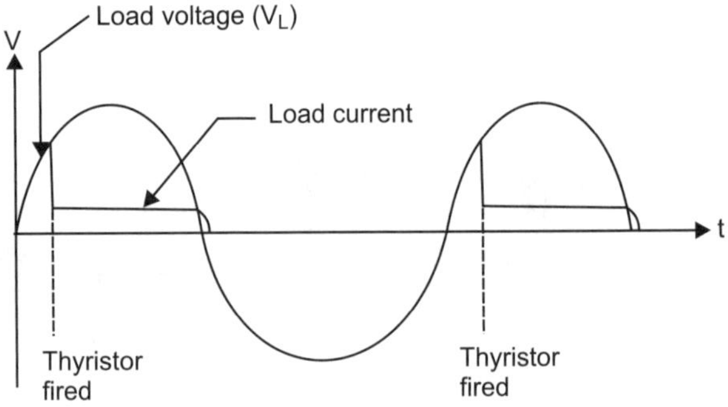 Load is connected to supply even when thyristor is not conducting.
(3) Inductive load 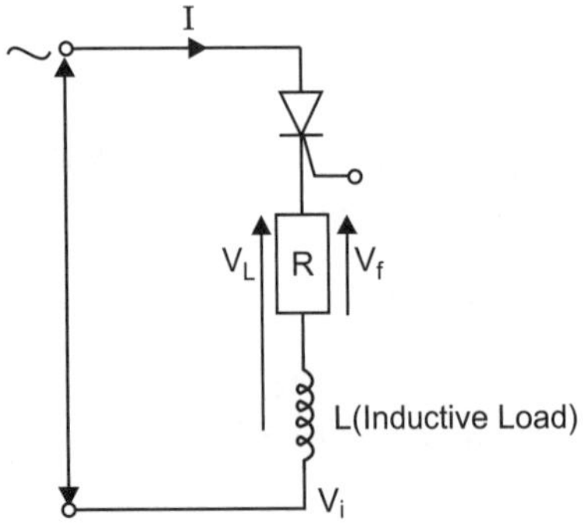 $V_f = (V_L - V_i)$	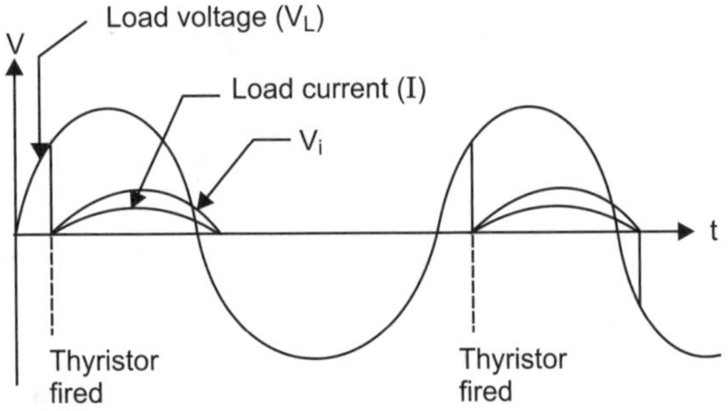 Used to control inductive loads such as motors. With the resistive load the load current follows the load voltage but with the inductive load the load current is sustained, hence, the thyristor current is beyond zero voltage.

4.7.3 Triacs

A triac is electrically equivalent to a pair of thyristors connected in reverse but parallel on the same chip. A triac is capable of turning on in both forward and reverse bias conditions. It can conduct in both directions under the appropriate gate current.

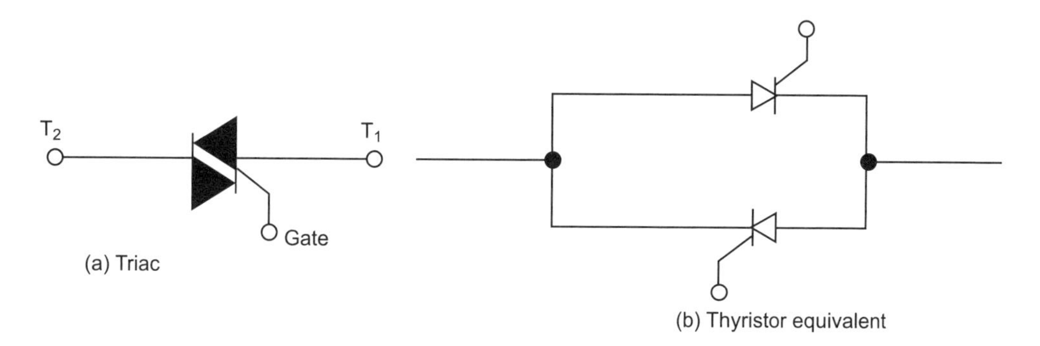

FIGURE 4.8 Thyristor equivalent of a triac.

When the terminal T_2 is positive the triac conducts and turns on with the application of forward current. When T_1 is positive with respect to terminal T_2, the triac allows negative current (in the reverse direction). Figure 4.8 shows the triac. Figure 4.9 shows the characteristic of the triac and the output of the triac. It provides an inexpensive method for control of A.C. power.

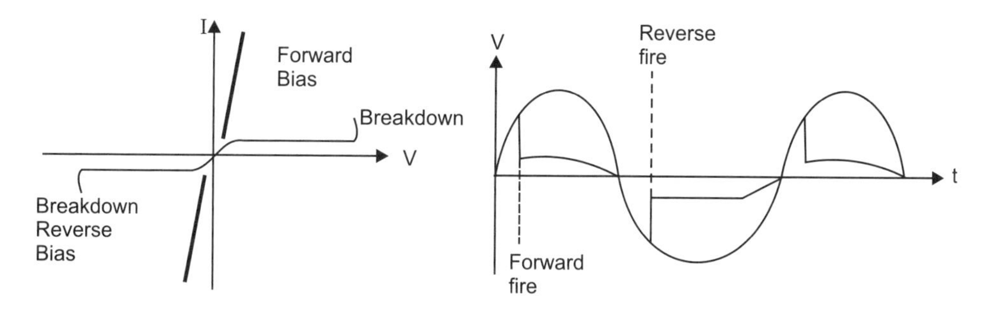

FIGURE 4.9 Triac characteristics.

4.7.4 Bipolar Transistors

In this section we discuss the characteristics of bipolar transistors.

TABLE 4.3 Bipolar Transistors

Transistor Type	Symbol	Feature
npn-transistor	Collector, I_B, Base, Emitter	In this type the current flows in from the collector and goes out at the emitter. At the base the current supplied acts as a controlling signal.
pnp-transistor	Emitter, I_B, Base, Collector	In a *pnp*-transistor the current flows in from the emitter and flows out at the collector. The controlling signal supplied at the base is opposite in direction.

Figure 4.10(*a*) shows the use of a *npn*-transistor in a common-emitter circuit. Figure 4.10(*b*) shows the characteristics depicting the relation between the collector current I_C and the potential difference between the emitter and collector V_{CE} for an increasing value of base current I_B. Beyond saturation the collector current does not increase with an increase in the base current and the base-collector junction becomes forward biased and acts as a switch.

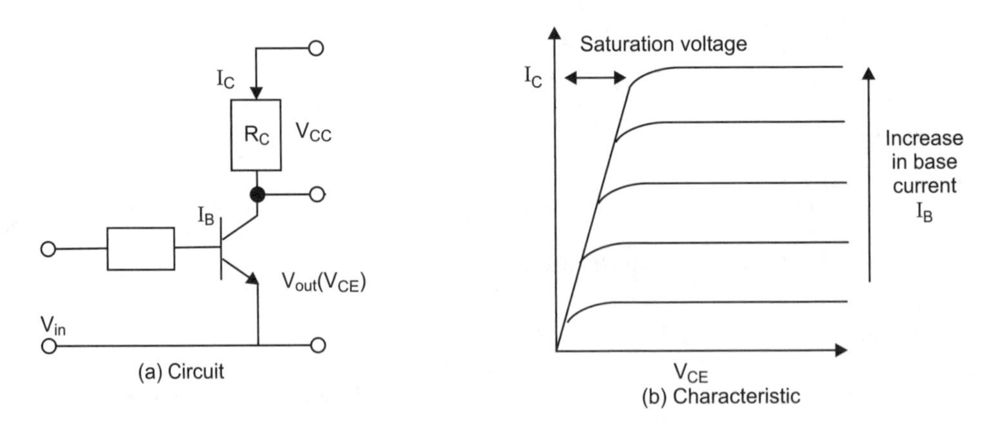

FIGURE 4.10 Common-emitter circuit.

4.7.5 Darlington Pairs

The process of the switching of high current with a small input current is facilitated by the combination of a pair of transistors known as a *Darlington Pair*. The pair of transistors are devised on a single chip.

TABLE 4.4 Darlington Pairs

npn-Darlington Pair	pnp-Darlington Pair
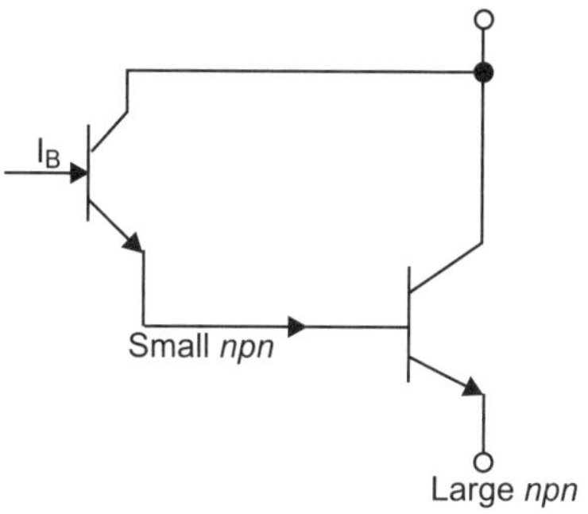 I_B Small *npn* Large *npn*	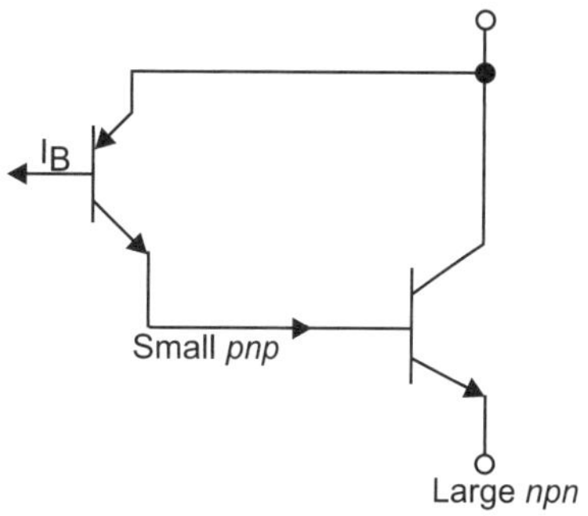 I_B Small *pnp* Large *npn*
Small *npn*-transistor in combination with a large *npn*-transistor forms a large Darlington *npn* pair. The amplification factor is large.	Small *pnp*-transistor in combination with a large *npn*-transistor forms a large *pnp*-Darlington pair.

Applications

- To control inductive loads.
- To switch off at large transient voltages.
- To switch the actuator with the microprocessor.
- Used to invert the output.
- For on/off switching of D.C. motors.
- High-frequency switching which is not possible by thyristors.
- Has less power handling capacity.

4.8 SOLENOIDS

4.8.1 Construction

A solenoid has a fixed ferrous body with a coil for excitation. Centrally positioned is the plunger which is a movable element. The body and the plunger (armature) are made of ferro-magnetic material for flux carrying. The tube carries a conical stopper plate at one end, which is also magnetic.

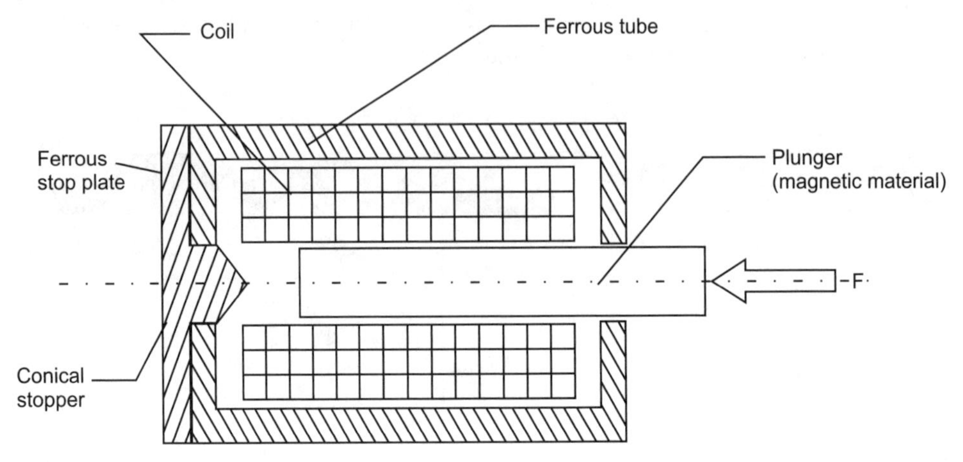

FIGURE 4.11 Construction of a solenoid.

4.8.2 Principle of Workings

Upon switching on the current, the body and the conical stopper get magnetized because the plunger is attracted to the stopper. Upon switching off the current to the coil the plunger returns back to the free position.

Solenoids are basically short stroke (up to 25 mm) unidirectional linear electrical actuators in which the action is always to pull the plunger into the coil irrespective of the polarity of the current. The force of actuation supplied by the simple solenoid of the type shown in Figure 4.11 is given by

$$F = \frac{1}{2}\left(\frac{N^2I^2}{x^2}\right)A\mu_0, \tag{4.1}$$

where

N = number of turns on the coil
I = the current through the coil
x = length of air gap
A = area of cross-section of air gap
μ_0 = permeability of air

4.8.3 Characteristics of Solenoids

The force exerted by the plunger is maximum in the beginning and varies non-linearly as the stroke progresses. The plunger force decreases with the stroke, as shown in Figure 4.12.

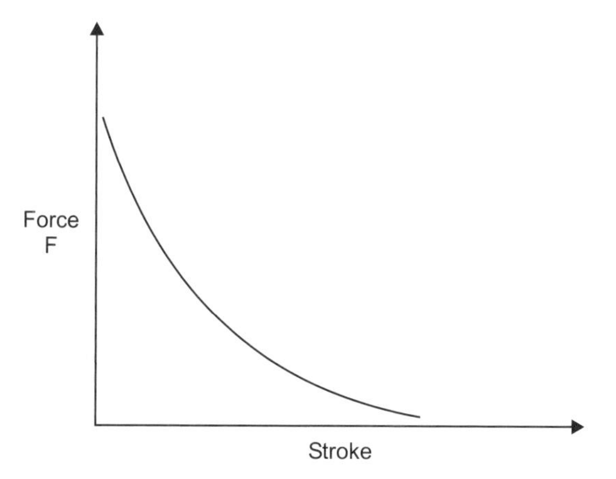

FIGURE 4.12 Force stroke current.

4.9 PULSE-LATCHING SOLENOIDS

4.9.1 Construction

In Figure 4.13, the moving element is a permanent magnet with actuator rods at two ends. When there is no current in the coils the plunger is latched to the left end and supplying current to coil B the magnet plunger is repelled to the right extreme position. Upon energizing coil A the current is pulsed in the opposite direction and the magnetic plunger is repelled to the left position.

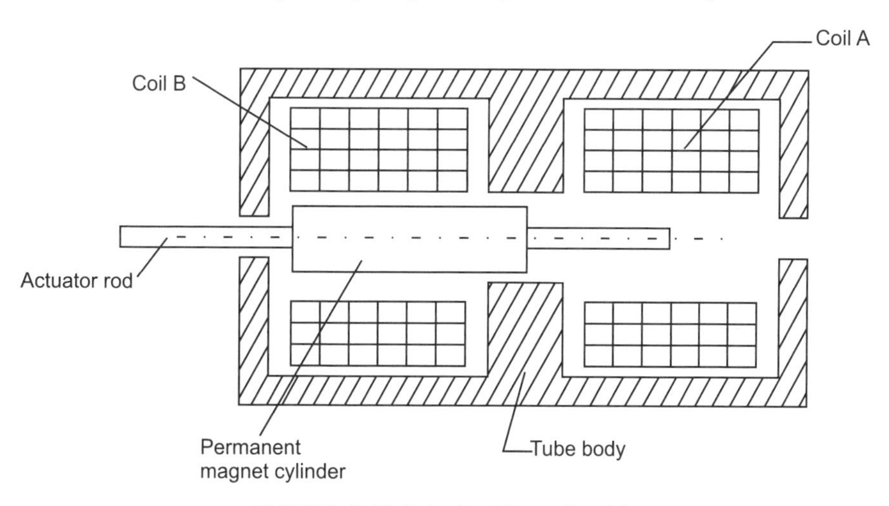

FIGURE 4.13 Pulse-latching solenoid.

4.9.2 Applications

- Flap valves of electrohydraulic systems.
- Pneumatic and hydraulic valves.
- Slide short stroke valves.
- Interlocks and brakes.
- Remote operation on/off valves.
- Central door-locking systems for cars.
- Electrohydraulic servovalves used in robots.

The non-linear force current characteristics have no particular importance in on/off applications. But linear relation is needed in a proportional solenoid, which is used in electrohydraulic servovalves. This is obtained by proper shape design of the armature, pole piece, and the core tube used in solenoids to produce constant force throughout the stroke length. With the use of a potentiometer and feedback loop it is possible to detect and control the position of the spool actuated by the solenoid. The stop plate is made conical to increase the flux density and reduce the current to avoid excessive heating effects. The non-magnetic spacer used in some designs prevents undesirable effects due to residual flux in the poles even after switch off of current that may result in a sticking effect of the plunger. The return of the plunger on de-energization may be supplemented by using a return spring.

4.10 RELAYS

Relays are basically switches that have electrical operation. By a change of the current in one leg of the circuit the other leg of the circuit is switched on or off the current. The schematic of a relay is in Figure 4.14.

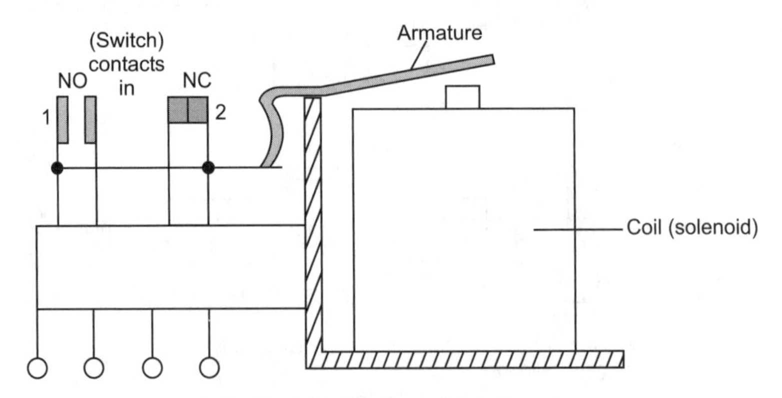

FIGURE 4.14 Configuration of a relay.

Relays in construction consist of a solenoid with a coil and armature. The armature is connected to switches with contact pairs. When there is no current in the coil of the solenoid the armature is not in touch with the plunger. The switch 1 is in the normally open (NO) state and switch 2 is in the normally closed (NC) state. The flow of current through the coil magnetizes the coil and the armature is pulled toward the plunger. As a result switch 1 attains the NC state and switch 2 attains the NO state.

4.10.1 Applications

- Operate controllers in a control system.
- Used in conjuction with transistors in temperature control systems.
- Control the operation of pneumatic and hydraulic valves.
- For delayed switching action time-delay relays are used.

4.11 ELECTRIC MOTORS

Rotational actuation is accomplished by electric motors. Based on the type of electrical energy domain (direct current or alternating current) they are classified as D.C. and A.C. motors. The principle behind the actuation (rotation) by an electric motor is depicted in Figure 4.15.

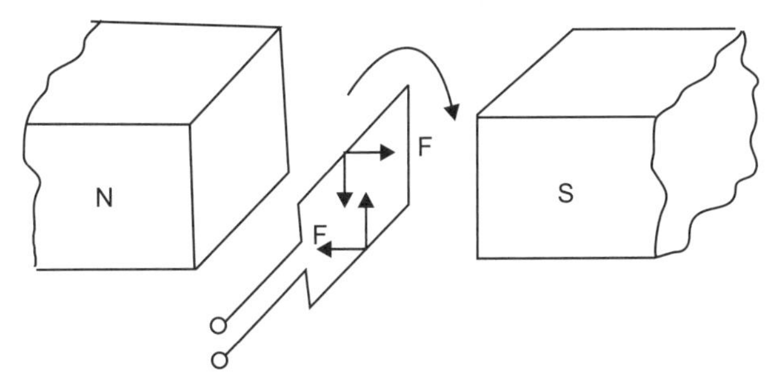

FIGURE 4.15 Principle of a motor.

A current carrying conductor kept in a magnetic field experiences a force which depends on the flux density of the magnetic field, the current flowing in the conductor, and the length of the conductor. The result of this force and the back emf produced is responsible for a torque-producing rotation.

Furthermore, based on the commutation (external or internal) and control methodology used, motors are classified in detail as given in Figure 4.15. Based on the synchronization of the speed of the rotor with the frequency of the current flowing motors are classified as asynchronous or synchronous motors. The position and speed control has led to the development of stepper motors and servomotors.

4.12 CLASSIFICATION OF ELECTRIC MOTORS

The following diagram shows the classification of electric motors.

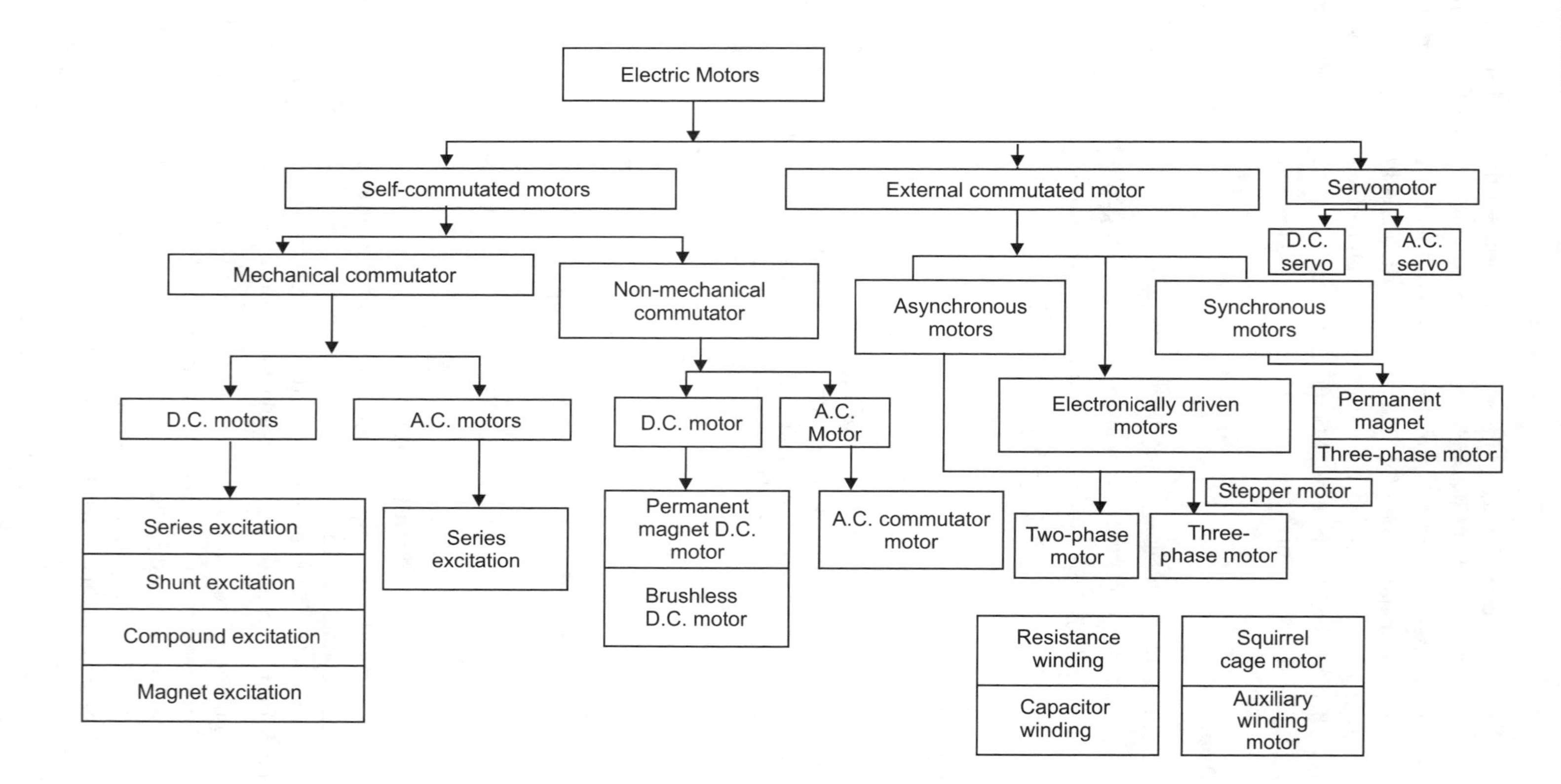

4.13 FOUR-POLE D.C. MOTORS

4.13.1 Construction

Figure 4.16 shows the construction of a four-pole D.C. motor. The coils of the main winding are wound first and slid on the yoke. The yoke is then moved into the stator packet. The rotor has an armature with squirrel cage windings.

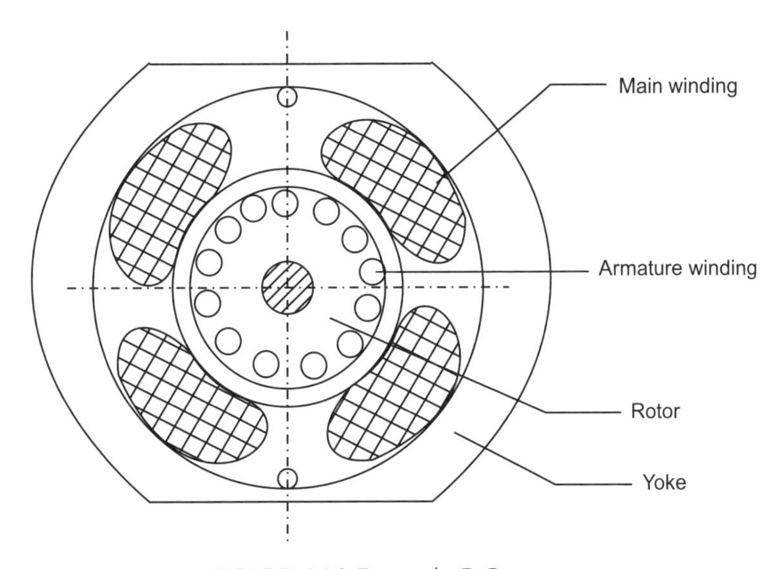

FIGURE 4.16 Four-pole D.C. motor.

4.13.2 Principle of Workings

The main field winding with D.C. current produces an elliptical static magnetic field. This field interacts with the current in the armature coil conductors. The torque produced by the interaction rotates the armature giving output to the shaft. The lag between the flux due to the auxiliary current and the main flux in the poles is responsible for the rotation of the armature.

4.13.3 Analysis

Let a voltage, V, be applied to the armature through a series connected resistance, R_a, with the armature. The back emf produced in the motor is

$$E_b = V - I_a R_a. \tag{4.2}$$

But for an angular speed, ω, and the torque produced, T, the mechanical power is

$$T\omega = E_b . I_a. \tag{4.3}$$

But from the motor configuration with torque constant, k, and flux, ϕ,

$$E_b = k\phi\omega. \tag{4.4}$$

From equations (4.3) and (4.4), the torque is given by

$$T = k\phi I_a. \tag{4.5}$$

This shows that T varies linearly with the armature current and hyperbolically with the speed of the motor (as $T\omega$ = constant).

Using equations (4.2) and (4.5) we get an expression for torque as

$$T = \frac{k\phi}{R_a}(V - E_b). \tag{4.6}$$

The arrangement of the armature resistance in the excitation of D.C. motors has lead to the following configurations:

1. Series excitation D.C. motor.
2. Shunt excitation D.C. motor.
3. Combination excitation D.C. motor.

In the first the resistance is connected in series. In the second the resistance is arranged in parallel. In the combination type one resistance is in series and one is in parallel form which results in the combined excitation.

4.14 SELF-COMMUTATED D.C. MOTORS

The following table shows the basic circuitry in self-commutated D.C. motors.

TABLE 4.5 Circuitry in D.C. Motors

Basic Circuitry	Torque-Speed Curve	Torque-Current Curve
Series Excitation 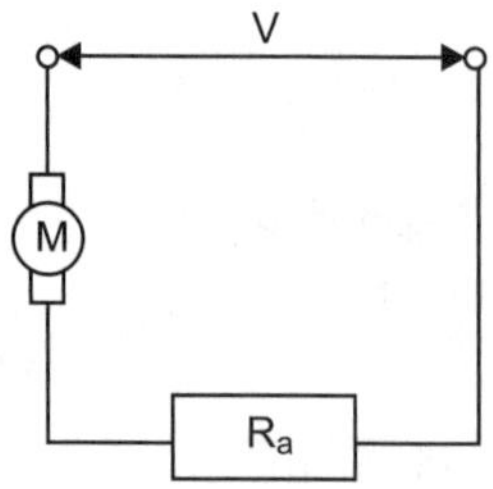 V M R_a	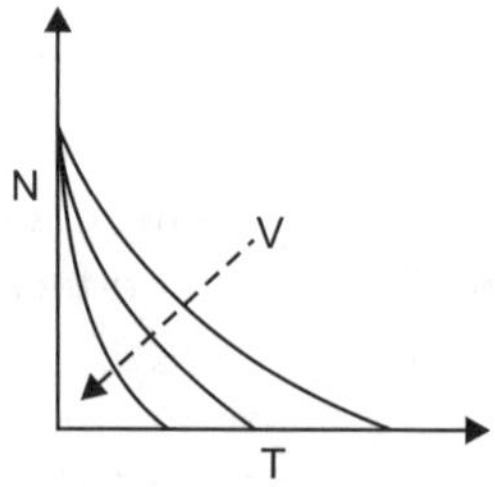 N V T	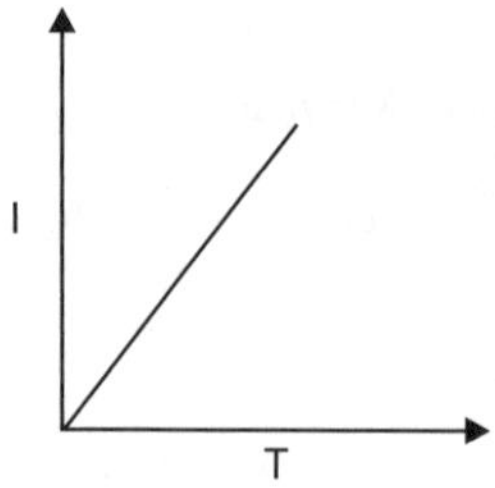 I T

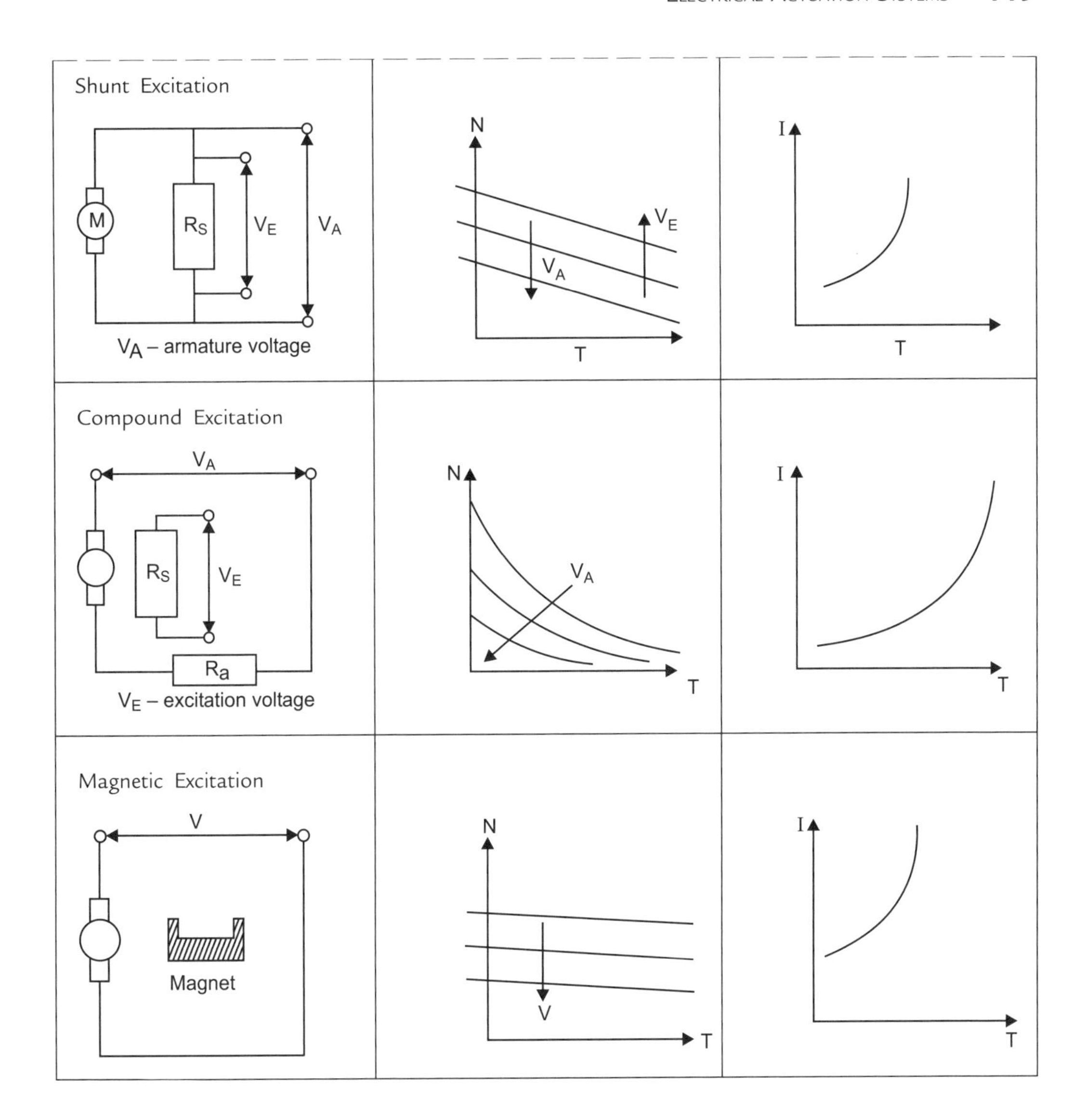

4.15 PERMANENT MAGNET D.C. MOTORS

4.15.1 Construction

Permanent magnet D.C. motors have magnets in the rotor structure to set up the required magnetic flux. They have a commutator winding in the stator structure. The current supplied to the stator coils sets up a torque in the rotor to produce rotational output. The permanent magnets are made of rare earth magnets such as SmCo, NdFeB, or AlNiCo. The different kinds, shapes, and mounting manners of permanent magnets offer a wide range of motor design possibilities.

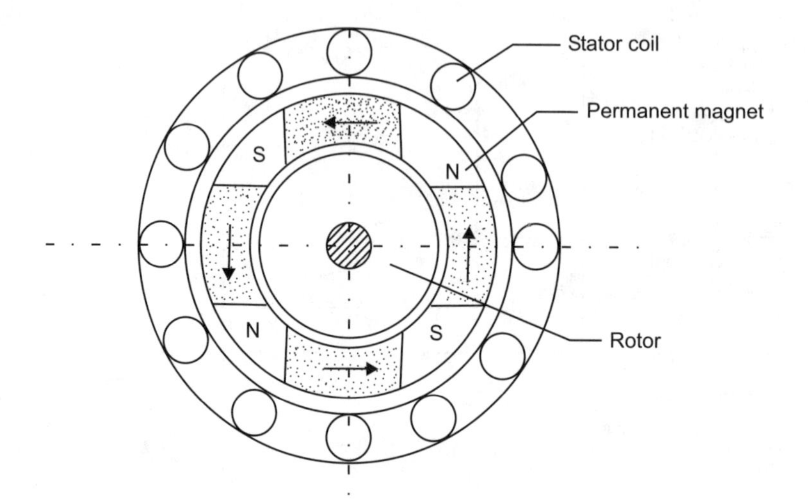

FIGURE 4.17 Permanent magnet D.C. motor.

4.15.2 Workings

The magnets in the rotor set up high magnetic flux, that is, they interact by the field set up by the D.C. current in the stator coil. This leads to the deflection of the rotor. The continuous interception of flux and the field results in the rotation of the rotor.

4.15.3 Analysis

The force, F, set up in the coil results in a torque, T, given by

$$T = F \times b. \tag{4.7}$$

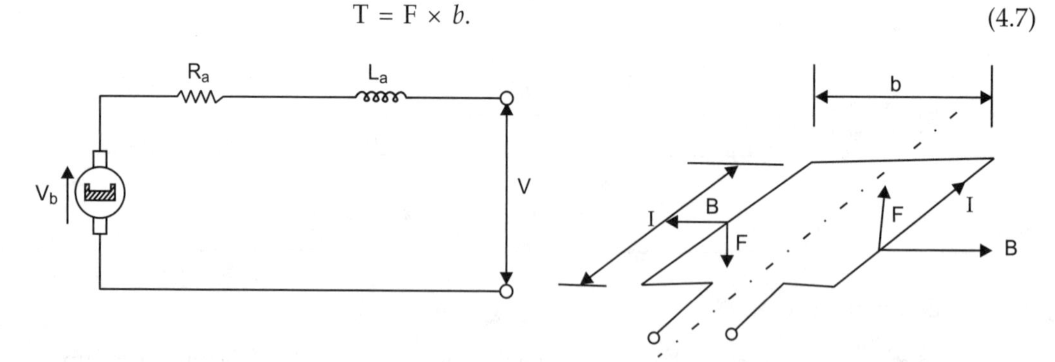

FIGURE 4.18 Analysis.

Let there be 'n' coils of total length 'L' in which current I is flowing. The flux density in the magnet, B, produces the force, F, given by

$$F = (n \perp BI). \tag{4.8}$$

From equations (4.7) and (4.8)

$$T = K_t.I, \tag{4.9}$$

where $K_t = (n\text{LB}.b)$.

But from the circuit analysis

$$I = \frac{V - V_b}{R_a}. \tag{4.10}$$

The back emf $V_b = k_b.\omega$, where k_b is the back emf constant and ω = angular velocity.

Hence, from equations (4.9) and (4.10) we have torque expressed by

$$T = \frac{K_t}{R_a}(V - k_b\omega). \tag{4.11}$$

4.15.4 Characteristics

The observation of equation (4.9) shows that the torque, T, varies linearly with current, I. The slope of the curve giving the torque constant K_t is specific for a given design. K_t is dependent on the coil dimension and numbers and the magnetic flux density as well. The torque also varies linearly with the speed of rotation as given by Eq. (4.11). It may be observed that as the speed increases the torque decreases linearly. The negative slope is given by the direct relation with the torque constant and the back emf constant, and the inverse relation with the series arranged armature resistance. Figure 4.19 shows the characteristics of torque variation with current and speed.

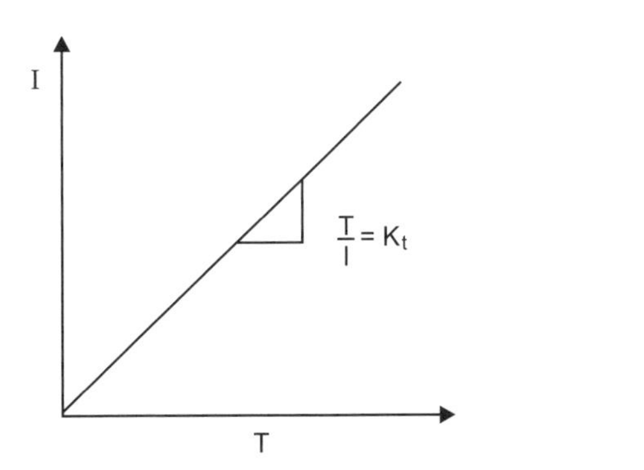

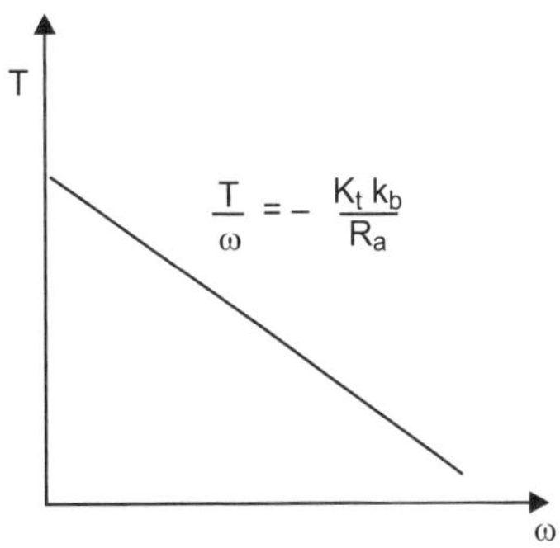

FIGURE 4.19 Characteristic curve.

4.15.5 Control

The motor is transistor driven in the on/off mode with Pulse-Width-Modulation

(PWM). An FD (free-wheeling diode) is connected in parallel with the motor that rectifies the current supplied.

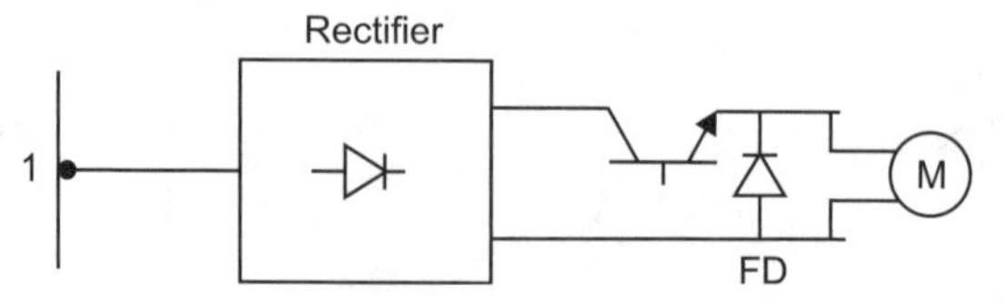

FIGURE 4.20 Control circuit.

4.16 FEATURE COMPARISON OF D.C. MOTORS

The following table compares the features of D.C. motors.

TABLE 4.6 Feature Comparison of D.C. Motors

Excitation Features	Series	Shunt	Compound	Magnet
• Speed torque characteristics	load-dependent	load-independent	load indepen-dent	load indepen-dent
• Locked rotor current	smaller	greater	greater	greater
• Speed range of same configurations	greater	greater	greater	smaller
• Efficiency with similar features	smaller (low)	smaller (low)	smaller (low)	greater
• Speed control capabilities	expensive	cheaper	cheaper	cheaper
• De-magnetizing risk	not applicable	not applicable	not applicable	exists
• Interference chances	simpler	more expensive	simpler	more expensive
• Cost component	medium	medium	greater	lower

4.17 BRUSHLESS D.C. MOTORS (BLDC)

4.17.1 Construction

Brushless D.C. (BLDC) motors have a poly-phase winding in the stator which is kept fixed with the yoke. With a small gap the stator coils are assembled concentric with the rotor fixed with a permanent magnet. The rotor is mounted on bearings at the extreme ends.

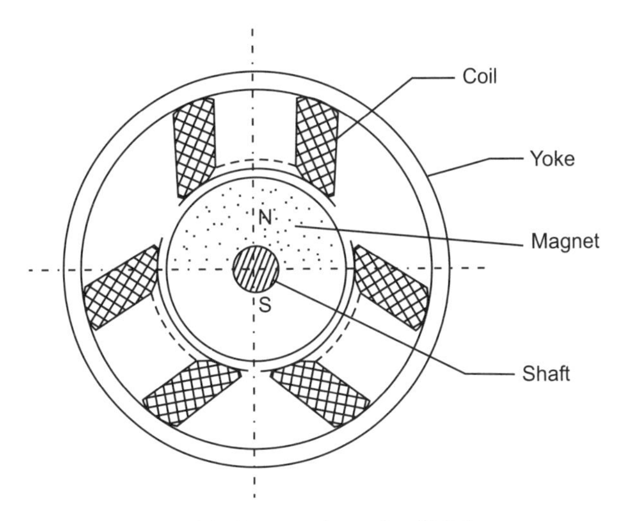

FIGURE 4.21 Construction of a BLDC motor.

4.17.2 Workings

An electronic control switches the stator phases of excitation in the cyclic order depending on the rotor position. A change of phase induces the torque on the rotor by the interaction with the permanent magnet attached to the rotor. The torque on the rotor results in rotation.

4.17.3 Features

- They cannot pull out of synchronism (starting problem, pull-out at overloads, oscillation on load variations).
- Speed is load-dependent and not constant.
- Electronic commutation makes it expensive.
- Vibration ability is decided by encoder resolution.

4.18 DISC-TYPE BLDC MOTORS

Figure 4.22 shows the constructional details of a disc-type BLDC motor. It has ironless self-supporting windings. The windings are photolithographic. The stator has three phases with two coils each. There are eight poles in the rotor. The disc motors with waves form coils which enable easy interconnection. The stator is divided into two parts on both sides of the rotor. Due to manufacturing difficulty of these motors they are expensive in cost.

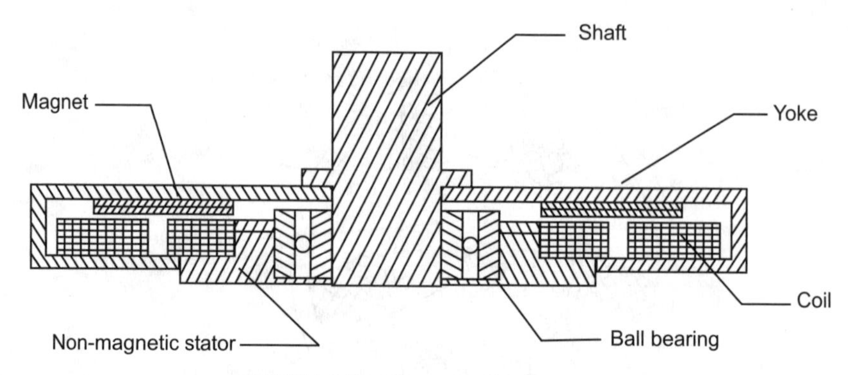

FIGURE 4.22 Disc-type BLDC motor.

4.18.1 Applications of BLDC Motors

- Tape and video recorders.
- Hard disc drives of computers.
- Digital cameras.
- Plotters and printers.
- Bar-code scanners.
- Robotics and feed drives.
- Electrocardiographs and dialysis apparatuses.
- Car auxiliary drives, small fans.
- Welding machines.
- Artificial limb control.

4.19 ASYNCHRONOUS MOTORS

Single-phase induction A.C. motors are known as *asynchronous motors*. They are mostly manufactured to operate with 230 V alternating (A.C.) voltage. They are characterized by the following features:

- Cost effective.
- Robust in design.
- Maintenance free.
- Low noise during operation.
- Low vibration and smooth running.
- No speed regulation is required.

They are widely used for the following applications:

- Washing machines and clothing dryers, dishwashers.
- Refrigerators and freezers and airconditioning.
- Garden chaff cutters and lawn mowers.
- Garage doors and fans.
- Typewriters and fax machines.
- Printers and paper shredders.
- Coin-counter machines.
- Machine tools such as circular saws and high-frequency tools.
- High pressure cleaners, compressors.
- Grinding machines and cement mixers.
- Stirring machines and analysis apparatuses.
- Heating pumps and centrifugal pumps.
- Oil and gas burners.

4.20 SINGLE-PHASE INDUCTION MOTORS

4.20.1 Construction

A squirrel cage with a slot for the windings is mounted on the rotor. It has two stator poles with the coils wound that have internal connections. The stator poles are integral with the yoke. The central rotor is mounted between the freely rotating ball bearings.

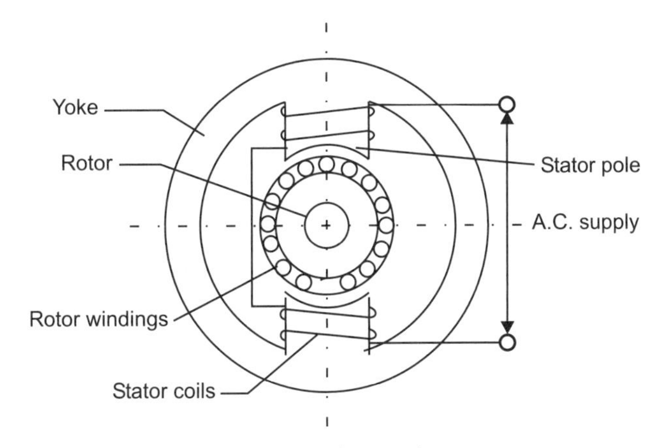

FIGURE 4.23 Single-phase induction motor.

4.20.2 Workings

The A.C. supply to the stator winding alternates the field in the poles. The induction effect of the poles' field with the current-carrying rotor coils set up the force on the rotor. The alternating field sets up a rotating torque in the rotor. But there will not be a starting torque due to which external arrangement starts initially.

4.20.3 Features

- The speed is not synchronous (speed of rotation of magnetic field) due to slip.
- The motor is not self-starting.
- The rotation can be reversed by interchanging lines.

4.20.4 Construction of Four-pole Asynchronous Motors

The stator and rotor are laminated to minimize eddy current losses. Figure 4.24 shows the internal rotor motor. The rotor has squirrel cage windings. The stator windings are, being two or three phases, distributed in slots.

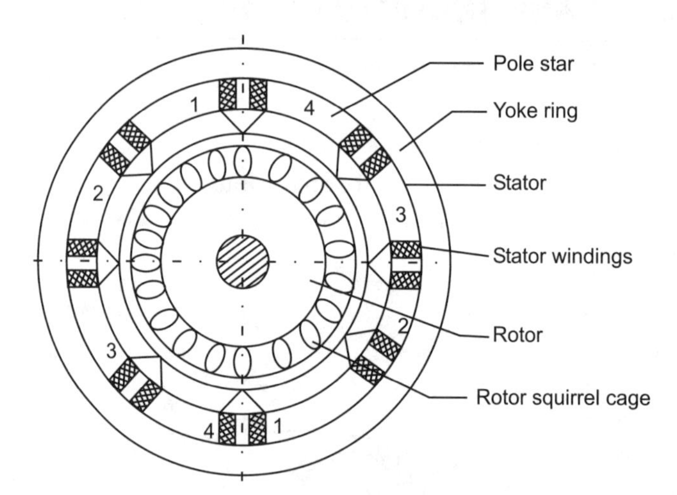

FIGURE 4.24 Four-pole asynchronous motor.

The revolving field induces voltage in the rotor bar with its current. The rotor current generates torque together with the rotary field. The stator current produces a magnetic field which rotates with the synchronous speed,

$$N_S = \frac{f}{p}, \tag{4.12}$$

where f is the line frequency and p is the number of pole pairs.

The characteristic plot showing the variation of torque (T) with the speed (N) of the rotor is given in Figure 4.25.

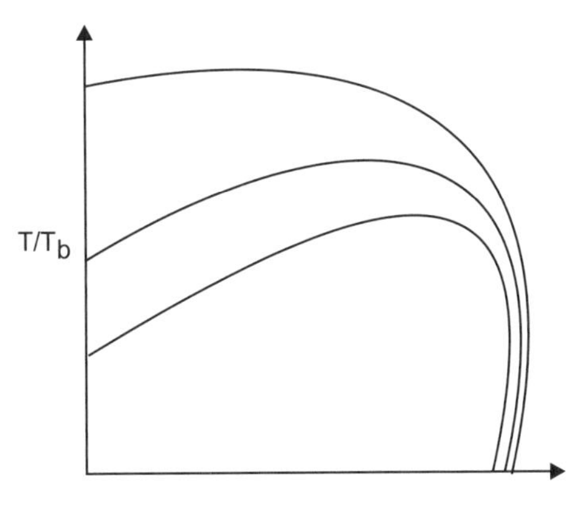

FIGURE 4.25 Char N/N_S.

4.21 SYNCHRONOUS INDUCTION MOTORS

4.21.1 Construction

The stator is a squirrel cage construction with the winding coils inserted in the slot of the stator. The cage of the stator is inserted into the yoke. This sub-assembly slides onto the rotor with a permanent magnetic cylindrical drum. The rotor is mounted onto free rotating bearings. They are not self-starters. Figure 4.26 shows the construction.

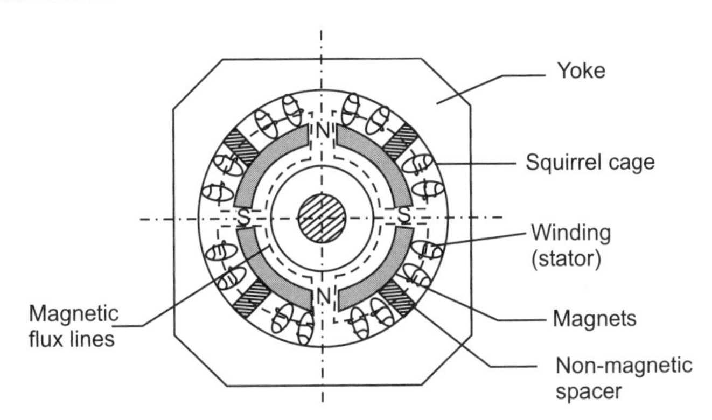

FIGURE 4.26 Synchronous motor.

4.21.2 Workings

The rotating electric field set up in the stator coil due to current interacts with the magnetic flux lines of the rotor-mounted permanent magnets to develop force that results in torque on the rotor to move the rotor with a speed equal to the speed of the rotating field. The frequency of the rotation of the rotor is proportional to the frequency of the current in the stator windings.

4.21.3 Applications

They are used in the following apparatuses: program controllers, clocks, time-delay relays, studio camcoders, measuring instruments, gyrodrives, fans and pumps, washing machines and dishwashers, can openers, and recorders.

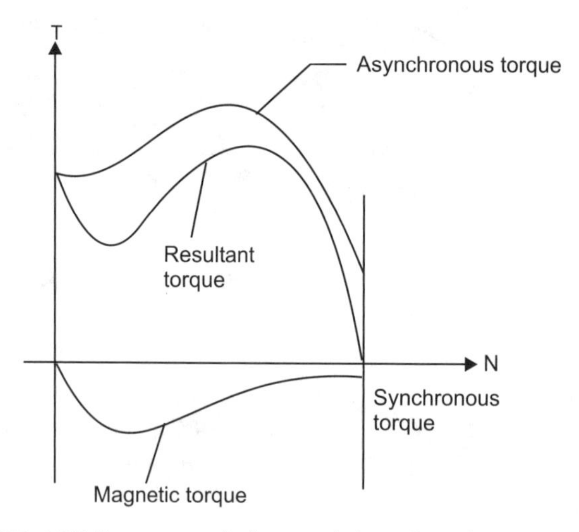

FIGURE 4.27 Torque-speed characteristics of synchronous motors.

The torque-speed characteristics of synchronous motors are given in Figure 4.27. The resultant torque is obtained by the combined effect of asynchronous torque and magnetic torque.

TABLE 4.7 Comparison of Synchronous and Asynchronous Motors

Synchronous Motor	Asynchronous Motor
• Starting problem, needs a starter in the beginning.	• No speed regulation is required.
• Speed does not vary with supply voltage.	• Speed varies with supply voltage.
• Simple in construction.	• Robust in construction.
• Efficiency is relatively high.	• Efficiency is not poor.
• Cost effective.	• Cost effective.
• Low noise and vibration.	• Low noise and vibration.
• Maintenance free.	• Maintenance free.
• Large part of the flux not entering the rotor gets wasted.	

4.22 COMPARISON OF D.C. AND A.C. COMMUTATOR MOTORS

The following table compares D.C. and A.C. commutator motors.

TABLE 4.8 Comparison of D.C. and A.C. Commutator Motors

Features	D.C. Motor	A.C. Motor
• Torque output	*identical*	*identical*
• Power losses	*identical*	*identical*
• Starting torque	*lower*	*greater*
• Speed control	*easier*	*more expensive*
• Speed range (at same cost)	*smaller*	*wider*
• Speed-torque characteristics	*load-independent*	*load-dependent*
• Commutation	*longer brush life*	*lower life*
• Rectification of the main's voltage	*necessary*	*not applicable*
• De-magnetizing	*preventive action needed*	*not applicable*
• Interference suppression	*more expensive*	*easier*
• Production cost component	*identical*	*identical*
Applications of D.C. and A.C. motors	• Household appliances • Hair dryers • Foil welding machines • Car auxiliary drives • Choppers • Windscreen wipers	• Portable appliances • Washing machines • Vacuum cleaners • Sewing machines • Drilling and grinding machines • Milling and planning machines • Power saws • Hot-air blowers • Mixer and coffee grinders

4.23 STEPPER MOTORS

The positioning of the rotatory movement in finite steps is required instead of continually rotating A.C. or D.C. motors. Such motors, called *stepper motors*, do not require rotor positioning encoders used in servodrives. Hence, they are cheaper as stepper motors operate under open loop control. But they do not have the check back on the position of the rotor to know whether the given position is really reached. In stepper motors it is the characteristic feature that for every electrical pulse there occurs a mechanical pulse rotation. In stepper motors electrical commutation of current is also responsible for rotation with a frequency of electronic control. Stepper motors are included in the synchronous motor class characterized by starting problems, pull-out at overload, and oscillations under sudden load variations. These problems should be solved in the design of the stepper motors so as not to disturb the condition mechanical pulses per electrical pulse of rotation.

4.23.1 Construction

Stepper motors are constructed in principle like BLDC motors. The stator has a winding made of concentrated coils on distinct poles. The rotor has a permanent magnet cylinder. The stator windings are inserted on the periphery of the yoke and this slides onto the permanent magnet pole cylinder. The rotor shaft is located at the center of the yoke, mounted on the bearings.

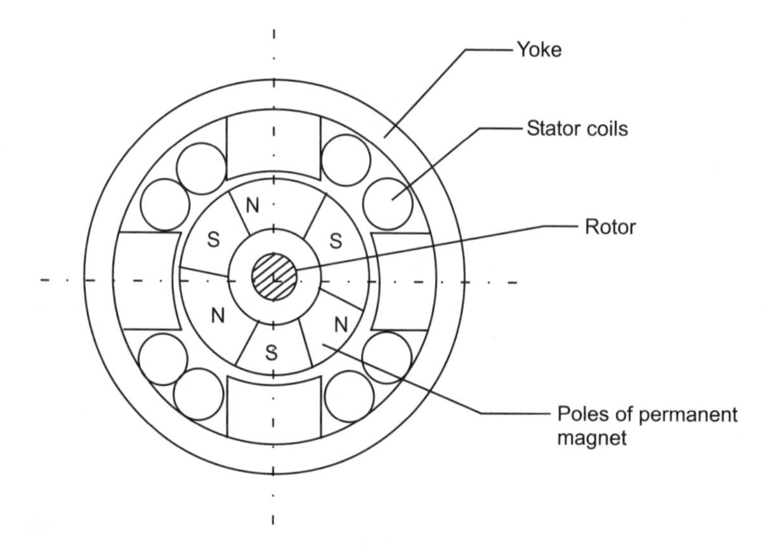

FIGURE 4.28. Stepper motor.

4.23.2 Workings

Pole 1 is energized and the N-pole of the magnet takes position as in Figure 4.29(*a*). The energization of pole 2 attracts the N-pole to rotate by 90° to the position in (*b*). The subsequent energization, poles 3 and 4, rotates the rotor further in steps of 90°. As the P1-P2-P3-P4 coils are energized in sequence one rotation takes place. The simultaneous energization of the two adjacent coils positions the rotor in between the coils, which is called the "half-step mode."

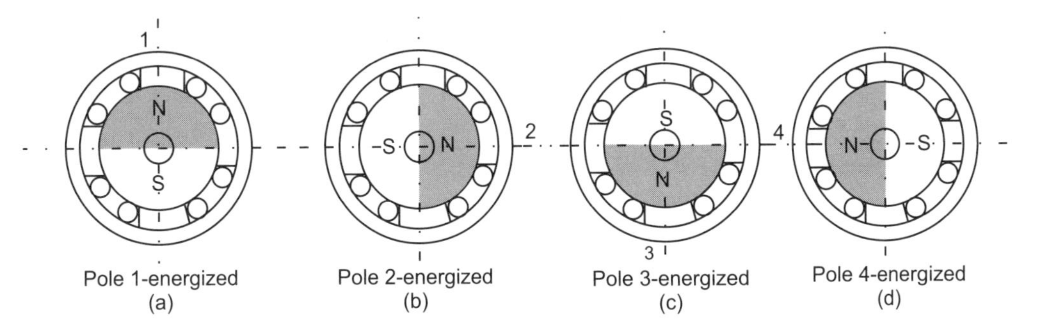

FIGURE 4.29 Workings of four-pole stepper motor.

The resolution or the number of steps, R, is given by

$$R = \frac{(2p)m}{F_B}, \tag{4.13}$$

where $2p$ = number of poles on the rotor

m = number of the system's energizing coil

F_B = operation mode factor

= 1 (for full-step operation)

= 1/2 (for half-step operation)

The step angle for each energization

$$\alpha = \frac{360°}{R}. \tag{4.14}$$

4.24 SINGLE-PHASE STEPPER MOTORS

A single-phase stepper motor has two poles with a single winding system. When the current in the coil is positive the flux ϕ, set up in the poles, keeps the shaft (rotor) in position 1. The opposite current in the coil leads to the reversal of the flux in the poles. This rotates the rotor to position 2 by 180°. Hence, the reversal of fluxes in the poles results in continuous rotation of the shaft in a clockwise rotation.

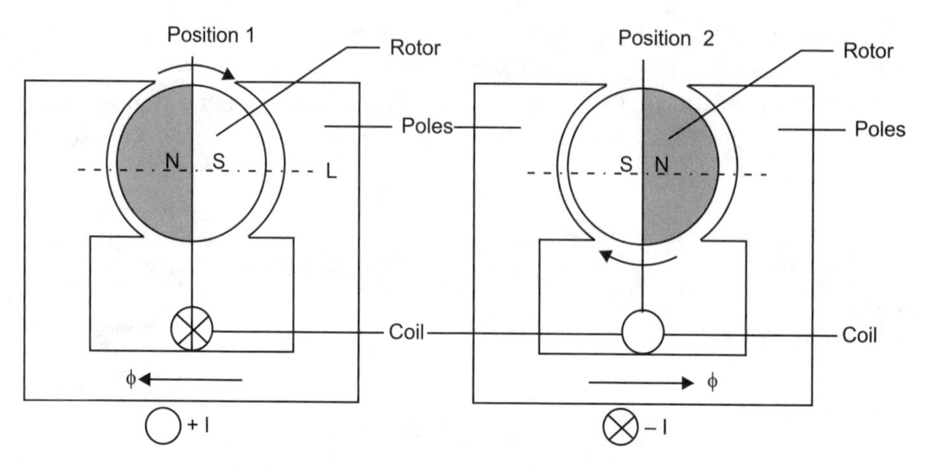

FIGURE 4.30 Single-phase motor.

All clockwork motors are 180° stepper motors rotating in only one direction. They are hardly used for other purposes than clockwork.

Figure 4.31 shows the torque-speed characteristics of a stepper motor. The characteristic shows the maximum starting torque and the maximum load torque that can be applied at a given speed of operation.

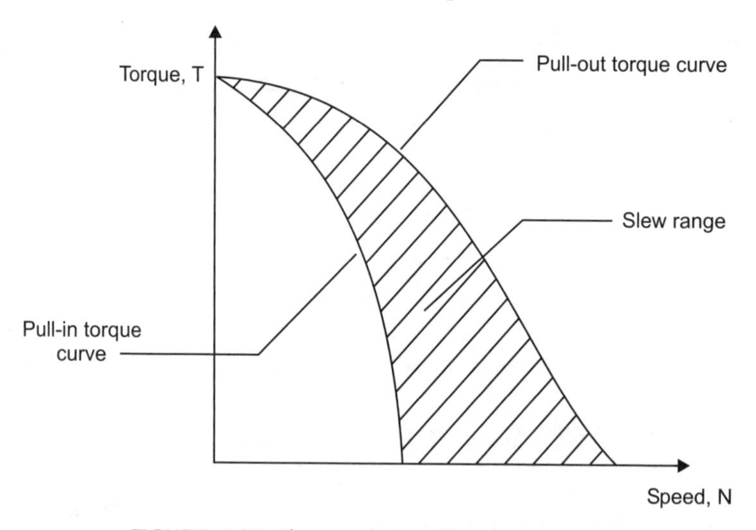

FIGURE 4.31 Characteristics of a stepper motor.

4.24.1 Specifications of Stepper Motors

A stepper motor is specified by the following:

1. *Phase.* Number of coils winding in the stator considered independently.
2. *Step angle.* The smallest angle that the rotor can rotate under a single pulse.

3. *Holding torque.* The maximum value of torque to be applied without producing rotation in the rotor.
4. *Pull-in torque.* The maximum torque required to start and reach synchronism.
5. *Pull-out torque.* The maximum load torque the motor can withstand without losing synchronism.
6. *Pull-in rate.* The maximum pulse rate required to start a loaded motor without missing a step.
7. *Pull-out rate.* The reduced switching pulse rate that retains synchronous speed.
8. *Slew range.* The range of the pulse rate between pull-in and pull-out without losing synchronism.

4.25 SERVOMOTORS: DEFINITION

The task of a servomotor is to position mechanical elements in a given position within a given time and with a given precision. These drives operate at a lower power range. These functions are required in feed screws and feed drives of machine tools, arm actuation, and positioning in robotic systems, positioning of conveyor belts, spool positioning of butterfly valves, orienting small antennas, and in the orientation of small telescopes. Several motors often have to run in synchronization. The factors to be considered in servo action are dynamics, positioning exactness, peak torque, rotational trueness at crawl speed, the speed regulation range and efficiency, explosion protection, and environmental influences. The components that are influenced by the above factors apart from servomotors are drive components, power amplifiers, sensors, electronic controls, and mechanical transfer elements. Optimal control is achieved through a low-time constant for the sub-assemblies participating in the dynamics of control. Electromagnetic actuators are advantageous in the lower power range over hydraulic and pneumatic actuators (see Table 4.11).

The easy adjustment capability in a D.C. commutator motor makes it more useful in servodrives in the earlier stages of development. Today, three-phase servodrives with electronic commutators are increasingly prevailing. In D.C. servomotors with constant excitation the torque depends linearly on the armature current. The regulation and control by current is easy. Hence, only the current is to be measured and controlled in D.C. servos which makes control electronics easy and simpler.

The rotors of three-phase motors with permanent magnet and squirrel cage windings have a greater moment of inertia compared to D.C. motors, with lower MI. In A.C. servomotors torque-causing current is generated inductively in the rotor. The speed is controlled close to the synchronous speed by electronic control and frequency of the pole number. The efficiency and power factors in A.C. motors are low which requires a costly supply unit. For low power output requirements A.C. servomotors are not used. The comparison of features of D.C. current servomotors with A.C. three-phase drives is given in Table 4.10. In Table 4.11 a comparison of three-phase drives with that of brushless D.C. (BLDC) servodrives are given to enable the easy selection of these drives to suit the application. A comparative analysis of the advantages and disadvantages of hydraulic servodrives with electromagnetic drives is provided in Table 4.12. BLDC motors are more advantageous in positioning accuracy. Higher speed range is given by A.C. servomotors. Both motors are dynamically equivalent. The applications of both BLDC and three-phase induction motors are equally favorable.

The workings of the servomotor through a block diagram and an explanation of D.C. and A.C. servocircuits is given in the following paragraphs.

4.26 SERVO-DRIVE CONTROL

In Figure 4.32 is an example of a control circuit for a servodrive.

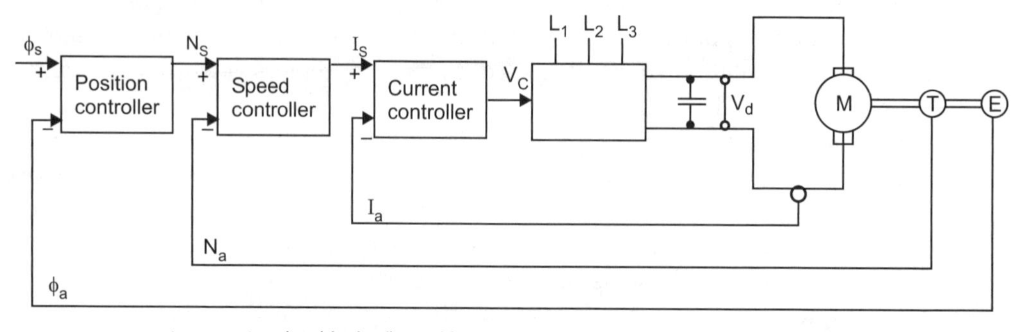

ϕ_s = set point (desired) position | ϕ_a = actual position
N_S = desired speed | N_a = actual speed
I_S = desired current | I_a = actual current
V_C = control voltage | V_d = direct voltage
L_1, L_2, L_3 = supply line current | M = motor
T = tachogenerator | E = encoder

FIGURE 4.32 Control circuit for a servodrive.

4.26.1 Workings of a Servomotor

The desired position, ϕ_s, is given as input to the system. The actual position, ϕ_a, measured by the encoder is given to the position controller as feedback. This results in the desired speed, N_S, which is compared with the actual speed, N_a, fedback by the tachogenerator, T. This comparison gives the output of the desired current, I_S, to be compared with the feedback of the actual current, I_a. The current controller gives the output of the control voltage to the motor. A six-pulse bridge supplied with a three-phase system gives the direct voltage, V_d, depending on the control voltage, V_C. The electronic commutated motor of A.C. is similar to this but requires the measurement of at least two currents.

4.27 CONTROL OF BLDC SERVOMOTOR

In Figure 4.33 is an example of a control circuit used in a D.C. servomotor.

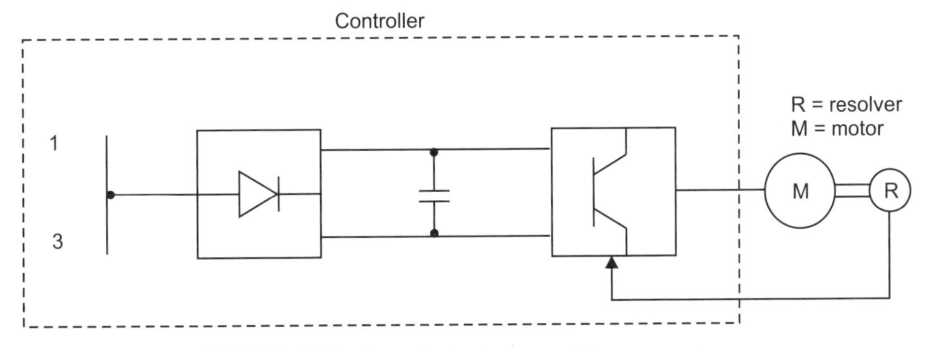

FIGURE 4.33 Control circuit for a D.C. servomotor.

Features

- Electronic switching by Hall elements.
- Sensorless switching.
- Transistor controller and diode rectifier.
- Resolver for speed control and position control and switching control.
- Resolver (R) performs the task of both the tachogenerator and the pulse generator (p-encoder).

4.28 A.C. SERVOMOTOR CONTROL

The position in a three-phase A.C. servomotor is accomplished by phase-angle control. The phase control is achieved by using a thyristor as shown in Figure 4.34.

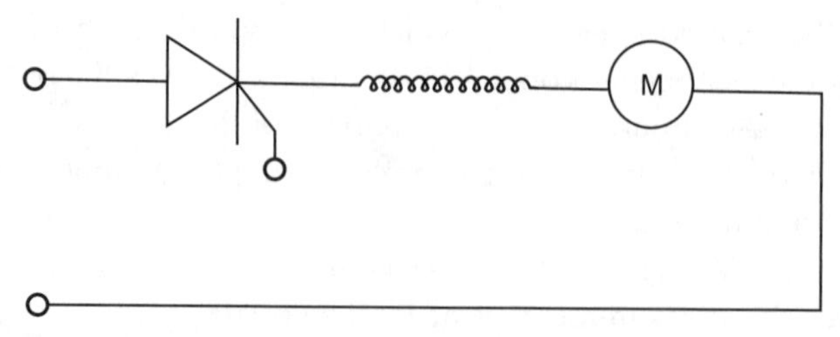

FIGURE 4.34 Thyristor control.

The phase angle can also be controlled by accommodating a triac in the control circuit of the A.C. servomotor as shown in Figure 4.35. The speed is controlled by the tachogenerator feedback to the transistor amplifier.

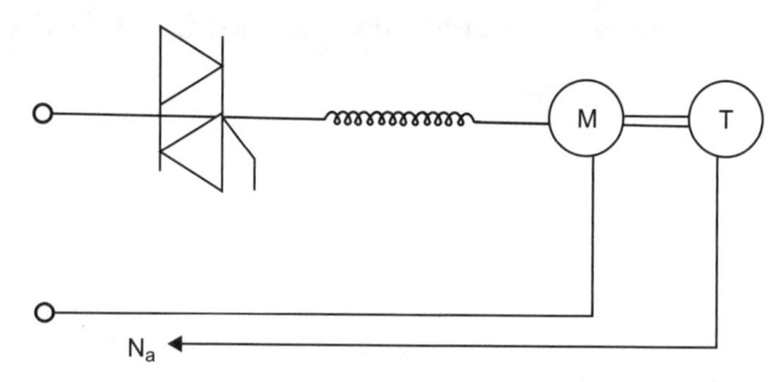

FIGURE 4.35 Triac control.

TABLE 4.9 Comparison of D.C. and A.C. Servomotors

D.C. Current Drives	A.C. Three-Phase Drive
• The power supply and control is easy as only one current is measured	• Control is by at least two currents and capture of rotor position
• Servo amplifier cost is low	• Expensive converter with higher switching capacity
• Higher noise and wear of brushes and commutator	• Maintenance free and low noise level
• Brush life is low (3000 *h*)	• Higher brush life (10000 *h*)
• Dynamics are better by low moment of inertia	• Higher MI effects dynamic of control
• Heat dissipation over the rotor shaft	• Higher protection grade
• A transformer necessary at voltages higher than 100 V	• Direction operation through the main's supply
• Bigger size and weight	• Lower weight and compact size
• Less expensive	• More expensive but robust

TABLE 4.10 Comparison of BLDC and Three-phase Servomotors

Comparative Features	BLDC Servomotor	Three-Phase A.C. Servomotor
• Robustness in design	• High	• Very high
• Control action	• Easy, cheaper	• Complex and expensive
• Overload capacity	• High	• Very high
• System cost	• High	• Very high
• Reactive power requirements	• Not applicable	• Necessary
• Ohmic (current) loss	• In the stator	• Instator as well as rotor
• Field weakening	• Can be restricted	• Occurs in a large range
• Size for the same power rating	• Smaller	• Higher
• Only motor costs	• Higher	• Lower

TABLE 4.11 Comparison of Hydraulic and Electromagnetic Servodrives

Hydraulic Drive	Electromagnetic Drive
• Higher power to weight ratio	• Lower energy density
• Smaller mounting space	• Better control response
• Lower cost	• Higher efficiency
• Higher accelerating ability	• Easier to maintain
• High torque at low speeds	• Easy adaption to diverse conditions
• Easy generation of linear motion	• Sensors, control electronics, and actuators operate under same type of energy domain
• Problem of leakage of hydraulic oil persists	

In fluid power drives, the force density is limited by the strength of the mechanical components. The force density ranges in hydraulic systems from 300 to 500 bars. The speed of operation is lower than electric drives. As the power-to-weight ratio is high in hydraulic drives they are commonly used in machine tools. In hydraulic drives the heat dissipated in actuation is carried away by oil to the tank, whereas in electrical machines the heat is dissipated by external cooling.

In electrical machines the power-to-weight ratio depends on the force density in the energy transformed. Because of the magnetic saturation in the armature and stator materials the force density is limited to a maximum of 15 to 16 bars. But in real-time applications it is limited to four bars only.

EXERCISES

1. Explain the working principle of a permanent magnet D.C. motor. How is it used for positive control devices?
2. Sketch and explain the workings of a stepper motor.
3. Differentiate between a diode, thyristor, and transistor.
4. Define:

   (*a*) Diodes (*b*) Thyristors
5. What is bouncing in mechanical switches? Show a transistor-based circuit used to prevent bouncing.
6. What is the role of a solenoid in a mechatronic system?
7. What does MOSFET mean? Illustrate how it can be used to control a D.C. motor.
8. How are D.C. motors classified? Illustrate how field windings and armature windings are connected in each case.
9. Explain the principle of operation of a VR stepper motor.
10. Draw the characteristics of a stepper motor and explain, in brief, the following:

    (*a*) Slew range (*b*) Pull-out torque (*c*) Holding torque
11. What is a triac? Sketch its characteristics and describe its operation.
12. Give the configurations of different types of mechanical switches.
13. Explain with symbols the features of three design varieties of mechanical switches.
14. With a sketch write the constructional and working details of a limit switch.
15. Explain the phenomenon of contract bounce in mechanical switches. Give an example.
16. How are the effects of contact bounce minimized in mechanical switches by mechanical ways?
17. How are the effects due to contact bounce eliminated by flip-flop use?
18. How does the Schmitt trigger function eliminate the effects of contact bounce?
19. Explain the functioning of a diode with the current voltage characteristics.
20. What is a thyristor? Explain its characteristics.
21. Explain the switching characteristics of a thyristor.
22. Explain the high side and low side switching arrangement of a thyristor.
23. How does a thyristor behave for an inductive load?
24. What is a triac? Give the thyristor equivalent of a triac.
25. Give the characteristic of a triac. How does it transform voltage?
26. What are the features of a bipolar transistor?
27. With a common-emitter circuit describe the characteristics of a transistor.
28. What is a Darlington pair? Explain two types.
29. Give the applications of a Darlington pair.
30. Give the construction and working principle of a solenoid. Give the force-displacement characteristics.

31. What is a pulsed latching solenoid? Explain.
32. Give the applications of a solenoid.
33. Give the configuration details of a relay.
34. Discuss the applications of a relay.
35. Explain the principle operation of an electric motor.
36. Give the detailed classifications of electric motors.
37. Explain the construction and workings of a four-pole D.C. motor.
38. Derive the torque-voltage relation for a D.C. motor.
39. Discuss the series, shut, and compound excitation characteristics of a D.C. motor.
40. Explain the constructional features and working principle of a permanent magnet D.C. motor.
41. Derive the torque-voltage relation for a permanent magnet motor. Discuss the characteristics.
42. Describe the current-control in a permanent magnet D.C. motor.
43. Compare the features of different excitation methods in a D.C. motor.
44. Enumerate the construction, workings, and features of a brushless D.C. motor.
45. Explain a disc-type BLDC motor.
46. List the applications of a BLDC motor.
47. What is an asychronous motor? What are its features and applications?
48. Give the arrangement of construction of a single-phase induction motor.
49. Explain with a sketch the workings of a four-pole asynchronous motor.
50. Describe a synchronous induction motor from the construction, workings, and application point-of-view.
51. Compare a synchronous motor with an asynchronous motor.
52. Compare A.C. and D.C. commutator motors.
53. Describe the construction of a stepper motor.
54. Explain the functioning of a stepper motor.
55. Discuss the workings of single-phase stepper motor.
56. List the specifications of a stepper motor. Explain.
57. Define and describe a servomotor.
58. How are the position, speed, and current control achieved in a servodrive?
59. Explain the workings of a servomotor (D.C.).
60. Explain the control circuit of a BLDC servomotor.
61. Explain the control circuits used in an A.C. servomotor.
62. Compare a D.C. servomotor with an A.C. servomotor.
63. Compare a BLDC servo with a three-phase A.C. servo.
64. Compare an electromagnetic drive with a hydraulic servodrive.
65. Explain the torque-speed characteristics of a stepper motor.

CHAPTER 5

SYSTEM MODELS

A mechatronic system consists of mechanical elements, electrical elements, hydraulic components, and thermal elements. The flow of energy through these systems experiences certain resistances. In mechanical systems the spring element represents resistance to deformation. The damping due to friction offers resistance to oscillation. The inertia provides resistance to acceleration. In mechanical systems it is the force input that causes the displacement output. Mechanical systems are analogous to electrical systems with the force compared to the voltage or the current. Hydraulic systems that carry energy through oil can be compared with electrical systems. The pressure difference and the volume flow rate are compared with the potential difference and the current, respectively. This chapter discusses the formulation of governing differential equations for the following elements:

- *Mechanical systems:* Spring-mass-damper representation.
- *Electrical systems:* Resistance—inductance-capacitance representation. D.C. motor models.
- *Thermal systems:* Thermal resistance and capacitance models.
- *Hydraulic systems:* Modeled elements are linear and rotary actuators, flow, pressure, and direction control valves.

5.1 ELEMENTS OF MECHANICAL SYSTEMS

Mechanical systems are made of elements such as levers, shafts, supports, frames, gears, flexible elements, and fasteners. The elements are characterized by flexibility, friction, and mass, which provide resistance to motion. The flexibility or stiffness provide resistance to deflection. The friction resulting in damping gives resistance to velocity. The mass representing inertia is responsible for resistance to acceleration. The properties of mechanical elements are represented by a spring dash pot and mass for stiffness, damping, and inertia, respectively. The modeling of these mechanical building blocks is presented in the following paragraphs.

5.1.1 Springs

Springs are characterized by stiffness which is the force per unit displacement either by extension or compression; a spring is a linear translational element. Figure 5.1 represents a spring symbolically and the block diagram represents the system.

The force, F, given as input to the spring of stiffness, k, results in the deflection, x. The extension or the compression of the spring is directly proportional to the applied force on the spring. The spring exerts an equal and opposite resistance to the applied force obeying Newton's Third Law of Motion. Hence, for a linear spring the output is proportional to the input, *i.e.*, $F \propto x$, leading to the equation

$$F = kx. \tag{5.1}$$

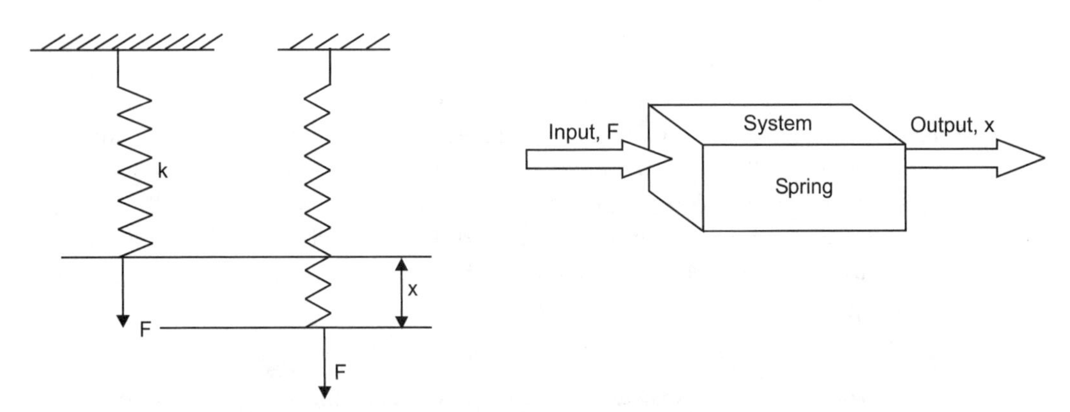

FIGURE 5.1 A spring and the system block diagram.

The proportionality constant obtained by the balance of resistive force in the spring and the applied force is known as the stiffness of the spring. The greater the stiffness the greater should be the input force to obtain the required deflection, x. For a bar of area of cross-section, A, and modulus of elasticity, E, and length, L, the stiffness is given by

$$k = \frac{AE}{L}. \tag{5.2}$$

For a shaft under torsion the stiffness is given by the torque per unit twist forming the relation

$$k_t = \frac{T}{\theta} = \frac{GJ}{L}, \tag{5.3}$$

where G is the rigidity modulus, J is the polar moment of inertia, and L is the length of the shaft.

The translatory and rotary element in the form of a bar and shaft, respectively, are shown in Figure 5.2.

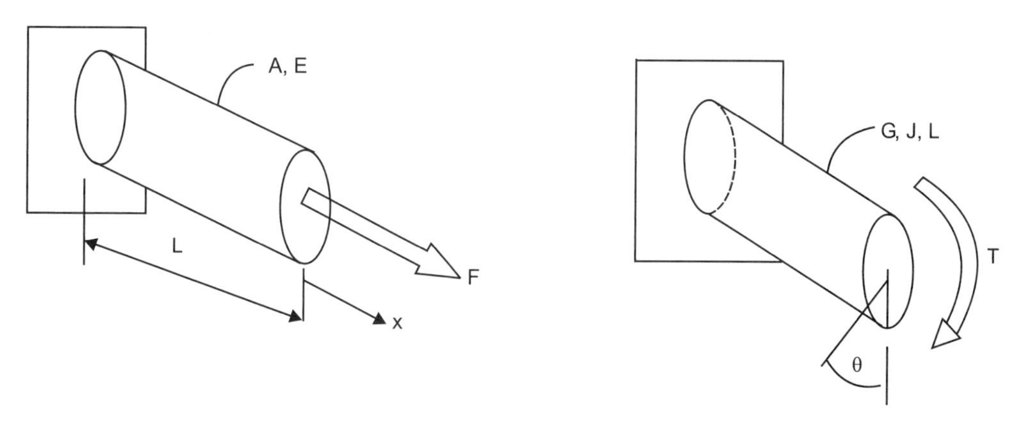

FIGURE 5.2 Bar and shaft.

The relation between the input and output is established by the following expressions:

$$F = \left(\frac{AE}{L}\right) x = kx \tag{5.4}$$

and for the shaft

$$T = \left(\frac{GJ}{L}\right) \theta = k_t \theta. \tag{5.5}$$

5.1.2 Transfer Function

If the input in Eq. (5.1) is a time-varying function, then the deflection (output) is also going to be a time-varying parameter leading to

$$F(t) = kx(t). \tag{5.6}$$

The transfer function is the ratio between the Laplace transform of the output to the input. Hence, by taking the Laplace transform of Eq. (5.6)

$$\frac{x(s)}{F(s)} = \frac{1}{k} = \text{Transfer function.} \tag{5.7}$$

The block diagram representation of Eq. (5.7) is given in Figure 5.3.

FIGURE 5.3 Block diagram of transfer function.

5.1.3 Position Feedback

In a practical situation the output, $x(s)$, desired would deviate from the actual output. Hence, by adjusting the input, F(s), suitably by a feedback signal control is achieved. This is known as positional feedback. The unity feedback block diagram is shown in Figure 5.4.

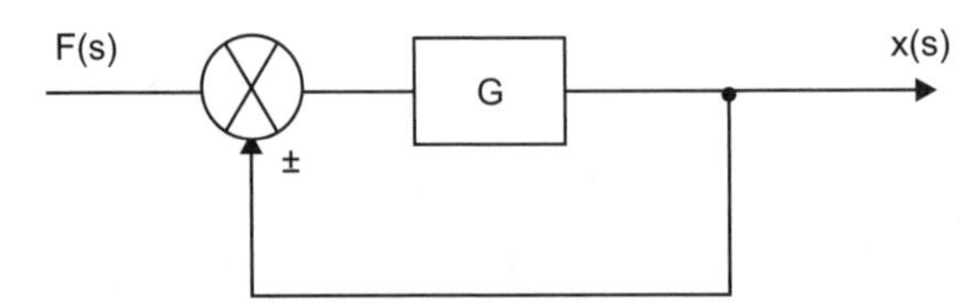

FIGURE 5.4 Equivalent unity feedback diagram.

By block diagram and reduction algebra

$$\frac{G}{1 \mp G} = \frac{1}{k}. \tag{5.8}$$

By simple manipulation

$$G = \frac{1}{(k \pm 1)}. \tag{5.9}$$

The G in Eq. (5.9) gives the gain of the unity feedback system that is equivalent to the open-loop transfer function ($1/k$) of Eq. (5.7) represented by Figure 5.3. The linear positional feedback provided by the incremental encoders finds its application in machine tools and robotics extensively, generally in cartesian robots.

5.1.4 Damping Element

In mechanical applications there are innumerable examples of relative motion between the parts of the system. This results in friction which is responsible for the damping action. The damping is explained by the Coulomb friction mechanism.

The dash pot represents the mechanism of damping as the mechanical element. In the dash pot it is the resistance offered by the fluid (gas or the oil) to the applied force that is modeled. The dash pot in construction consists of a piston and a cylinder arrangement with fluid on one side of the cylinder. The flow of oil past the piston during the motion results in the resistive force. A dash pot symbolically is shown in Figure 5.5.

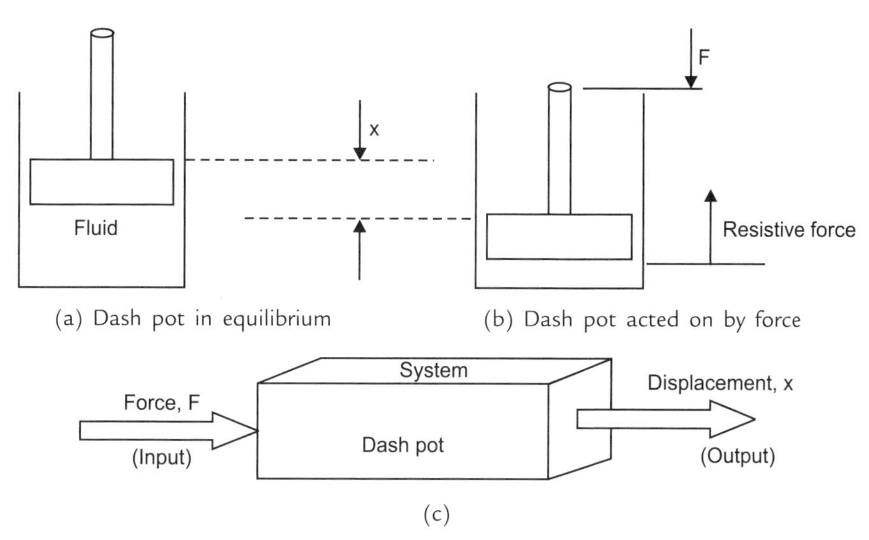

FIGURE 5.5 Dash pot element.

The resistive force developed to the applied force, F, is directly proportional to the velocity of the piston, v. The balancing action of the dash pot thus gives

$$F = cv, \tag{5.10}$$

where 'c' known as the damping coefficient is the constant of proportionality and v is the first time derivative of the output, x $\left(i.e., \ v = \frac{dx}{dt}\right)$, then

$$F = c.\frac{dx}{dt}. \tag{5.11}$$

Thus, the damping force bears a direct relationship with the rate of change of displacement, x, caused by the applied force, F.

5.1.5 Transfer Function

The variability of force, F, and the displacement, x, in the time domain sets up the relation as

$$F(t) = c.v\ (t) = c.\ \dot{x}(t). \tag{5.12}$$

The transfer function, which is the ratio of the Laplace transform of the output to the input, is thus obtained by taking the Laplace transform of Eq. (5.12), *i.e.*, $$F(s) = c.s\ x\ (s)$$

or $$\frac{x\ (s)}{F\ (s)} = \frac{1}{cs.} \tag{5.13}$$

In Figure 5.6, we see the block diagram representation of the transfer function (Eq. 5.13).

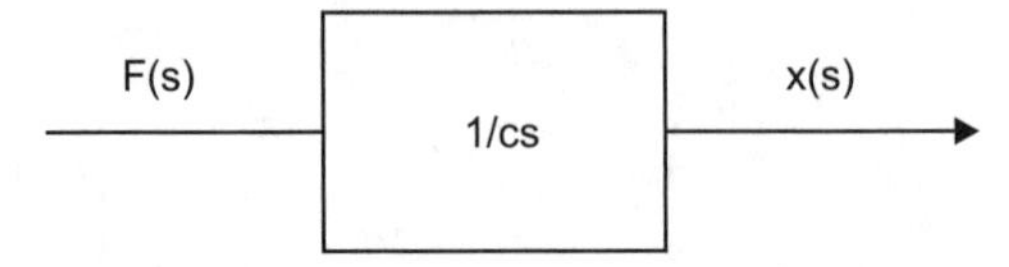

FIGURE 5.6 Transfer function (open loop) of dash pot.

5.1.6 Velocity Feedback

In a time-varying practical situation the desired output velocity may not be the same as the actually obtained velocity output. This error can be balanced by suitably tuning the input based on the velocity feedback arrangement in the control circuit.

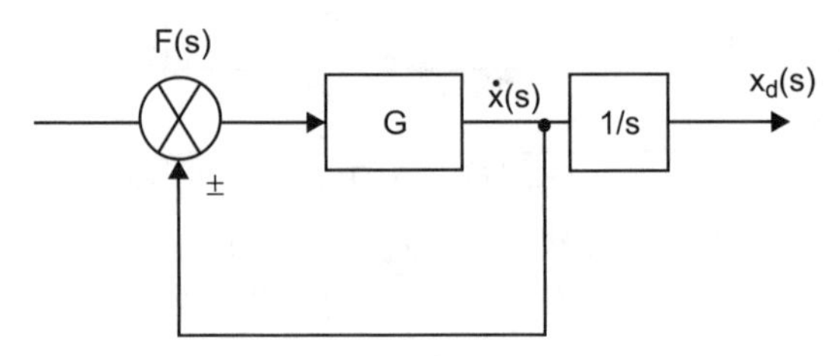

FIGURE 5.7 Equivalent unity feedback diagram.

Figure 5.7 is the equivalent unity feedback circuit for the velocity feedback. Hence, G, the derivative gain, is given by

$$G = \frac{1}{c \pm 1}. \tag{5.14}$$

The sign is indicative of whether the feedback is positive or negative.

For a rotational system the torque input balances with the resistive force proportional to the angular velocity given by

$$T = c.\frac{d\theta}{dt}, \tag{5.15}$$

where $d\theta/dt$ is the angular velocity due to twist.

The velocity feedback is provided, in a rotational system, by the tachogenerator connected at the output, and the controller is designed for the gain given by Eq. (5.14) which depends on the damping coefficient of the system.

5.1.7 Mass

Mass contributes to the inertial effect to the motion of the elements of a mechatronics system. The force acting on a mass ensuring motion is resisted by the opposite inertial force which is proportional to the acceleration.

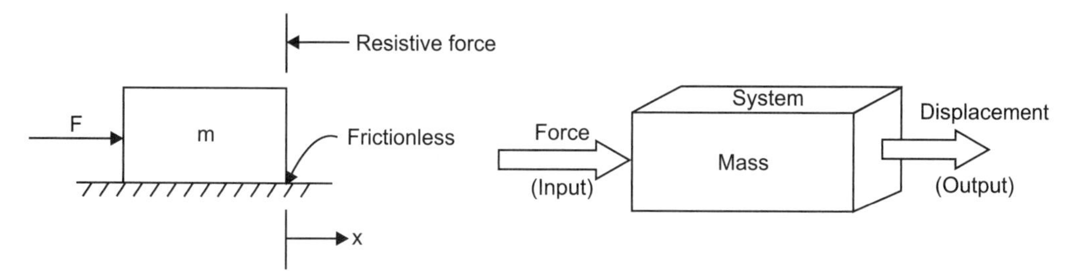

FIGURE 5.8 Inertial effect of mass.

The mass, m, kept on a frictionless surface is acted upon by the force, F, and results in a displacement, x. The inertial effect due to mass balances the force, F. The resistive force is given by the product of mass and the acceleration in motion, *i.e.*,

$$\text{F} = m.a. \tag{5.16}$$

The acceleration is given by the second time derivative of displacement x, hence,

$$\text{F} = m\frac{d^2x}{dt} = m\ddot{x}. \tag{5.17}$$

Figure 5.8 shows the mass kept on a frictionless surface, acted upon by force, F. Here, the input is the force and the output is displacement.

5.1.8 Transfer Function

The time-varying force, F, and the time-dependent displacement are considered and Eq. (5.17) takes the form

$$\text{F}(t) = m\ddot{x}(t). \tag{5.18}$$

The transfer function is obtained by the Laplace transform applied to Eq. (5.18) as

$$\frac{x(s)}{\text{F}(s)} = \frac{1}{s^2m}. \tag{5.19}$$

Eq. (5.19) is represented by the block diagram as shown in Figure 5.9.

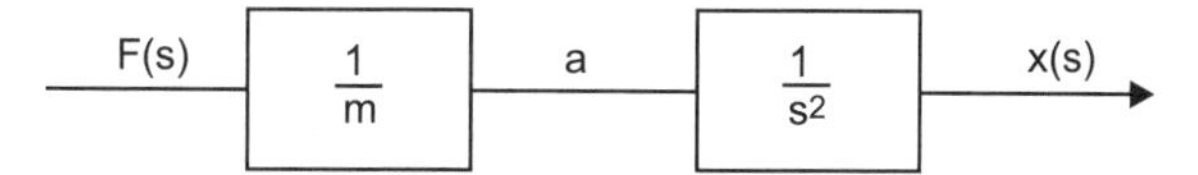

FIGURE 5.9 Open-loop system.

5.1.9 Acceleration Feedback

The feedback of acceleration actually obtained in a practically feasible system is required to control the system to achieve the desirable output. This can be achieved through the unity feedback system as shown by the block diagram in Figure 5.10.

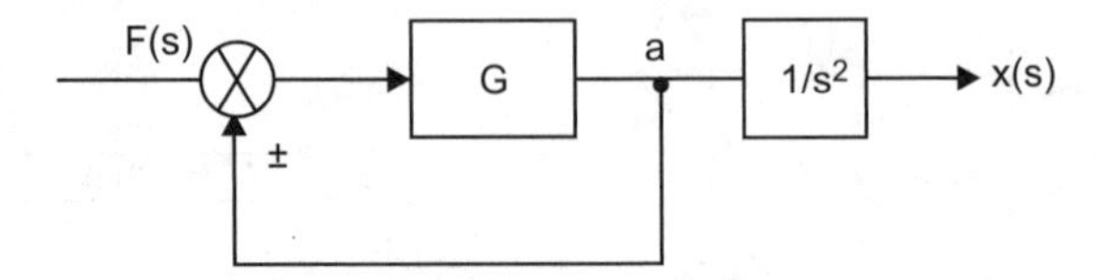

FIGURE 5.10 Unity feedback system.

The gain, G, is obtained by the block diagram algebra, by comparing the closed-loop transfer function with that of the open-loop transfer function as

$$G = \left(\frac{1}{m \pm 1}\right). \tag{5.20}$$

Figure 5.10 is the equivalent feedback circuit of Figure 5.9 given by the open-loop circuit. Eq. (5.20) shows that the gain is dependent on the mass of the system.

5.2 SPRING-MASS-DAMPER SYSTEM

A practical situation in mechanical design is the combination of all three elements: spring, damper, and mass. Figure 5.11 shows the schematic of such a system.

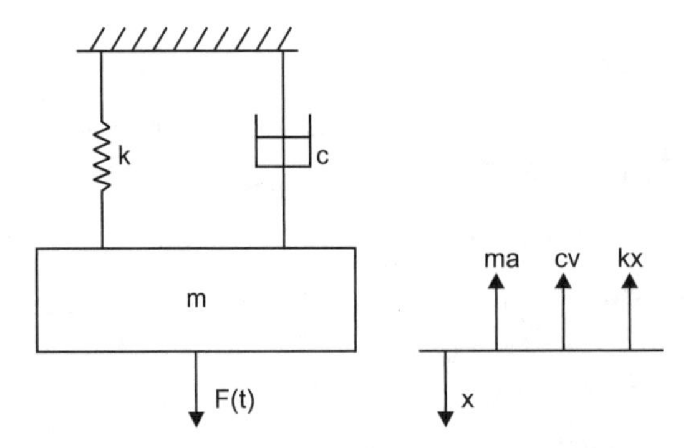

FIGURE 5.11 Spring-mass-damper system.

The force, F, acting on mass, m, has to balance the inertial force, the damping force, and the spring force.

$$\begin{Bmatrix}\text{Applied} \\ \text{force}\end{Bmatrix} = \begin{Bmatrix}\text{Inertial} \\ \text{force}\end{Bmatrix} + \begin{Bmatrix}\text{Damping} \\ \text{force}\end{Bmatrix} + \begin{Bmatrix}\text{Spring} \\ \text{force}\end{Bmatrix}$$

i.e.,
$$F(t) = ma + cv + kx \tag{5.21}$$

or
$$F(t) = \frac{m.d^2x}{dt^2} + \frac{c.dx}{dt} + kx. \tag{5.22}$$

By taking the Laplace transform of Eq. (5.22) we get the transfer function as

$$\frac{x(s)}{F(s)} = \frac{1}{s^2m + sc + k}. \tag{5.23}$$

Eq. (5.23) represents the second-order system. The response of such a system is given by considering the characteristic equation in the form

$$s^2m + sc + k = 0, \tag{5.24}$$

which is a quadratic equation in 's'. The roots are given by

$$s_{1,2} = \frac{-c \pm \sqrt{c^2 - 4mk}}{2m}. \tag{5.25}$$

Depending on the value of the discriminant $\sqrt{c^2 - 4mk}$, the responses are as shown in Figure 5.12.

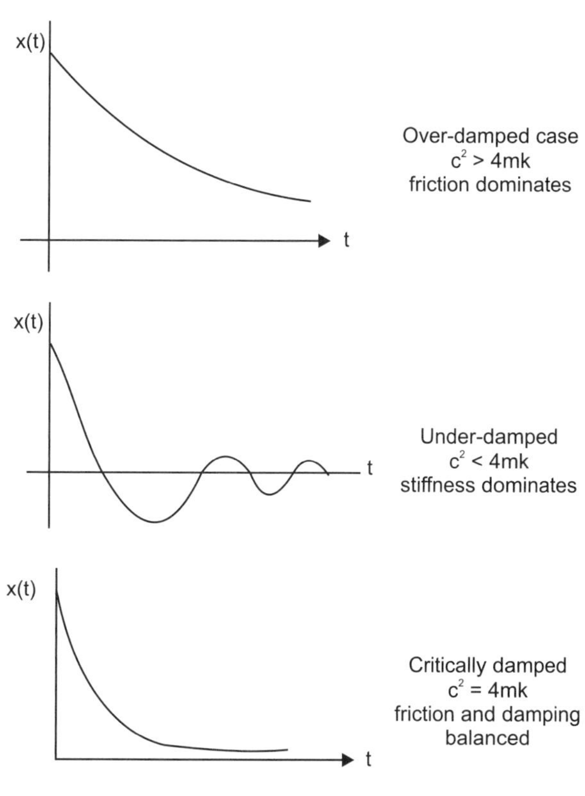

FIGURE 5.12 Responses.

The block diagram representation of the unity position and velocity feedback for a spring-mass-damper system is depicted in Figure 5.13.

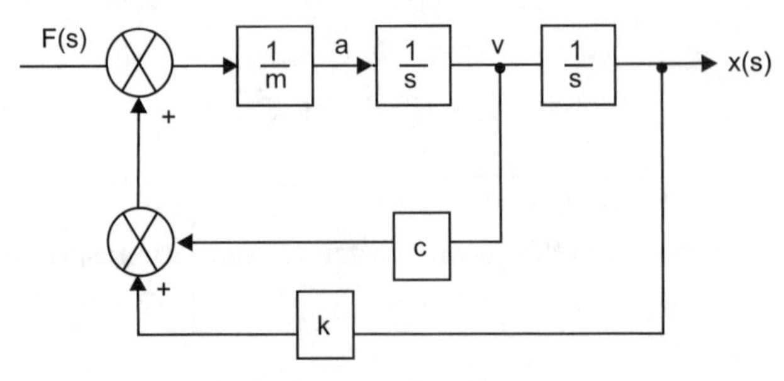

FIGURE 5.13 Block diagram of spring-mass-damper system.

5.3 AN UNCONVENTIONAL APPROACH

The second-order differential equation (5.22) can be solved unconventionally giving the solution in the form

$$a = \frac{F(t)}{k}$$

$$v = \left(1 - \frac{m}{c}\right)\frac{F(t)}{k}$$

$$x = \left(1 - \frac{c}{k}\right)\frac{F(t)}{k}. \tag{5.26}$$

In block diagram form it is shown in Figure 5.14, and the unity feedback system for position and velocity feedback is shown by the block diagram in Figure 5.15.

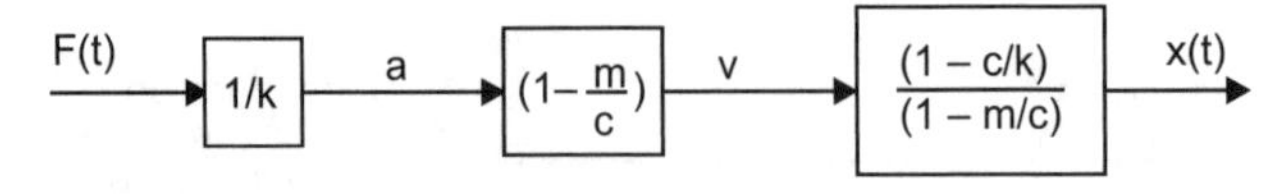

FIGURE 5.14 Open-loop system.

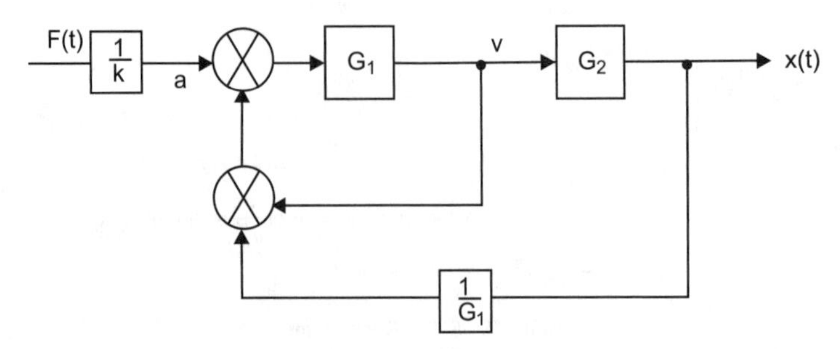

FIGURE 5.15 Closed-loop system.

5.3.1 Reduction of Figure 5.15

$$G_1 = (1 - m/c),$$

where $G_2 = (1 - c/k)$

and the ratio of output to input

$$\frac{x(t)}{F(t)} = \left(1 - c/k\right) \cdot \frac{1}{k} \cdot \tag{5.27}$$

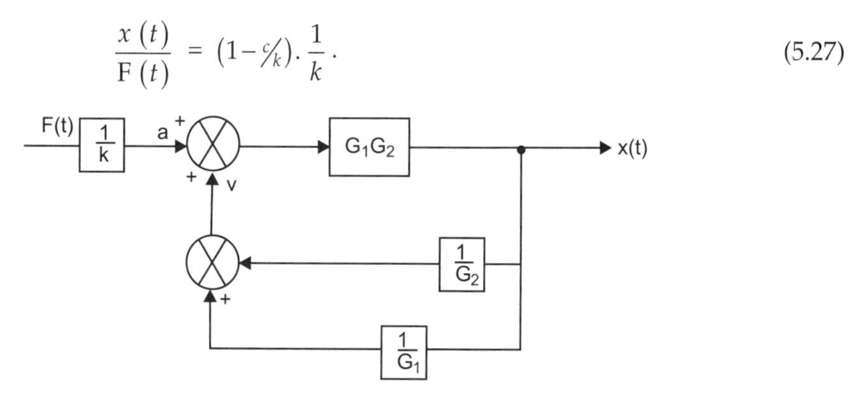

FIGURE 5.16 Reduction of closed-loop system.

5.4 ARRANGEMENT OF MECHANICAL ELEMENTS

5.4.1 Series Arrangement

Mechanical elements may be connected in series and represented by springs of different stiffness arranged in series. Such a system symbolically is shown in Figure 5.17.

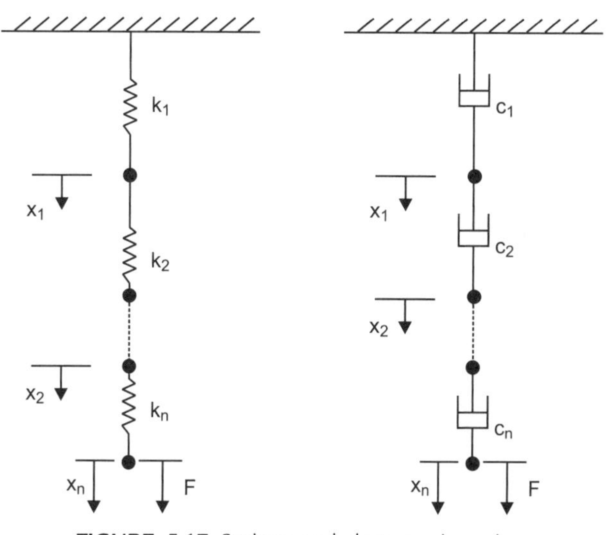

FIGURE 5.17 Springs and dampers in series.

The effective stiffness of such an arrangement is given by

$$\frac{1}{k_{eff}} = \frac{1}{k_1} + \frac{1}{k_2} + \ldots\ldots + \frac{1}{k_n}. \tag{5.28}$$

The dampers arranged in series as shown in Figure 5.17 can be represented by an equivalent damper with an effective damping coefficient given by

$$\frac{1}{c_{eff}} = \frac{1}{c_1} + \frac{1}{c_2} + \ldots\ldots + \frac{1}{c_n}. \tag{5.29}$$

5.4.2 Parallel Arrangement

There are mechanical systems which have to be represented by the springs and dampers arranged in parallel as shown in Figure 5.18. The analysis is done by obtaining the equivalent effective stiffness and damping coefficients.

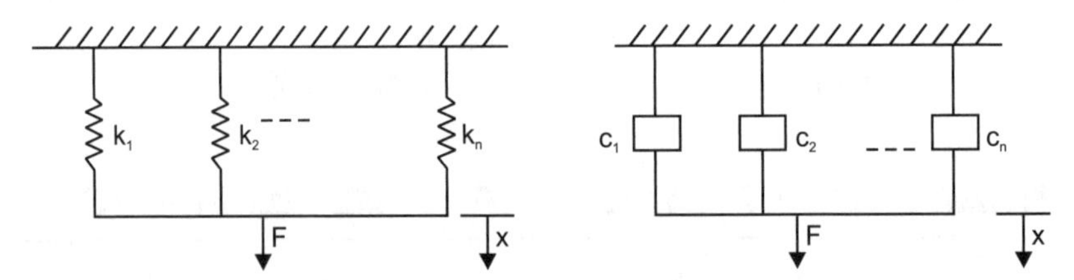

FIGURE 5.18 Springs and dampers arranged in parallel.

The effective spring stiffness can be given by

$$k_{eff} = k_1 + k_2 + k_3 + \ldots + k_n. \tag{5.30}$$

The effective damping coefficient is written as

$$c_{eff} = c_1 + c_2 + \ldots + c_n. \tag{5.31}$$

5.5 APPLICATION: RACK-AND-PINION ARRANGEMENT

The force, F, acting on the rack is the input that results in the rotation of the pinion with an angular velocity, ω, of the pinion with a torque, T. The rack-and-pinion arrangement is shown in Figure 5.19. If m is the mass of the rack moving with a velocity, v, the force, F, is given by

$$F = m\frac{dv}{dt}. \tag{5.32}$$

The force, F, is consumed in the friction and development of torque, T, in the pinion, hence,

$$F = \frac{T}{r} + c.v. \tag{5.33}$$

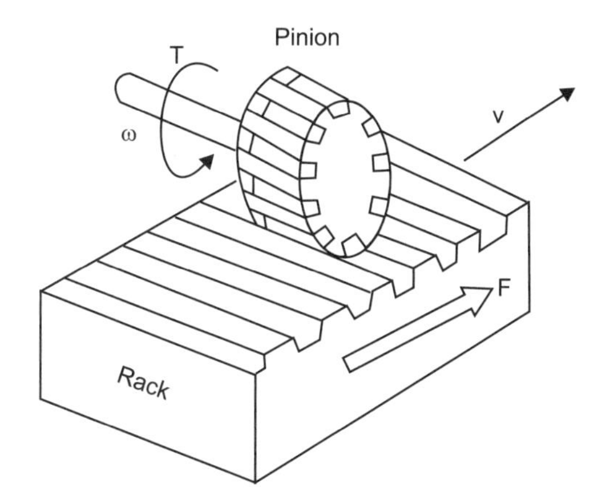

FIGURE 5.19 Rack-and-pinion system.

From Eqs. (5.32) and (5.33),

$$\frac{T}{r} + c.v = m\frac{dv}{dt}. \tag{5.34}$$

But
$$T = I\frac{d\omega}{dt} = \frac{I}{r}\frac{dv}{dt}. \tag{5.35}$$

Eqs. (5.34) and (5.35) yield

$$\frac{I}{r^2}.\frac{dv}{dt} + c.v = m\frac{dv}{dt}$$

or
$$\frac{dv}{dt}\left(m - \frac{I}{r^2}\right) - c.v = 0. \tag{5.36}$$

Eq. (5.36) is the governing differential equation for the rack-and-pinion arrangement. By transforming Eq. (5.36) into

$$F - \frac{I}{r}\frac{d\omega}{dt} - cr.\omega = 0. \tag{5.37}$$

Here, F is the input and ω is the output. The block diagram representation of the system can be obtained by taking the Laplace transform of Eq. (5.37)

i.e.,
$$F(s) = \left(\frac{sI}{r} + cr\right)\omega(s)$$

$$\frac{\omega(s)}{F(s)} = \frac{1}{\left(s\,{}^{I}\!/_{r} + cr\right)}. \tag{5.38}$$

The block diagram is given in Figure 5.20 and represents an open-loop system. This represents a first-order system.

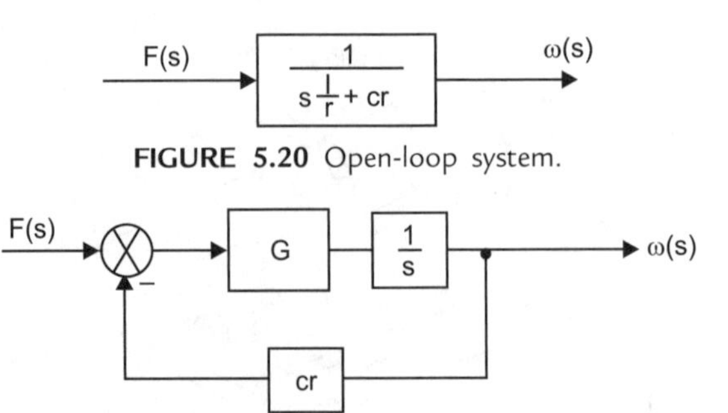

FIGURE 5.20 Open-loop system.

FIGURE 5.21 Closed-loop system.

The closed-loop system giving the velocity feedback is done by the circuit shown in Figure 5.21. The controller is designed by knowing the friction damping, *c*, the moment of inertia, I, and the radius, *r*, of the pinion. The angular velocity is measured and fed back to the controller using a tachogenerator.

Certain systems such as machine slides that get translational movement by a shaft and lead-screw arrangement can also be modeled in the same fashion. Other examples are a pulley and belt arrangement, the gear trains in a gearbox, and suspension and type arrangements. Before designing the controller for these systems to control the output, the modeling of the system with various elements into a governing differential equation is a necessity, and further Laplace transformation gives the transfer function.

5.6 ELEMENTS OF AN ELECTRICAL SYSTEM

An electrical system is identified by basically three elements—resistors, inductors, and capacitors. The voltage across a resistor as shown in Figure 5.22 is directly proportional to the current flowing through it. Here, the system is the resistor, R. The input is the potential difference, V, and the output is the current, I.

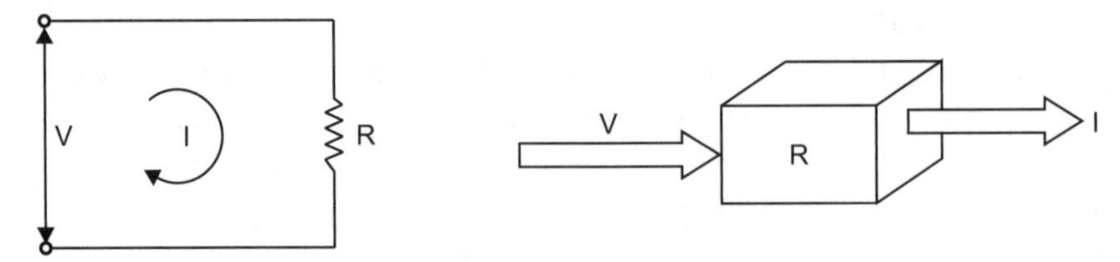

FIGURE 5.22 Resistor.

The governing equation is

$$V = I R. \tag{5.39}$$

In a time-varying circuit where the input and the output vary with time,

$$V(t) = R I (t). \tag{5.40}$$

By taking the Laplace transform of Eq. (5.40) the transfer function is given by

$$\frac{I(s)}{V(s)} = \frac{1}{R}. \tag{5.41}$$

The transfer function in Eq. (5.41) can be represented by an open-loop circuit as in Figure 5.23.

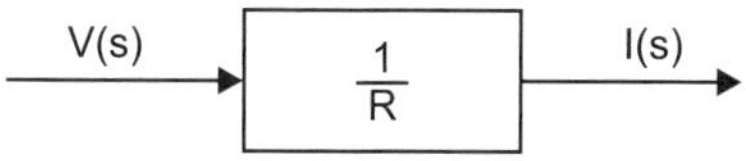

FIGURE 5.23 Open-loop function.

The unity feedback current control block diagram can be represented by Figure 5.24.

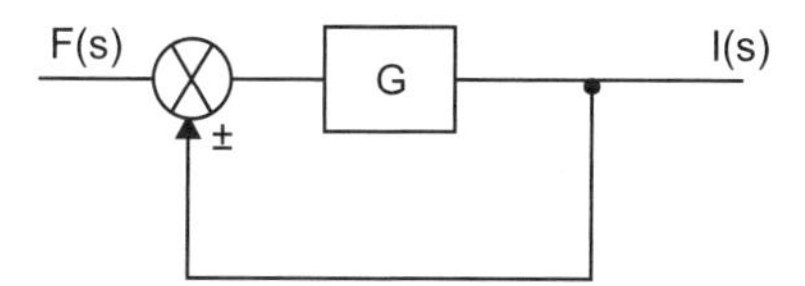

FIGURE 5.24 Closed-loop function.

To obtain G from the block diagram algebra

$$\frac{G}{1 \mp G} = \frac{1}{R}. \tag{5.42}$$

By simple manipulation of Eq. (5.42)

$$G = \frac{1}{(R \pm 1)}. \tag{5.43}$$

G is the closed-loop gain determined by the value of the resistor, R. The sign indicates the type of feedback, either positive or negative.

The inductor element connected in series with the resistor as shown in Figure 5.25 produces a current output to the voltage input governed by the relation

$$V(t) = RI(t) + L\frac{dI}{dt}. \tag{5.44}$$

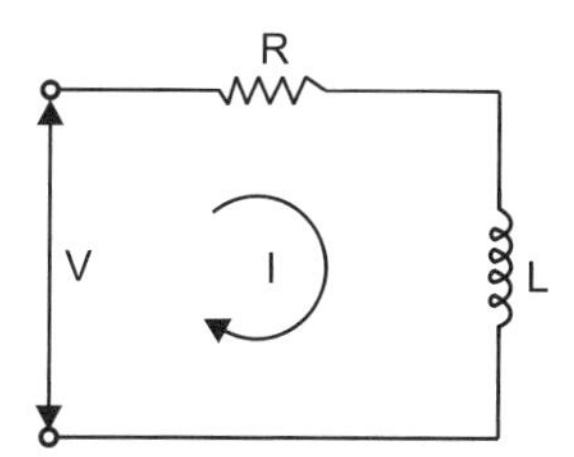

FIGURE 5.25 R-L circuit.

By the Laplace transformation, Eq. (5.44) yields

$$V(s) = RI(s) + sL\,I(s). \tag{5.45}$$

The transfer function is given by

$$\frac{I(s)}{V(s)} = \frac{1}{sL+R}. \tag{5.46}$$

Eq. (5.46) represents a first-order system for which the feedback control circuit is given by the closed-loop circuit in Figure 5.26.

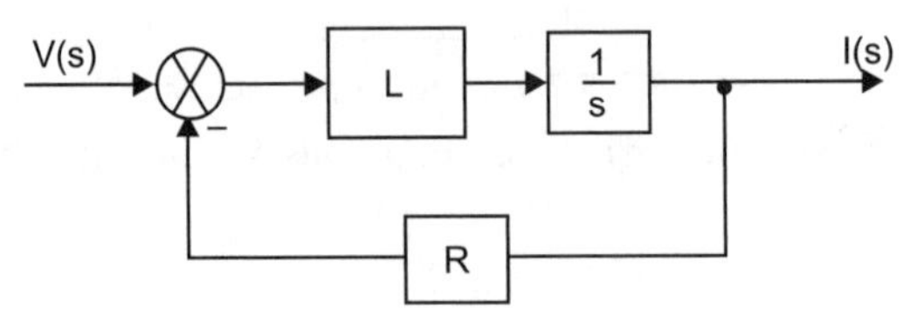

FIGURE 5.26 Closed-loop circuit.

The output current is controlled by the negative feedback circuit shown in Figure 5.26.

5.6.1 R-L-C Circuit

The capacitor arranged in series with the resistor and the inductor constitute a dynamic circuit as shown in Figure 5.27. The voltage input given across an R-L-C circuit results in an output current in a time-varying domain.

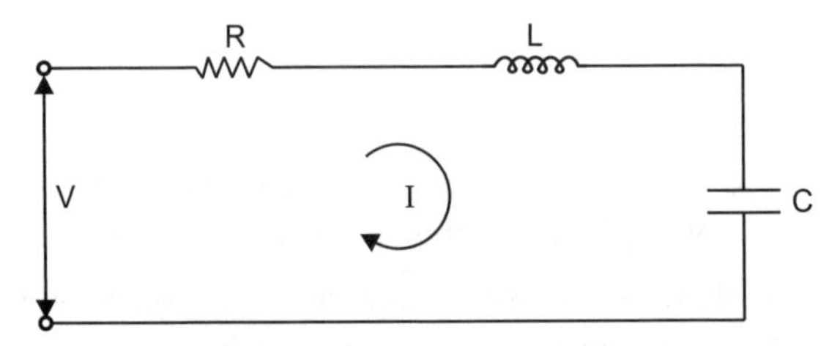

FIGURE 5.27 R-L-C circuit.

The governing equation that relates the output with the input can be written as

$$V(t) = RI(t) + L\frac{dI}{dt} + \frac{1}{C}\int I(t)\,dt. \tag{5.47}$$

Subjecting Eq. (5.47) to the Laplace transformation leads to

$$V(s) = RI(s) + sL\,I(s) + \frac{1}{sC}\,I(s). \tag{5.48}$$

The transfer function can be obtained by

$$\frac{I(s)}{V(s)} = \frac{sC}{sRC + s^2LC + 1} \tag{5.49}$$

$$= \frac{sC}{1+s(RC+sLC)}.$$

The closed-loop block diagram representation of Eq. (5.49) is shown in Figure 5.28.

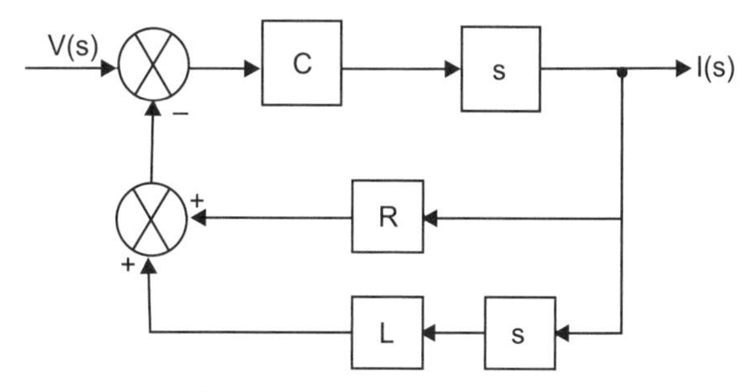

FIGURE 5.28 Closed-loop circuit.

The R-L-C circuit is a second-order system where the responses depend on the value of the resistor, inductor, and capacitor. The R-L-C parameters can be tuned suitably to get critical, over-damped, and under-damped responses to suit the application.

5.7 UNCONVENTIONAL SOLUTION TO THE R-L-C CIRCUIT

The dynamic Eq. (5.47) relating the input, V(t), and the output, I(t), is solved by an unconventional method to yield

$$\left.\begin{aligned} I(t) &= CV\,(t) \\ \frac{dI}{dt} &= \left(1-\frac{R}{L}\right) C.V\,(t) \\ \int I\,(t)\,dt &= (1-LC).\ C.V\,(t) \end{aligned}\right\}. \tag{5.50}$$

The block diagram representing Eq. (5.50) is presented in Figure 5.29.

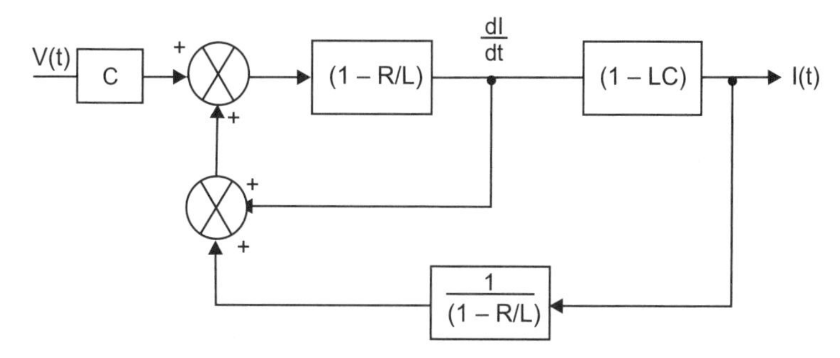

FIGURE 5.29 Unconventional block diagram.

The advantage of the unconventional approach is evident from the fact that there is no need for the Laplace transform, and the control circuit can be designed by playing with the R-L-C values.

5.8 APPLICATION TO D.C. SERVOMOTOR

A D.C. servomotor with inertia, I_m, transmits power with a torque, T_m. The output shaft of the motor drives a load of inertia, I_L, through a gear train of gear ratio, N. Let the damping in the motor and load be represented by the coefficients, c_m and c_L, respectively. Figure 5.30 shows a D.C. motor with the relevant circuit and load.

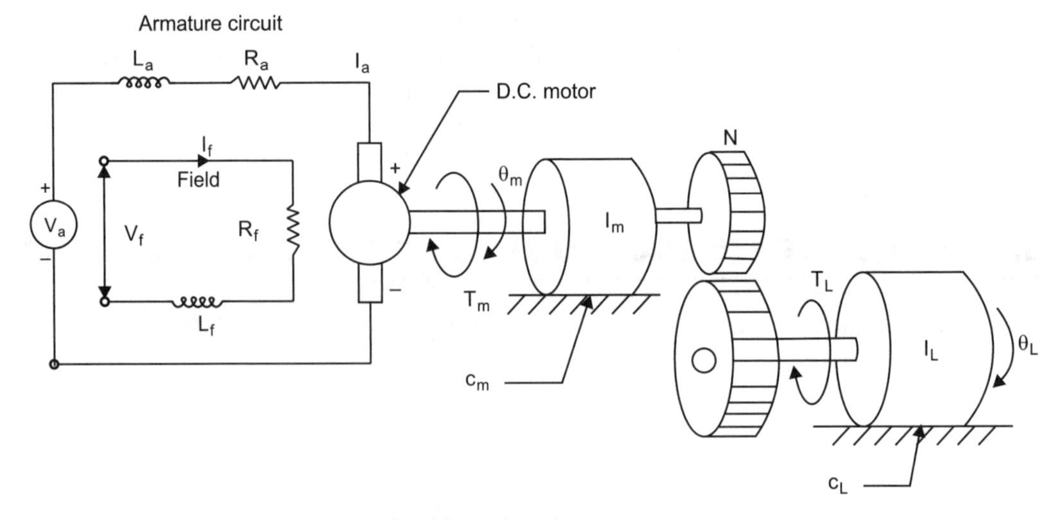

FIGURE 5.30 D.C. servomotor.

By the reduction model of the gear train we establish

$$\left.\begin{aligned} \theta_m &= N\,\theta_L \\ \dot{\theta}_m &= N\,\dot{\theta}_L \\ \ddot{\theta}_m &= N\,\ddot{\theta}_L \end{aligned}\right\}. \tag{5.51}$$

The torque equation governing the input torque and the output rotation can be derived in the following manner:

$$\begin{Bmatrix}\text{Torque developed}\\ \text{by the motor shaft}\end{Bmatrix} = \begin{Bmatrix}\text{Torque dissipated}\\ \text{on the motor}\end{Bmatrix} + \begin{Bmatrix}\text{Torque consumed by load}\\ \text{referred to the motor}\end{Bmatrix}$$

$$\begin{aligned} T &= T_m + (T_L)_m \\ &= \left[I_m\ddot{\theta}_m + c_m\dot{\theta}_m\right] + \left[I_L\ddot{\theta}_L + c_L\dot{\theta}_L\right]_m. \end{aligned} \tag{5.52}$$

But $\qquad N(T_L)_m = T_L$

and also from Eq. (5.51)

$$T = N\left[I_m\ddot{\theta}_m + c_m\dot{\theta}_m\right] + \left[I_L\ddot{\theta}_L + c_L\dot{\theta}_L\right]$$

$$= N^2\left[I_m\ddot{\theta}_L + c_m\dot{\theta}_L\right] + \left[I_L\ddot{\theta}_L + c_L\dot{\theta}_L\right]$$

$$T = \left(I_L + N^2 I_m\right)\ddot{\theta}_L + \left(c_L + N^2 c_m\right)\dot{\theta}_L$$

$$T = I_{eff}\,\ddot{\theta}_L + c_{eff}\,\dot{\theta}_L, \tag{5.53}$$

where $\qquad I_{eff}$ = effective inertia = $(I_L + N^2 I_m)$

c_{eff} = effective damping = $(c_L + N^2 c_m)$.

By taking the Laplace transform of Eq. (5.53)

$$T(s) = (s^2 I_{eff} + s\, c_{eff})\, \theta_L(s). \tag{5.54}$$

Electric modeling

$$V_a = R_a i_a + L_a \frac{di_a}{dt} + e_b.$$

The back emf e_b is

$$e_b = k_b\,\dot{\theta}_m, \text{ where } k_b = \text{constant of back emf.}$$

Hence, $$i_a(s) = \frac{V_a(s) - sk_b\,\theta_m(s)}{R_a + sL_a}$$

$$= \frac{V_a(s) - sN\,k_b\,\theta_L(s)}{R_a + s\,L_a}. \tag{5.55}$$

The torque developed by the motor is directly proportional to the armature current, i_a,

$$T = k_t.\, i_a$$

or $$T(s) = k_t\, i_a(s). \tag{5.56}$$

Using Eqs. (5.54), (5.55), and (5.56)

$$\left(s^2[I_{eff}] + s\,c_{eff}\right)\theta_L(s) = \frac{k_t\,[V_a(s) - sN\,k_b\theta_L(s)]}{R_a + sL_a}.$$

Rearranging terms

$$\frac{\left[(R_a + sL_a)\left\{s^2 I_{eff} + s\,c_{eff}\right\} + k_t\,s\,N\,k_b\right]}{k_t} = \frac{V_a(s)}{\theta_L(s)}.$$

Hence, the transfer function of output, θ_L, to input, V_a, is

$$\frac{V_a(s)}{\theta_L(s)} = \frac{k_t}{(R_a + sL_a)\{s^2 I_{eff} + s\, c_{eff}\} + sN\, k_t\, k_b}. \tag{5.57}$$

The open-loop transfer function is given in Figure 5.31.

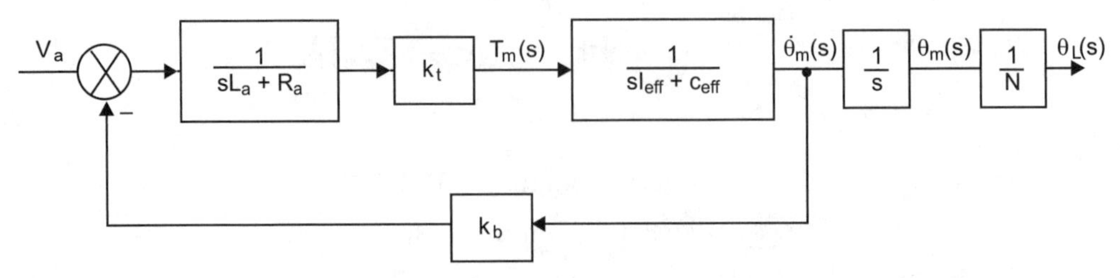

FIGURE 5.31 Open-loop transfer function.

The unity feedback positional control diagram is given in Figure 5.32.

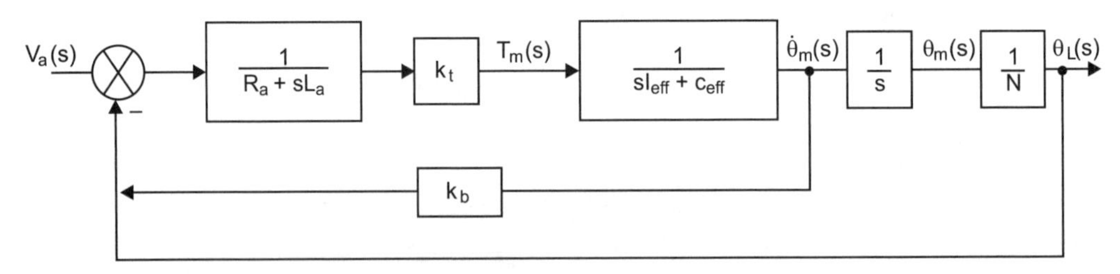

FIGURE 5.32 Unity feedback positional control.

5.9 HYDRAULIC SYSTEM MODELING

The hydraulic system components during transmission of energy through pressurized oil offer the following three types of resistances:

1. Resistance to motion
2. Resistance to acceleration
3. Resistance to deformation

5.9.1 Resistance to Motion

The oil flowing through hydraulic components such as valves, cylinders, and hoses with a flow rate, Q, experience the resistance to motion, R, that results in a change in pressure, Δp. The parameters Δp, Q, and R are related by the expression

$$R = \frac{\Delta p}{Q^n}. \tag{5.58}$$

For $n = 1$, Eq. (5.58) is linear exhibiting analogy to Ohm's law in electrical engineering with Δp comparing with voltage, V, and Q bears analogy to current, I. But in a mechanical situation the non-linearity exists with the index $n = 2$.

5.9.2 Resistance to Acceleration

Consider a mass, m, attached to a cylinder piston of area, A, as shown in Figure 5.33. The change of flow from Q_1 to Q_2 into the cylinder results in a change of pressure from p_1 to p_2.

Let V be the velocity with which the piston moves the mass. The work done in the process is the kinetic energy,

$$W_p = \frac{1}{2} m v^2. \tag{5.59}$$

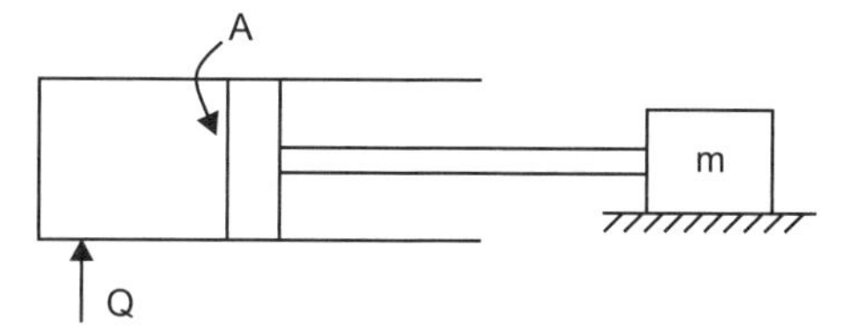

FIGURE 5.33 Resistance to acceleration.

But for continuity, Q = AV, and from Eq. (5.59)

$$W_p = \frac{1}{2} \frac{m}{A^2} Q^2. \tag{5.60}$$

The term $\frac{m}{A^2}$ = H is the resistance to acceleration.

Hence,
$$W_p = \frac{1}{2} HQ^2. \tag{5.61}$$

By definition, the power P is the change of work with respect to time, giving the relation,

$$P = \frac{dW_p}{dt} = HQ. \frac{dQ}{dt}. \tag{5.62}$$

But from hydraulic fundamentals

$P = \Delta p.\, Q$, which on comparison with Eq. (5.62) yields

$$\Delta p = H \frac{dQ}{dt}, \tag{5.63}$$

where $\frac{dQ}{dt}$ is the volumetric acceleration. Eq. (5.63) compares well with the relation in electrical sciences as

$$V = L.\frac{di}{dt}, \text{ where L is the inductance.}$$

Hence, the resistance to acceleration, H, is analogous to the inductance of the electrical circuit.

For a pipe of cross-section, A, and length, L, the resistance to acceleration to the oil of density ρ,

$$H = \frac{m}{A^2} = \frac{(Ah)\rho}{A^2} = \frac{h\rho}{A}. \tag{5.64}$$

For a rotary actuator of moment of inertia, I, and rotating with angular velocity ω,

$$P = \frac{d}{dt}\left(\frac{1}{2} I\, \omega^2\right).$$

If v is the volume displaced with flow rate Q,

$$\omega = \left(\frac{2\pi}{V}\right) Q$$

$$P = Q\, p = \left(\frac{2\pi}{V}\right)^2 IQ.\frac{dQ}{dt}.$$

Hence, the resistance to acceleration is

$$H = \left(\frac{2\pi}{V}\right)^2 .I. \tag{5.65}$$

5.9.3 Resistance to Deformation

If the change in pressure, Δp, in a hydraulic component results in a change in volume, Δv, due to deformation, then the resistance to deformation is defined by the ratio of Δp with that of Δv.

Hence, $$D = \frac{\Delta p}{\Delta v} \tag{5.66}$$

or $$D\, \Delta v = \Delta p$$

$$\Delta p = D.\, Q\, dt.$$

By integrating, $$p = D \int_0^t Q\, dt. \tag{5.67}$$

By comparing this with the relation in the electrical circuit

$$v = \frac{1}{C} \int i\, dt.$$

The resistance to deformation bears analogy to the reciprocal of the capacitance.

The resistance to deformation of certain hydraulic components is given in Table 5.1.

TABLE 5.1 Resistance to Deformation

Component	Symbol	D-resistance
1. Single-acting Spring return	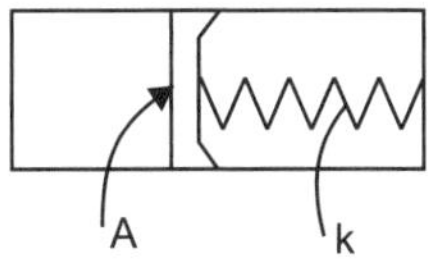 A k	$D = \frac{k}{A^2}$ k = spring rate
2. Linear actuator	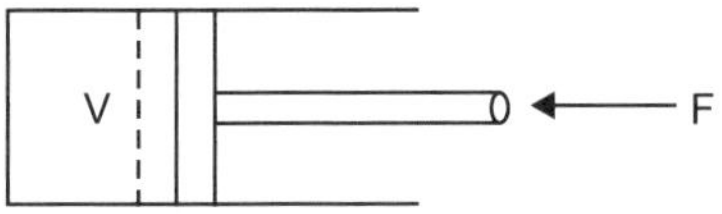 V F	$D = \frac{K}{V}$ K = bulk modulus
3. Accumulator	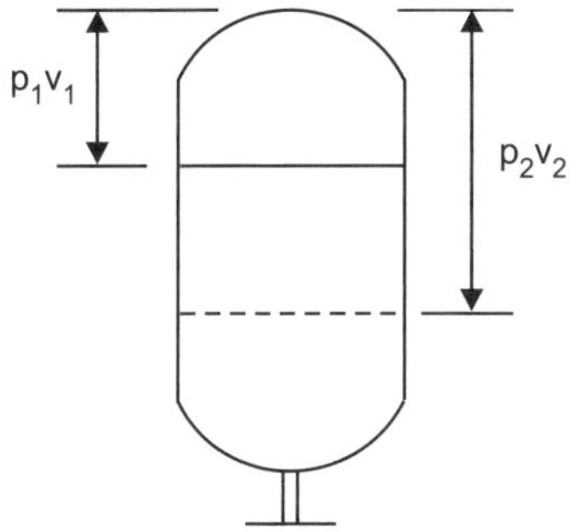 p_1v_1 p_2v_2	$D = \frac{p_2}{V_1}$

5.10 MODELING OF ACTUATORS

In this section the linear and rotary actuators are modeled for the transfer function that controls the output for the given input. Here, the input is the pressure energy and the output is the displacement.

5.10.1 Linear Actuator

The linear actuator represented by the hydraulic cylinder (double-acting) is modeled as shown. Figure 5.34 shows the linear actuator assumed.

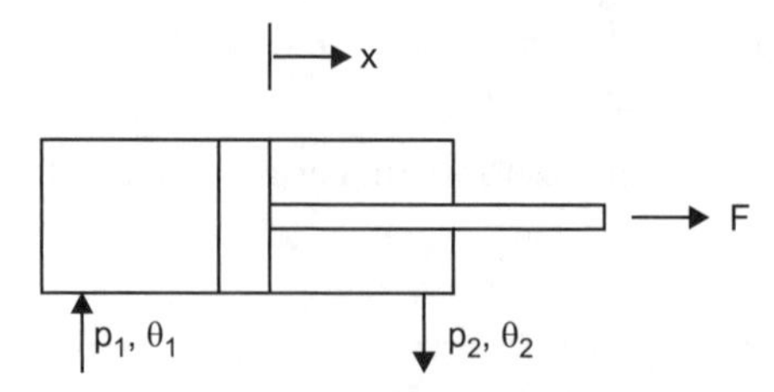

FIGURE 5.34 Linear actuator.

$$\text{Pressure force} = \begin{Bmatrix}\text{Force exerted} \\ \text{by the piston}\end{Bmatrix} + \begin{Bmatrix}\text{Force required to} \\ \text{overcome resistance}\end{Bmatrix}$$

Here, the cylinder is considered to be rigid and the deformation is neglected,

$$(p_1 - p_2) = \frac{F}{A} + (\Delta p)_R + (\Delta p)_H$$

$$\Delta p = \frac{F}{A} + R\Delta Q + H.\frac{d\theta}{dt}$$

$$= \frac{F}{A} + R.A\frac{dx}{dt} + H.A\frac{d^2x}{dt^2}. \tag{5.68}$$

Taking the Laplace transform of the differential equation (5.68),

$$\Delta p\ (s) = \frac{F(s)}{A} + sRA\ x(s) + s^2\ HA\ x(s)$$

$$\frac{x\,(s)}{\Delta p(s)} = \frac{1}{\dfrac{F\,(s)}{A\,x(s)} + sRA + s^2HA}. \tag{5.69}$$

Eq. (5.69) gives the transfer function, where x displacement is the output and the change in pressure, Δp, is the input.

5.10.2 Rotary Actuator

In a rotary actuator the input pressure, p, is responsible for the development of torque against the resistance motion and acceleration. The pressure consumed to overcome resistance to deformation is considered negligible. Hence, the input is the pressure force and the output is the rotation.

By balance of difference in pressure with that of the torque developed and the resistive forces, we write

$$\Delta p = \frac{T}{V_m} + (\Delta p)_R + (\Delta p)_H$$

$$= \frac{T}{V_m} + R.\Delta Q + H.\frac{d\theta}{dt}$$

$$= \frac{T}{V_m} + R.V_m\frac{d\theta}{dt} + HV_m\frac{d^2\theta}{dt^2},$$

where V_m is the volume swept by the motor.

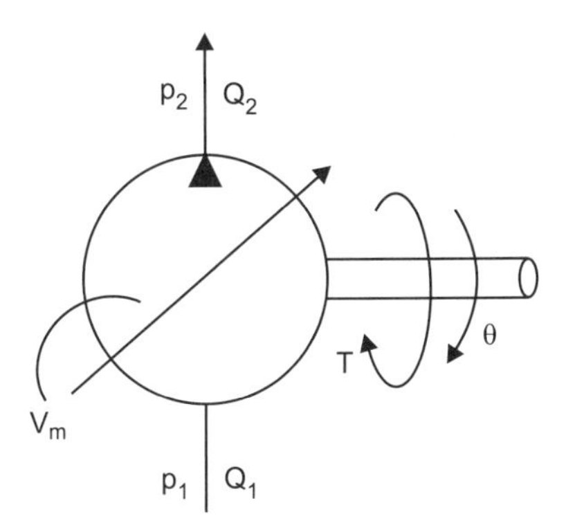

FIGURE 5.35 Rotary actuator.

By applying the Laplace transform,

$$\Delta p\ (s) = \frac{T(s)}{V_m} + sRV_m\ \theta(s) + s^2\ HV_m\theta\ (s).$$

The transfer function

$$\frac{\theta\ (s)}{\Delta p(s)} = \frac{1}{\dfrac{T(s)}{V_m\theta(s)} + sRV_m + s^2\ HV_m}. \tag{5.70}$$

Eq. (5.70) represents a transfer function for a second-order system.

5.11 MODELING OF CONTROL VALVES

In general, there are three types of control valves classified based on the parameters they control. The throttle valve controls the flow of oil, the relief valve controls the pressure in the circuit, and the spool valve controls the direction of flow. The modeling of these valves plays an important role in the design of the hydraulic circuit leading to the development of the power pack for various applications. The following paragraphs deal with the modeling of control valves.

5.11.1 Flow Control Valves

A throttle valve giving a variable flow rate of oil in the circuit is shown in Figure 5.36. If Q is the volume rate of flow and Δp is the pressure drop across the circuit, then the variable resistance, R_n, is given by Eq. (5.58) as

$$R_n = \frac{\Delta p}{Q^2}. \tag{5.71}$$

If Ω is the fixed resistance of design and ϕ is the design characteristic parameter, then the variable resistance

$$R_n = \left(\frac{\Omega}{\phi^2}\right). \tag{5.72}$$

From Eqs. (5.71) and (5.72), by manipulation, we get

$$\Delta p = \left(\frac{\Omega}{\phi^2}\right) Q^2, \tag{5.73}$$

where Δp is the input and Q is the output that both vary with time. By applying the Laplace transform,

$$\Delta p\ (s) = \left(\frac{\Omega}{\phi^2}\right) [Q(s)]^2$$

$$\frac{\Delta p\ (s)}{Q\ (s)} = \left(\frac{\Omega}{\phi^2}\right) Q(s). \tag{5.74}$$

Eq. (5.74) is the transfer function needed for control of the output.

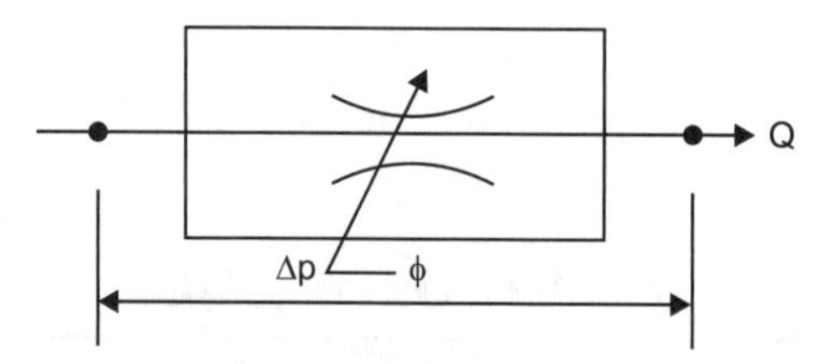

FIGURE 5.36 Throttle valve.

Flow control valves with variable flow are characterized by two specific parameters:

1. Flow sensitivity defined by

$$C_Q = \frac{dQ}{d\phi} = \left(\frac{\Delta p}{\Omega}\right)^{0.5}. \tag{5.75}$$

It is the square root of the ratio of the pressure drop for a fixed resistance of the valve.

2. Pressure sensitivity by definition is $\frac{d(\Delta p)}{d\phi}$. From Eq. (5.73),

$$C_P = \frac{d(\Delta p)}{d\phi} = \frac{-2Q^2\,\Omega}{\phi^3}. \tag{5.76}$$

The pressure sensitivity is directly proportional to the fixed resistance and the square of the flow at that resistance, but inversely proportional to the cube of the characteristic parameter, ϕ, which is design specific. The negative sign signifies a decrease in ϕ and results in an increase in pressure drop.

5.11.2 Relief Valves

The pressure control carried out by relief valves is modeled to predict the performance of the hydraulic circuit. A relief valve is schematically shown in Figure 5.37.

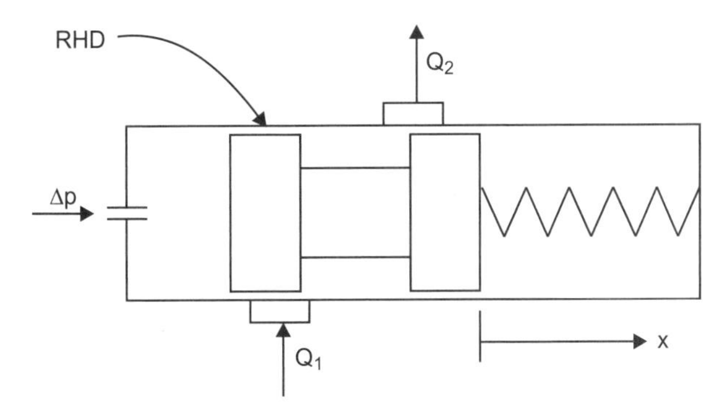

FIGURE 5.37 Relief valve schematic.

The pilot pressure balances the spool in the value against the spring force. Let ΔQ be the change in flow and Δp be the change in pressure and x the spool displacement. The flow change is effected by the three hydraulic resistances to the flow:

$$\left\{\begin{matrix}\text{Drop in flow rate}\\ \text{through valve}\end{matrix}\right\} = \left\{\begin{matrix}\text{Drop due to resis-}\\ \text{tance to motion}\end{matrix}\right\} + \left\{\begin{matrix}\text{Drop due to resista-}\\ \text{nce to acceleration}\end{matrix}\right\} + \left\{\begin{matrix}\text{Drop due to resista-}\\ \text{nce to deformation}\end{matrix}\right\}$$

$$\Delta Q = (\Delta Q)_R + (\Delta Q)_H + (\Delta Q)_D. \tag{5.77}$$

By manipulation of Eqs. (5.58), (5.63), and (5.67),

$$(Q_1 - Q_2) = \frac{\Delta p}{R} + \frac{\Delta p}{H}.\,dt + \frac{1}{D}\frac{\Delta p}{dt}$$

or $$\frac{dQ}{dt} = \frac{1}{R}\frac{d}{dt}\Delta p + \frac{1}{H}\Delta p + \frac{1}{D}\frac{d^2}{dt^2}\Delta p. \tag{5.78}$$

By applying the Laplace transform to Eq. (5.78)

$$sQ(s) = \frac{s}{R}\Delta p(s) + \frac{1}{H}\Delta p(s) + \frac{s^2}{D}\Delta p(s).$$

The transfer function is given by

$$\frac{Q(s)}{\Delta p(s)} = \frac{1}{s}\left(\frac{s}{R} + \frac{1}{H} + \frac{s^2}{D}\right) = \frac{sHD + RD + s^2RH}{SRHD}. \tag{5.79}$$

It is evident from Eq. (5.79) that the output, Q, and the input of valve, Δp, represent a second-order system.

The values of R, H, and D can be tuned to a critical or over-damped situation to prevent oscillation in flow.

5.11.3 Direction Control Valves

The direction control valve shown in Figure 5.38 executes the function of forward, reverse, and the idling operations of the actuator movements. The connected actuator can be linear or rotary. The motion is effected by the position of the spool operated by the solenoid or pilot line.

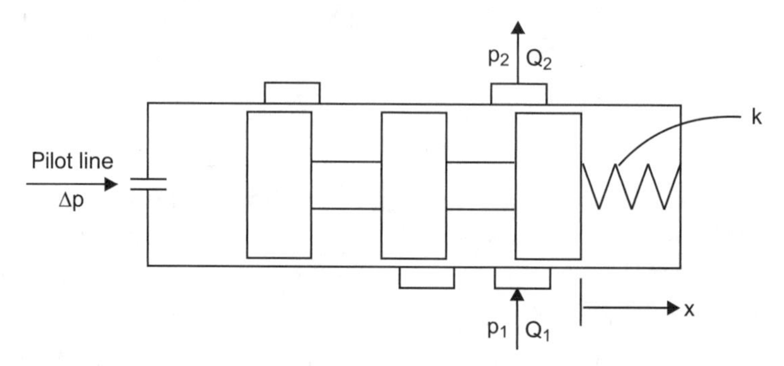

FIGURE 5.38 Direction control valve.

The various resistances offered by the spool result in a drop in pressure and flow rate, considering the change in flow rate the input and the spool displacement, x, the output against the spring force, F, with a spring rate, k.

The spring force, F, and the deflection, and spring rate, k, are related by

$$F = kx. \tag{5.80}$$

But $$F = \Delta pA.$$

$$kx = [(\Delta p)_R + (\Delta p)_H + (\Delta p)_D]A. \tag{5.81}$$

From Eqs. (5.58), (5.63), and (5.67)

$$kx = \left[R\Delta Q + H\frac{dQ}{dt} + D\int \Delta Q\,dt\right]A. \tag{5.82}$$

By taking the Laplace transform of Eq. (5.82),

$$\frac{k}{A}x(s) = RQ(s) + sHQ(s) + \frac{D}{s}Q(s).$$

$$\text{The transfer function} = \frac{x(s)}{Q(s)},$$

$$\frac{x(s)}{Q(s)} = \frac{A}{sk}\left[sR + s^2H + D\right]. \tag{5.83}$$

Eq. (5.83) is the transfer function representing the second-order system. The oscillation in movement of the actuators can be prevented by choosing tuned parameters for R, H, and D that play an important role in the design of direction control valves.

5.12 THERMAL SYSTEMS

Elements of mechanical systems that transfer heat are made of solids, liquids, and gases. Solids transfer heat by conduction, and the liquids and gases transfer heat by convection. These media offer resistance to the flow of heat. They are also responsible in the process of storage of heat. Since the heat lacks the property of inertia there is no resistance to acceleration in thermal systems. Hence, thermal systems are characterized by thermal resistance and thermal capacitance.

5.12.1 Thermal Resistance

Consider a metallic rod as shown in Figure 5.39 which is heated at end, A, to a temperature, T_1. Heat is conducted through the rod to the other end, B, to maintain a temperature, $T_2 < T_1$. The drop in temperature is attributed to resistance, R, offered by the rod.

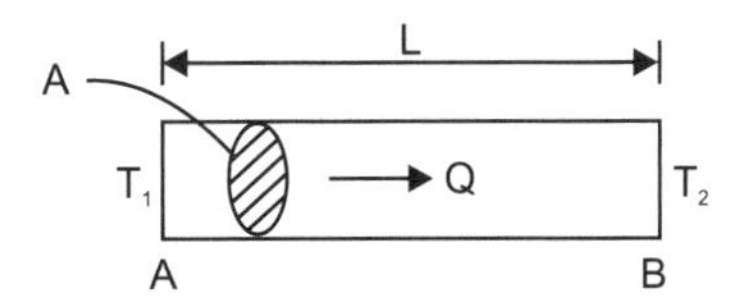

FIGURE 5.39 Circular rod.

If Q is the heat transfer rate, it bears an analogy to the current of the electrical system and the temperature difference compares with the potential difference, then we can relate it to resistance by

$$Q = \frac{(T_1 - T_2)}{R}. \tag{5.84}$$

But the rate of heat transfer is given by

$$Q = kA\,\frac{(T_1 - T_2)}{L}, \tag{5.85}$$

(through solids)

where k is the thermal conductivity, A is the area of cross-section, and L is the length of the rod. Comparison of Eqs. (5.84) and (5.85) yields,

thermal resistance,

$$R = \frac{L}{kA}. \tag{5.86}$$

But the heat transfer through fluids has a different mechanism, *i.e.*, convection. Hence, the governing equation is

$$Q = hA\,(T_1 - T_2). \tag{5.87}$$

Comparison of Eq. (5.87) with Eq. (5.84) gives

$$R = \frac{1}{hA}, \tag{5.88}$$

where h is the heat transfer coefficient for convection.

5.12.2 Thermal Capacitance

Thermal systems have a capacity to absorb heat with an increase in internal energy. This capability defines the thermal capacitance as a ratio of change in the heat transfer rate to the rate of change of temperature. Hence,

$$\Delta Q = C\,\frac{dT}{dt}, \tag{5.89}$$

where C is the thermal capacitance.

The process of heat absorption can be analyzed to formulate

$$\Delta Q = \text{rate of change of internal energy}$$

$$(mc) = \text{rate of change of temperature.}$$

Hence,

$$\Delta Q = mc\,\frac{dT}{dt}. \tag{5.90}$$

Comparing Eq. (5.90) with Eq. (5.89) the thermal capacitance

$$C = mc, \tag{5.91}$$

where m is the mass and c is the specific heat capacity.

5.13 MODELING OF THERMAL SYSTEMS

Consider a thermometer dipped in a container with liquid of temperature, T_1. The temperature indicated by the thermometer is T_2. The rate of heat transfer

$$Q = \frac{T_1 - T_2}{R}. \tag{5.92}$$

But the amount of heat absorbed in the process is given by

$$Q = C\frac{dT}{dt}. \tag{5.93}$$

Equating Eqs. (5.92) and (5.93),

$$C\frac{dT}{dt} = \frac{T_1 - T_2}{R}.$$

Hence,

$$RC\frac{dT}{dt} + T_2 = T_1. \tag{5.94}$$

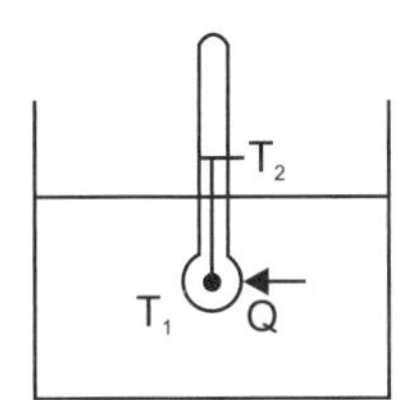

FIGURE 5.40 Thermometer system.

Eq. (5.94) governs the heat transfer process. To design a controller for temperature control, by taking the Laplace transform of Eq. (5.94),

$$sRCT_1(s) + T_2(s) = T_1(s).$$

Hence,

$$\frac{T_2(s)}{T_1(s)} = (1 + sRC). \tag{5.95}$$

Eq. (5.95) is the transfer function with T_1 as the input and T_2 as the output, which represents a first-order system.

5.13.1 Temperature Controller

Consider a bath of liquid with temperature, T_1, and the liquid in the bath is heated by a coil that heats it with a rate of heat transfer, Q_1. The rate heat is transferred to an ambient of temperature, T_0, is Q_2. The bath is shown in Figure 5.41.

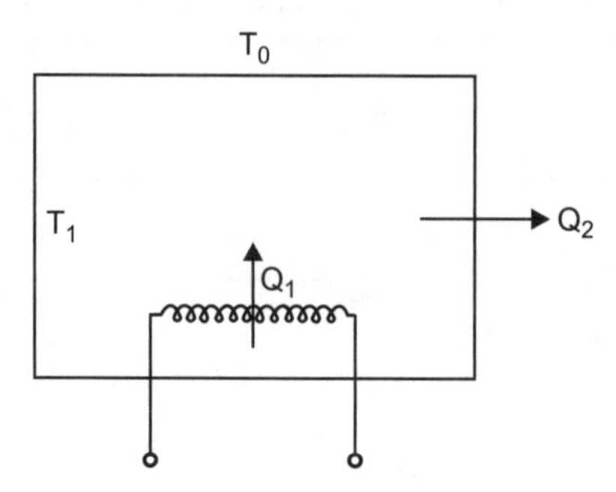

FIGURE 5.41 Heating of a liquid bath.

By definition of resistance,

$$Q_2 = \frac{T_1 - T_0}{R}. \tag{5.96}$$

Heat absorbed by liquid,

$$(Q_1 - Q_2) = C\frac{dT}{dt}. \tag{5.97}$$

Substituting Eq. (5.96) into Eq. (5.97),

$$Q_1 - \left(\frac{T_1 - T_0}{R}\right) = C\frac{dT}{dt}. \tag{5.98}$$

Rearranging Eq. (5.98),

$$RC\frac{dT}{dt} + dT = Q_1 R. \tag{5.99}$$

Taking the Laplace transform of (5.99),

$$sRC\ T(s) + T(s) = RQ_1(s).$$

The transfer function,

$$\frac{T(s)}{Q_1(s)} = \frac{R}{1 + sRC}. \tag{5.100}$$

Equation (5.100) is the transfer function of a first-order system. The closed-loop circuit for the transfer function is given by

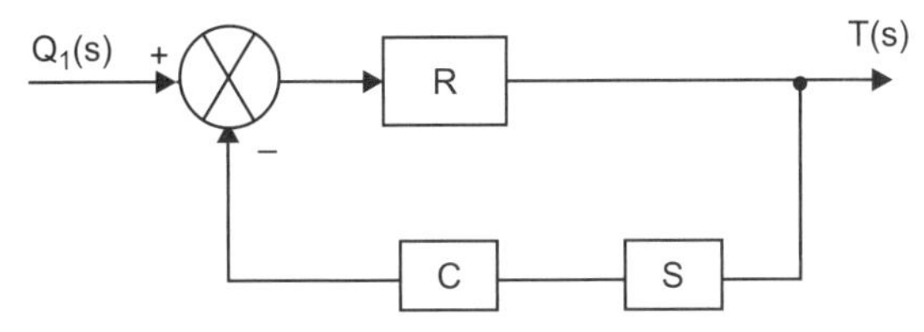

FIGURE 5.42 Feedback block diagram.

In the transfer function of Eq. (5.100) the input is the heat, Q_1, from the coil and the output is the temperature rise of the bath with respect to the ambient. The desired temperature is obtained by the heating rate change by feedback of the actual input. The content of the bath and dimensions of the bath give the values R (resistance) and capacitance (C).

EXERCISES

1. In terms of input and output, model the mechanical elements that provide:
   (*a*) Resistance to deformation
   (*b*) Resistance to velocity
   (*c*) Resistance to acceleration
2. Obtain the transfer function for a spring element. Explain how it can be used for position feedback.
3. Model the damper element for velocity feedback.
4. How does the mass of mechanical elements contribute to motion analysis?
5. Discuss the response of a spring-mass-damper system as a mechanical element.
6. Give a block diagram to show the velocity and position feedback of a mechanical system. Explain the transfer function.
7. Obtain the expression for effective stiffness of (*a*) springs in series and (*b*) springs in parallel. Give the electrical analogy.
8. Obtain the expression for effective damping of (*a*) dampers in series and (*b*) dampers in parallel.
9. Derive the governing equation of motion for a rack-and-pinion arrangement. Obtain the transfer function as a ratio of angular velocity of pinion as output and force on pinion as input.
10. What are the elements of an electrical system? Model them individually in terms of current and voltage.
11. Model the first-order system of an electrical circuit.
12. Discuss the R-L-C circuit for transient response. What is the order of the system?
13. Write the transfer function and block diagram for an R-L-C circuit.

14. Model a D.C. motor to control the rotation for an input voltage to the armature.
15. Give a block diagram showing the position and velocity feedback of a D.C. servomotor.
16. How is the resistance to motion defined in a hydraulic system? Give an electrical analogy.
17. Derive the expression for resistance to acceleration in a hydraulic system. Give the electrical analogy to it.
18. Establish an electrical analogy to the resistance to deformation in hydraulic systems.
19. Give (derive) the expression that governs the input and output relation in a linear hydraulic actuator.
20. Establish a relation between rotation and fluid pressure in a rotary hydraulic actuator.
21. Obtain the transfer function to model the flow through a throttle valve.
22. Define flow sensitivity and pressure sensitivity related to flow control valves.
23. Derive the governing equation for a relief valve and obtain the transfer function.
24. How is a directional control valve modeled to get the transfer function?
25. Define thermal resistance and obtain the expression for the thermal resistance of a
    (*a*) Conducting system
    (*b*) Convective system
26. How is thermal capacitance defined? Obtain the expression relating heat, thermal capacitance, and temperature.
27. Obtain an equation that governs the heat transfer process in a thermometer.
28. Give an example for a first-order thermal system and obtain the transfer function for the temperature controller.

CHAPTER 6

ELEMENTS OF MACHINE TOOLS

Machine tools, in general, have tool-holding and driving systems, work- (job) holding and driving systems, and support structures to rigidly accommodate the two systems. The translatory lines and rotary motions of tools and workpieces are common in machine tools. For guidance and positioning, linearly moving systems require guide ways. The guides can be the sliding type or rolling type. The rotary motions are provided by machine spindles supported on bearings. There are various types of spindles designed to carry static, dynamic, and thermal loads. Bearings can be sliding type journal bearings or rolling type anti-friction bearings. The conversion of rotary motion to linear motion is accomplished by screws and not arrangements. Especially in CNC machines that have very high precision, ball or roller type recirculating screws and nuts are used. Each part type used in machine tools has its own advantages and disadvantages. The selection of specific types of guides, spindles, bearings, or screws/nuts depends on parameters such as load, speed, size, and accuracy of rotation and position. Frictionless systems have application in limited motion devices. Automation in machine tools, for better dimensional control over machined parts, has lead to unconventional type systems, replacing conventional systems. The design for static and dynamic stiffness of parts has lead to accurate control that favors using the mechatronic principle in the making of sophisticated machine tools that are interfaced with computers.

The content of this chapter covers the following topics:

- Types of machine tool structures and design requirements.
- Types of loads on structures.
- Types of guide ways—slide ways, anti-friction ways, and their classification, principles, and configurations.
- Stick-slip phenomena.
- Hydrostatic and hydrodynamic slide ways.
- Comparison of slide ways and anti-friction ways.
- Recirculating ball screws and nuts.
- Concept of pre-loading of ball nuts.
- Roller screws and planetary roller screws, advantages.
- Spindles—functions, types of noses and loads, and selection criteria.
- Spindle bearings—types, classification.
- Principle of sliding bearings. Hydrostatic and hydrodynamic bearings.
- Bearing materials.
- Anti-friction bearings—different configurations and applications, design considerations.
- Comparison of sliding and anti-friction bearings.
- Pre-loading of anti-friction bearings—methods.
- Advantages and disadvantages of anti-friction bearings.
- Frictionless bearings—types.
- Advantages and disadvantages of frictionless bearings.

6.1 STRUCTURES

A machine tool is characterized by three types of functions:

- Work- (job) holding and driving
- Tool-holding and driving
- Rigidly supporting

The tables and their mechanisms, the chucks and their actuation, are the work-holding and driving structures. The spindles, arbors, carriages, turret heads, etc., are the tool-holding and driving structures. The beds, columns, frames, bases, etc., are the support structures. Based on the above functions machine tool structures are grouped into the following categories:

Category 1: beds, bases, columns, frames on which different sub-assemblies are mounted.

Category 2: box-type housings in which the gear box and spindle drives are housed.

Category 3: table, carriage, knee, and tailstock that serve to hold and move the workpiece and the cutting tool.

The following requirements should be kept in mind while designing and manufacturing machine tool structures:

- The major, relatively moving mating surface should be machined to a high degree of accuracy.
- The geometrical accuracy of a structure should be maintained throughout its life.
- The shape and size should ensure safe operation and maintenance/repair.
- The stresses and deformations in the elements of mechanisms should be within acceptable limits.

6.2 DESIGN CONSIDERATIONS OF STRUCTURES

The important design considerations to be understood by machine tool designers (for design of structural parts) are:

- High wear resistance of guiding and guided surfaces.
- Low thermal resistance to dissipate heat generation and minimize the effects of thermal loading.
- Static stiffness should be high enough to limit deformation within a permissible value.
- High fatigue strength to withstand the cyclic loading on the structure.
- High dynamic stiffness to withstand vibratory forces encountered during operation.
- Corrosion resistance should be high for long life.
- Dimensioning and tolerances should favor an easy process plan and manufacture to keep costs low.
- Shape and size should enable easy assembly and disassembly for maintenance and repair.

6.3 LOADS ON STRUCTURES

- *Static loads:* The loads that can be expected to act on the elements of machine structures do not vary with time:
  - The thrust loads act axially on the members.
  - The bending loads act transversely.
  - The torsion loads act on rotating members that tend to twist the members.
- *Dynamic loads:* Time-varying loads, either axial, bending, or torsion, lead to cyclic loads that can be reversed, repeated, or fluctuating. The effect of these loads results in a decrease in strength at which the structure may fail. Such strength is called *endurance strength*. The machine tools in operation are generally subjected to fatigue loads that can develop fatigue cracks at certain cycles of operation. Furthermore, when the natural frequencies of the elements match with the frequency of the dynamic loads this may result in resonance of vibration with large amplitude leading to a breakdown of parts. Under dynamic loads, the frames, beds, and columns can vibrate by the action of cutting forces.
- *Thermal loads:* The heat generated by friction in relative moving parts results in expansion of the parts. Generally, machine tool parts are made of metallic parts that expand when heated. The deformation due to thermally induced loads results in thermal stresses. The selection of materials of low thermal expansion result in high thermal resistance. Because of this the dissipated heat should be removed by a proper cooling system. In unavoidable circumstances, compensatory allowances in the dimensions to cater to the expansion due to thermal loads can be provided. Ignorance of the thermal effect can seriously affect the accuracy and performance of machine tools that are sophisticated.

6.4 GUIDE WAYS

The systems carrying the tool or the structures supporting the workpiece should be moved systematically and accurately. There is the need for a guide way to move in a predetermined path.

Examples include:

- The carriage moving on the bed of the lathe.
- The knee in a milling machine moving vertically.
- The drill head moving radially on the arm of a radial drilling machine.

- The spindle head of a grinding machine (vertical) moving along the column.
- The boring head moving vertically in the guide ways of a column.
- The turret with tools moving circularly on circular guide ways.
- The broaching gripper guided by vertical columns.
- The boring tool guided in the quill of a boring machine.

6.5 CHARACTERISTICS OF GOOD GUIDE WAYS

- Surface of guide ways should have a good finish along with controlled dimensional accuracy.
- Minimum deviation in the travel about the actual path from the predetermined path.
- Initial accuracy of manufacture to determine reliability and durability.
- Guide ways should ensure minimum friction and wear to prevent power loss and ensure long life.
- Variation in the coefficient of friction results in a jerky motion due to stick-slip phenomena which should be avoided.
- Bending and torsional rigidity should be sufficiently high to minimize deflection and twist.
- High damping ability aids in the dynamic stability of guide ways.
- Thermal effects should be minimized by dimensional compensation and cooling by lubrication.
- Wear-compensating elements can help reconditioning and preventive maintenance.
- Design for load distribution and transfer can minimize guide way deflection.

6.6 CLASSIFICATION OF GUIDE WAYS

The first level of classification is based on the friction component. The second level of classification is based on design configuration. The third level is based on geometric configuration.

Guide ways with sliding friction are called *slide* ways. The rolling elements, balls and rollers, incorporated in guide ways make *anti-friction* ways. Slide ways can be geometrically symmetrical, asymmetrical, open or closed according to which the classification is further accomplished. The basis of exposure of rolling elements and circulation leads to a third level of classification of anti-friction ways. Guide ways designed to offer no friction are called *frictionless* guide ways. The details of the principle of hydrostatic and hydrodynamic slide ways are provided in Sections 6.13 and 6.14. Ball type anti-friction ways use spherical balls and roller type anti-friction ways use cylindrical rollers.

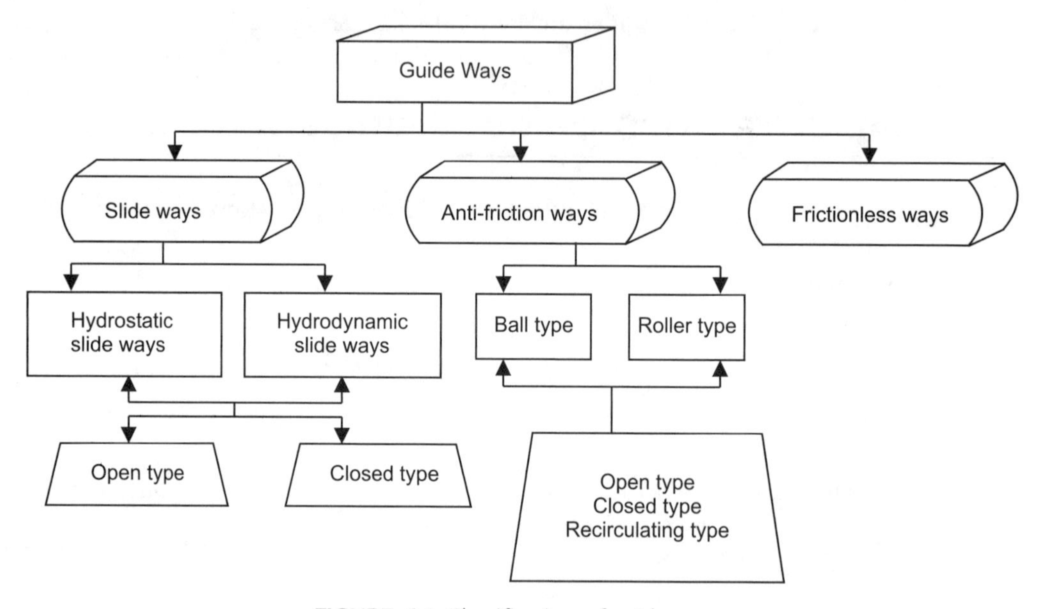

FIGURE 6.1 Classification of guide ways.

6.7 PRINCIPLE OF SLIDE WAYS

The friction between the relatively moving sliding surfaces in slide ways is characterized by the type of lubricating media, *i.e.*, solid, liquid, or semi-liquid. Solid friction is rarely encountered in machine tools as the frictional force offered is high. The other two types of friction can be explained by the following principle. (The simplest way to show the slider is given in Figure 6.2.)

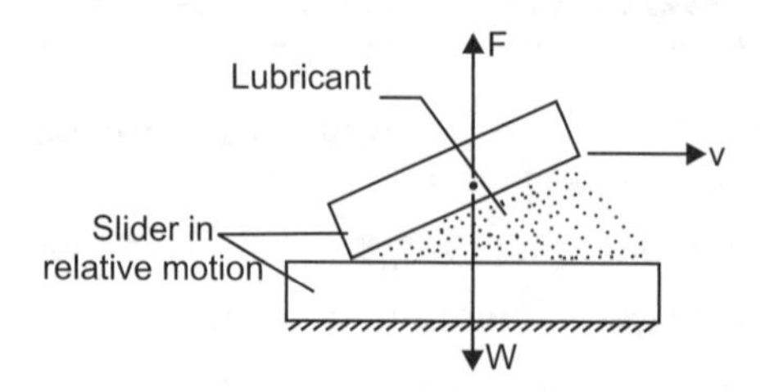

FIGURE 6.2 Schematic of a slider.

The Principle: The hydrodynamic force due to the presence of lubricant between the surfaces in relative motion velocity, v, is written as

$$F \propto v. \tag{6.1}$$

If W is the weight of the sliding body, the resultant force that tends to lift the body, R, is given by

$$R = F - W. \tag{6.2}$$

It is evident from Eqs. (6.1) and (6.2) that the resultant force, R, depends on the slide velocity for a given set of design parameters and lubricating conditions. For $R < 0$ metal-to-metal contact exists. When $R > 0$ there exists a thin film of lubricant that reduces the friction by floating action. When the velocity of motion is small a film of lubricant can be forced to exist between the sliding surfaces by a supply of oil under pressure. Such a slide way is called a *hydrostatic* slide way. Readers are encouraged to refer to Section 6.13 for delibration on hydrostatic slide ways.

6.8 STICK-SLIP PHENOMENA

The beginning of sliding includes the problem of jerky motion. This is known as *stick-slip phenomena*. This fact is attributed to the variation of the friction coefficient during the commencement of motion, but not to the low or high value of friction. The variation in μ is caused by:

- The change in sliding speed.
- The elapse of time.

A typical graph showing the variation of μ with sliding velocity, v, is shown in Figure 6.3. When the velocity $v < v_0$, the coefficient of friction, μ, decreases with an increase in velocity. But beyond v_0 the value of μ increases with velocity.

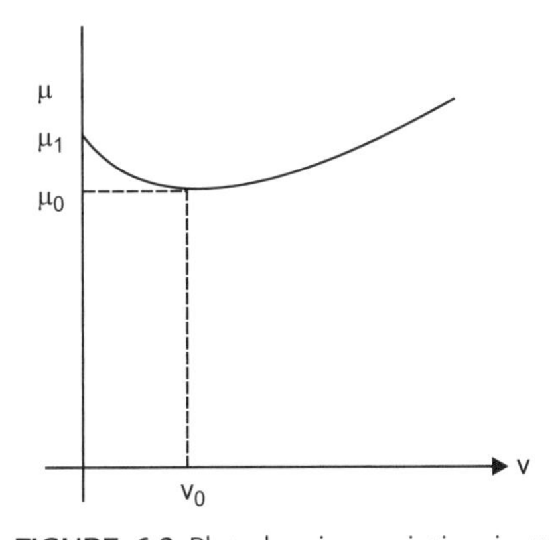

FIGURE 6.3 Plot showing variation in μ.

The coefficient friction decreases from a value μ (static friction) to a minimum value μ_0 as the sliding velocity increases. Hence, to start the motion a higher force should be applied. But with the motion, the friction is less and the excessive force applied is released to produce a jerky motion exhibiting the "stick-slip" phenomena. With the squeezing of oil with time, the actual coefficient of friction varies. Hence, the time lapse between two oiling replenishments also contributes to stick-slip phenomena.

6.9 PRINCIPLE OF ANTI-FRICTION WAYS

Anti-friction guide ways make use of either spherical balls or cylindrical rollers in between the relatively moving surfaces of the guide ways. Point contacts or line contacts keep the friction to a minimum. The selection of a ball or roller is decided by the load capacity of the anti-friction guide way. The contact strength of the element (ball or roller) is the basis for the determination of load capacity of the anti-friction way. For ball type ways, the load capacity

$$L = Kd^2, \tag{6.3}$$

where K is the load coefficient (N/mm^2) and d is the diameter of the ball in mm. The roller type of ways have the load capacity relation written by

$$L = Kb.d, \tag{6.4}$$

where b is the length (mm) and d is the diameter (mm) of the roller. It is clear from Eqs. (6.3) and (6.4) the high load capacity for roller ways is achievable with designs of $b > d$.

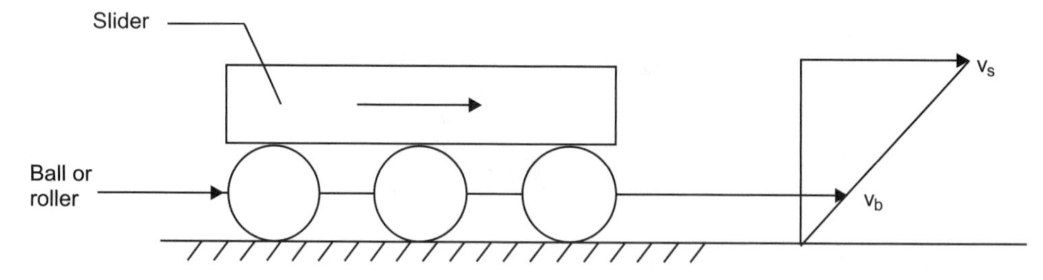

FIGURE 6.4 Schematic of anti-friction way.

Anti-friction ways are shown in principle by Figure 6.4. The problem faced in this type is the balls or rollers lag behind the slider as ball speed, v_b, and slider speed, v_s, are related by

$$v_b < \frac{1}{2} v_s . \tag{6.5}$$

The frictional force acting on the surfaces is virtually independent of speed. But the frictional force depends on the weight of the slider and table, and also the vertical component of the cutting force. The pulling force needed by the anti-friction way is given by

$$R = \mu(W + P_V) + P_x,$$

where R = resultant pulling load,

W = weight of table and guide,

P_V = vertical component of cutting force,

P_x = axial component of cutting force, and

μ = coefficient of friction.

Anti-friction ways are characterized by low friction, uniform motion, high stiffness, and high speed operation.

6.10 DESIGN SHAPES OF SLIDE WAYS WITH APPLICATIONS

The following table shows slide way designs and their applications.

TABLE 6.1 Design Shapes of Slide Ways with Applications

Shapes	Features	Applications
1. Flat slide ways	• Easy to manufacture • Accumulate dirt • Poor in retaining lubricant	• General purpose lathes • Boring machines • Plano milling machines
2. V slide ways	• Automatic clearance adjustment • Greater accuracy • Less effect of wear • Difficult to produce	• Planing machines • Precision lathers • Turret lathes • Surface grinding machines
3. Dovetail slide ways	• Compact and easy to adjust clearance • Difficult to manufacture and inspect	• Knee-type milling machines • Drilling machines • Cross slides and compound rests

4. Cylindrical slide ways	• Easy to manufacture • Low rigidity • High tolerance • Difficult to adjust	• Columns • Over arms • Tail stock • Sleeves

6.11 SHAPES OF ANTI-FRICTION WAYS

The following table shows the various shapes of anti-friction ways.

TABLE 6.2 Shapes of Anti-friction Ways

Shapes	Features	Applications
1. Ball Type (open)	• Open type • Accumulates dust • Uses flat and V profiles	• Used when dead weight is higher • Resultant load does not vary
2. Roller Type	• Open type • Accumulates dirt • Uses flat and V profiles • Takes more load	• Used in heavy duty applications • Non-varying load applications
3. Ball Type (closed)	• Closed balls • Lubrication easy • Pre-loading possible • High stiffness	• Automatic machines • Heavy duty machines • Precision machines
4. Roller Type	• Closed type rollers • Retains lubricant • Takes more load • High stiffness • Pre-loading easy	• Automatic machines • Heavy duty machines • Precision machines

6.12 RECIRCULATING TYPE OF ANTI-FRICTION WAYS

The schematic of a recirculating type of anti-friction way is shown in Figure 6.5. The rolling elements enclosed in grooves circulate with the movement of the guide way providing unlimited travel. The rolling elements are free to move along the recirculation path without being enclosed in the cage. Among the rolling elements some are made smaller in diameter and act as separators between effectively functioning elements. Such a guide way is characterized by high stiffness and low friction. They are generally used in precision automatic machine tools. The continuously packed rolling elements in the groove recirculate giving sufficient balls between the stationary surface and the moving surface.

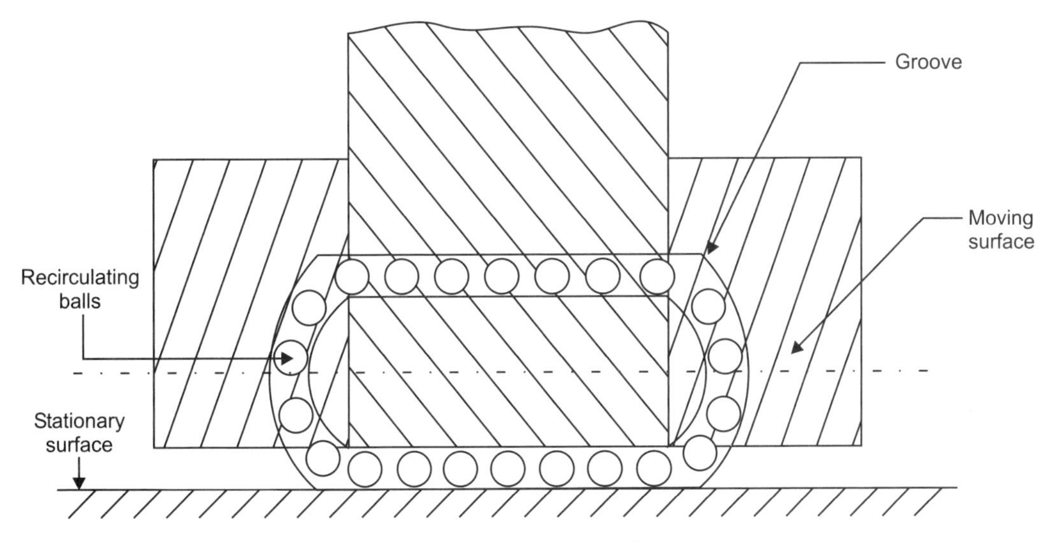

FIGURE 6.5 Recirculating anti-friction way.

6.13 HYDROSTATIC SLIDE WAYS

Hydrostatic slide ways provide liquid friction in motion. The interface between the sliding surface and the stationary surface is supplied with lubricant under pressure. The pressure in the lubricant supplied should be high enough to float the sliding body and prevent metal-to-metal contact. A typical hydrostatic slide way is shown in Figure 6.6.

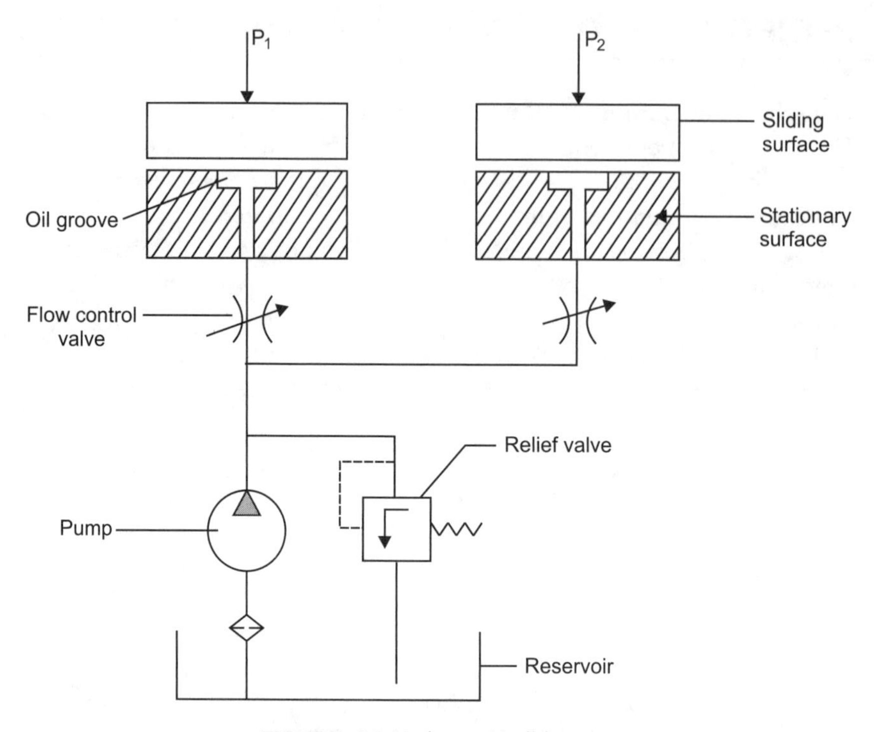

FIGURE 6.6 Hydrostatic slide way.

Hydrostatic slide ways have certain advantageous features including:

- High load bearing capacity at all speeds.
- No starting friction and low running friction.
- Wear is almost negligible.
- Stiffness is very high.
- Vibration resistance is high.
- Highly accurate and uniform feed possible.

The principle of operation can be explained by pad bearings. The oil grooves and the path are connected to a pump that supplies oil at high pressure. The pressure of oil depends on the maximum load acting on the pad. The pressure of the oil should be sufficient enough to raise the sliding surface by the development of a thin oil film between mating surfaces.

6.14 HYDRODYNAMIC SLIDE WAYS

In hydrodynamic slide ways the surfaces are inclined to each other and slide with relatively high speed and the hydrodynamic action can be observed. The liquid friction condition developed between the sliding surfaces of slide ways is achieved due to hydrodynamic action of the lubricating oil film. The oil film formed between the sliding surfaces by the action of the hydrodynamic effect at high sliding speed is capable of lifting the guided member.

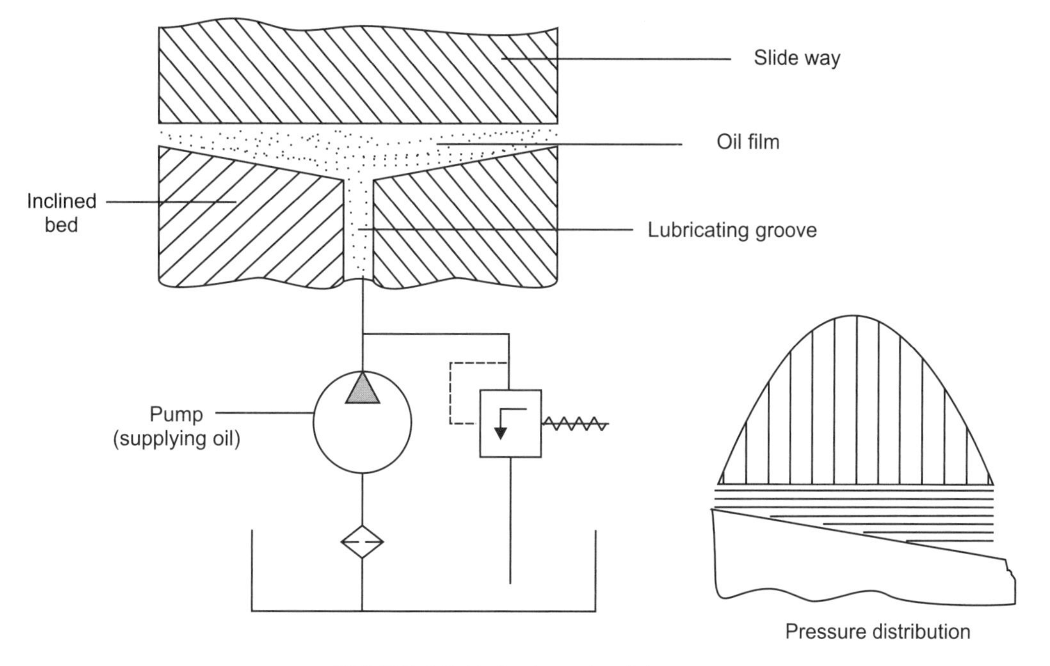

FIGURE 6.7 Hydrodynamic slide way.

Hydrodynamic slide ways are characterized by:

- High load bearing capacity.
- High stiffness.
- Low running friction at high speeds.
- High friction at low and breaking speeds.
- High damping capability.
- Excessive wear at wedge corners.
- High accuracy and uniform feed.

The hydrodynamic bearing surfaces are generally used in heavy duty precision machine tools such as vertical boring machines, turning mills, planing machines, floor boring machines, etc. Machines with hydrostatic and hydrodynamic slide ways are equipped with a power pack to supply and control the pressurized lubricating oil through the grooves. The schematic showing a hydrodynamic slide way is shown in Figure 6.7.

6.15 SLIDE WAYS AND ANTI-FRICTION WAYS

The following table compares slide ways and anti-friction ways.

TABLE 6.3 Comparison of Slide Ways and Anti-friction Ways

Comparison Features	Slide Ways	Anti-friction Ways
• Frictional forces	• Higher friction exists in this type due to surface contact between sliding surfaces. Effort is made to reduce it by hydrostatic or hydrodynamic action.	• Lower friction as the relatively moving surfaces are separated by balls or rollers that establish point or line contact, respectively, but relative motion exists between slider and rolling elements.
• Stiffness to deflection	• High stiffness, low deflection attributed to direct contact.	• High stiffness is developed by pre-loading of the members.
• Uniformity in motion	• Stick-slip phenomena attributed to variation in friction coefficient which gives non-uniform motion.	• Meritted by uniform motion with the absence of stick-slip phenomena. The rolling action is responsible for this.
• Wear of the surfaces	• Higher wear because of direct contact.	• Lower wear because of indirect contact.
• Velocity of motion	• Used in machine tools operated at low speeds because of demerits adversely effecting it.	• Machine tools operated at high speed and precision favor its use.
• Load bearing capacity	• Higher load retaining capacity.	• Load bearing capacity is low but can be improved.

6.16 RECIRCULATING BALL SCREW-AND-NUT ARRANGEMENT

The ball recirculating screw-and-nut arrangement as shown in Figure 6.8 is the rolling friction power screw. In between the threads of the screw and nut there are spherical balls that establish contact which is not direct. The balls move in the grooves between the threads of the screw and nut resembling ball bearings with balls between the outer and inner races. Such power screws are characterized by the recirculation of the balls.

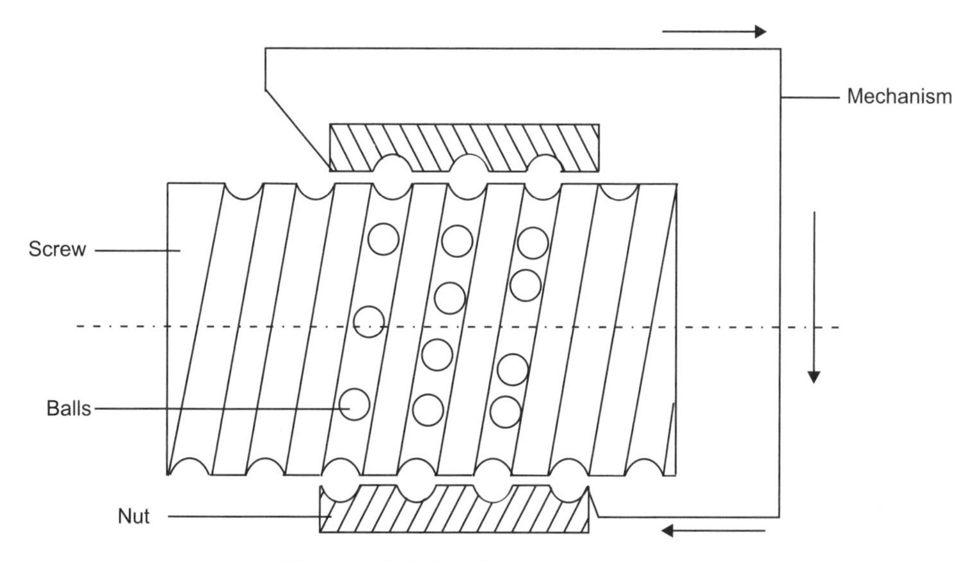

FIGURE 6.8 Ball recirculating power screw.

- *Application:* Anti-friction power screws find application in precision machine tools such as:
  - Grinding machines.
  - Boring and jig-boring machines.
  - Special purpose machine tools.
  - Numerically controlled machine tools.

The thread profiles can be square and trapezoidal for low load carrying capacity. A commonly used profile is semicircular that provides higher stiffness giving a higher load-carrying capacity. The design requirements of anti-friction power screws are: consideration of stiffness for low deflection, consideration of static strength, and fatigue strength for different cyclic loading. The material of the nut and the screw is selected for heat treatment to give a higher hardness that reduces wear.

6.17 ADVANTAGES AND DISADVANTAGES OF ANTI-FRICTION POWER SCREWS

The dominating advantages of anti-friction power screws over conventional types justify their application in high precision machine tools.

- **Advantages**
  - Low coefficient of friction compared to sliding friction power screws.
  - Characterized by high transmission efficiency.
  - Improvement in stiffness requires pre-loading of power screws because of which clearance is also eliminated.
  - Uniformity of motion at all velocities because of the absence of stick-slip phenomena.
  - High reliability and less wear.
- **Disadvantages**
  - Bigger than the sliding type which are compact in size.
  - Run-out is higher.
  - Less stability under axial, tensile, or compressive loads.
  - Long and heavy loaded applications cannot use this type.
  - Residual stress needs to be minimized for accurate transmission.
  - Material and manufacturing costs are high.
  - Maintenance is costly and difficult.

The screw and nut in anti-friction power screws are made of low carbon alloy steels that are nitrided, and high carbon steels that are fully hardened to improve wear and strength properties.

6.18 PRE-LOADING BALL NUTS

Ball screws and nuts are invariably used in CNC machines because of the favorable features over conventional screws and nuts. The advantages of ball screws and nuts used in machine tools are:

- Low frictional loss.
- Elimination of stick-slip effect.
- High wear resistance and high reliability.
- Better efficient operation.
- Effect of low temperature increases.

- Low power consumption.
- Adjust to higher speed of motion.

The pre-load in ball screws and nuts for elimination of backlash includes two types:

1. ***Tension Pre-loading:*** On a screw and nut with balls the screw will have gaps that must be kept to a minimum by developing the tensile force between the screw and nut. This is achieved by assembling a spacer (shim) plate of thickness slightly more than the width of the groove in which the balls circulate in between the two nuts. (Refer to Figure 6.9.)

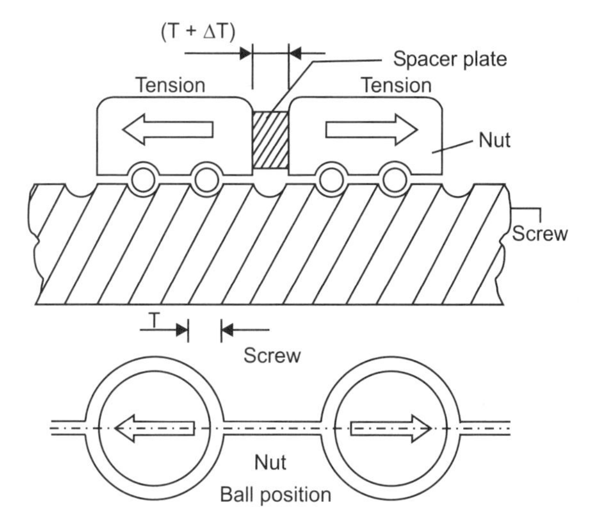

FIGURE 6.9. Nut pre-loading (tension).

2. ***Compression Pre-loading:*** The compressive load directed inward on the nuts tends to reduce the backlash between the screw and nut. The compressive load is developed by locating the shim plate between the nuts. The shim is a thickness less than the width of the ball groove.

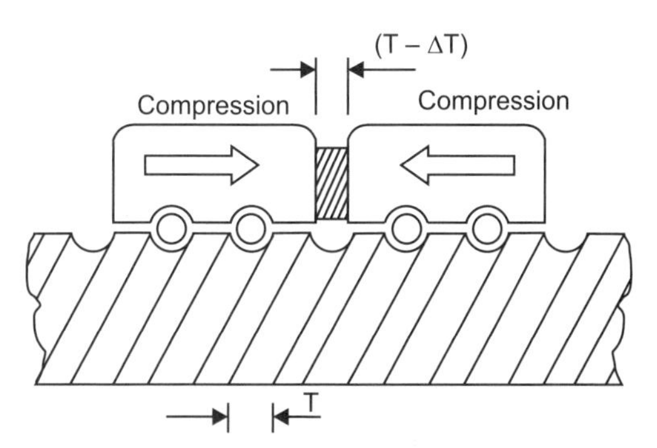

FIGURE 6.10 Nut pre-loading (compression).

6.19 PLANETARY ROLLER SCREWS

Figure 6.11 shows the arrangement of a recirculating type roller screw. The rollers are located in the circular groove along the length of the screw shaft. The rollers are equally spaced and kept in the circumferential position by the cage. With the rotation of the nut the rollers also move axially along the length of the screw shaft. Such screws are capable of taking greater loads with higher accuracy.

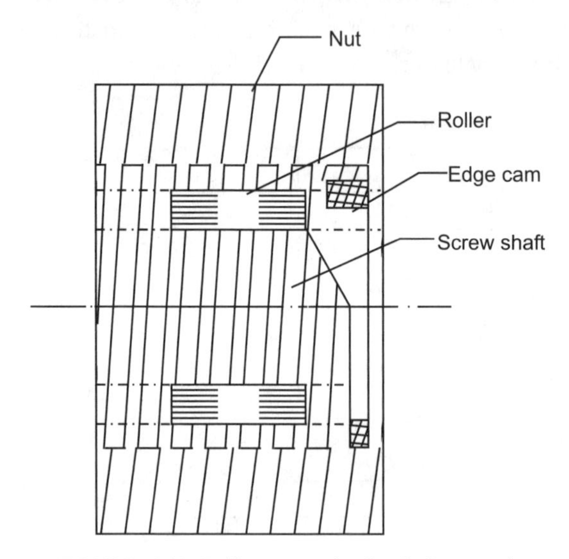

FIGURE 6.11 Roller screw (recirculating type).

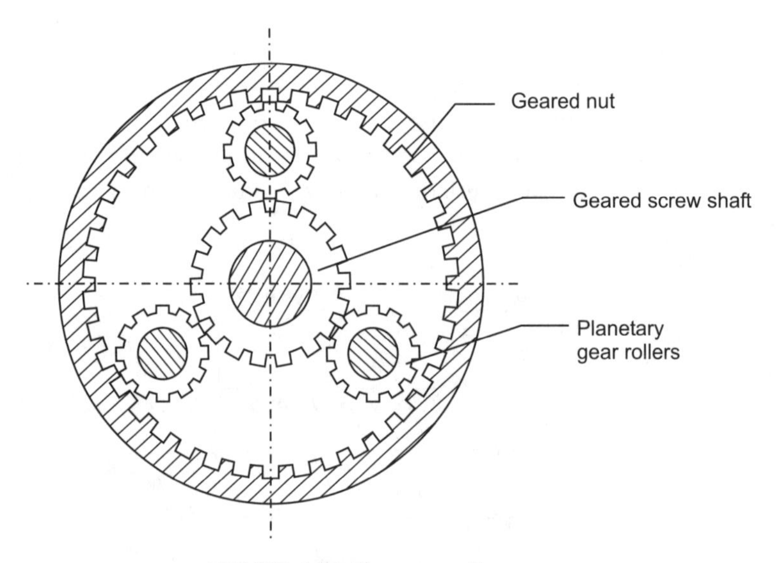

FIGURE 6.12 Planetary roller screw.

Figure 6.12 shows a planetary roller screw. The rollers are threaded in the middle and cut with gear teeth at the ends. The rollers from the planets which engage with the nut have internally cut gear teeth. The threads in the rollers engage with the screw shaft. The equally spaced rollers are mounted to a spacer ring with the spigot provided at the end of the rollers. The nut drives the rollers to provide a rolling motion between the nut and the screw. The rollers do not have axial motion in the rolling action.

Advantages of Roller Screws Over Ball Screws

- Backlash-free movement with equal efficiency.
- Pitch is less, hence, the movement (axial) per rotation of the nut is smaller.
- Provide very accurate positional control.
- Can operate at higher shaft speeds.
- Capable of transmitting higher loads.
- Axial and radial stiffness is higher.
- Do not require pre-loading.

6.20 SPINDLES AND SPINDLE BEARINGS

A spindle is a hollow shaft mounted on the bearing (supports) which has features to locate and clamp work-holding and tool-holding attachments. It is the spindle that receives power from transmitting systems to impart rotary and translatory motion to the job or cutting tool. The cutting action in machine tools requires certain degrees of motion depending on the type and shape of the surface generated. The spindle is one of the elements of a transmission system that acts as a link between the driving system and the work/tool fixtures. Spindles perform the following functions:

- Locating and centering of the work/tool attachments (*e.g.*, the chucks in the lathe and turrets, and drill chucks or bits in drilling machines).
- Clamping of work/tool holding systems (such as face plate and chucks). They have clamping holes or internal threads for this purpose.
- Importing motions (rotary motion as in lathes, and translatory motion as in drilling machines).

The design of spindles must satisfy the following requirements:

- Radial and axial run-outs should be within permissible limits for a high degree of rotational accuracy.

- Static stiffness such as axial bending and torsional stiffinesses should be high for good machining accuracy.
- Damping and dynamic stiffness should be reasonably high for dynamic stability and minimum vibrational effects for a better surface finish.
- High-wear resistance for mating surfaces that can be accomplished by surface hardening.
- Thermal loads due to heating of bearing, cutting tool, and work can cause thermal deformation which should be kept to a minimum by selection of material with low thermal expansion.
- Clamping and fixturing arrangement must be easy and reliable for easy clamping and disassembly of attachments.

Depending on the accuracy requirements and the size of the spindles, the following materials are recommended:

Requirements	Materials
• For normal accuracy	C 45 and C 59 (Hardened and tempered) HR_C = 30
• For higher accuracy	Hardened low alloy steel (40 Cr|Mn 60 Si 27 Ni 25) HR_C = 55 - 60
• Precision machine tool spindles	Hardened alloy steels (20 Cr|Mn 60 Si 27 Ni 25) HR_C = 56 - 60
• Hollow heavy duty spindles	Gray cast iron, spheroidal graphite iron

The strength of the material does not have a big part in the design of the spindle. The parameter of importance is the stiffness and rigidity that controls the deflection of the spindle. A careful look into the properties of steel reveals that it more or less has the same modulus of elasticity, the determining factor of stiffness. Hence, there is no real advantage derived from using high grade costly steel as the material for spindles. For spindles of a diameter less than 150 mm, rolled steel stock is used. For diameters greater than 150 mm, centrifugally cast CI is used.

Types of Spindle Noses

1. *Type A spindle nose:* An external taper spigot acts as a locator for the chuck and a tennon acts as the stopper. The chuck is mounted to the spindle nose by clamping cap screws. Lathe chucks and face plates are located and mounted in this type of spindle nose.

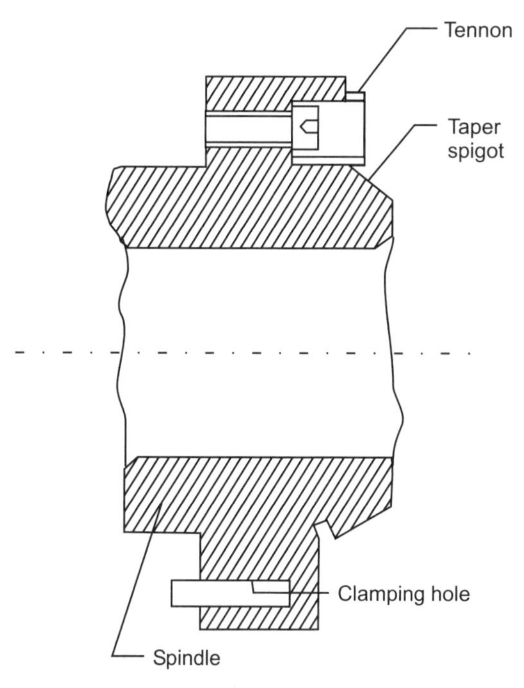

FIGURE 6.13 Type A spindle nose.

2. ***Steep taper nose:*** The nose is a steep taper 7:24 conical hole provided in the milling arbor. The cutter is located and clamped to the nose to withstand the vibrating intermittent cutting force. A conical hole provides the location for alignment, and the tapped holes accommodate clamping screws.

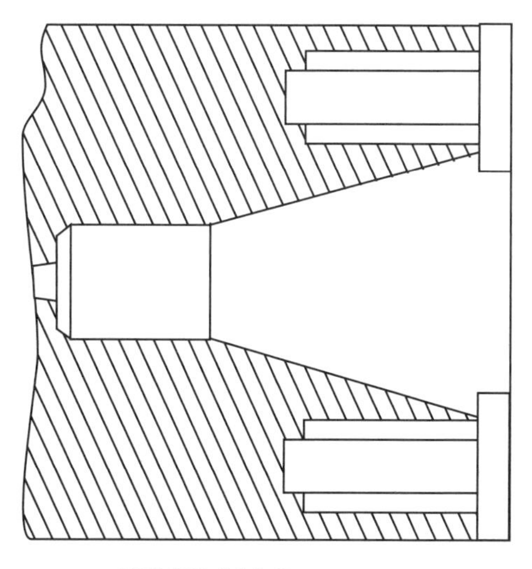

FIGURE 6.14 Steep taper.

3. ***Morse taper:*** A long, narrow taper provided in the vertical spindle adapts to tools with tapering shanks. This type is generally used in drilling, boring, and grinding machines.

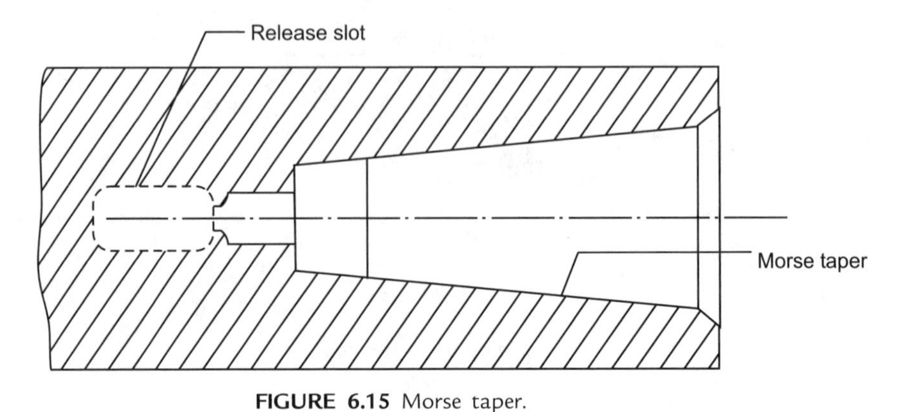

FIGURE 6.15 Morse taper.

A release slot is provided to unlock the tool held by compression.

6.21 TYPES OF LOADS ON SPINDLES

The spindle, being the power transmitting element and supporting system for the work piece and/or tool, is subjected to the following types of loads:

- ***Static loads:*** Loads that are time independent and do not vary with time are static loads. All types of static loads, axial, bending, and torsional loads act on the spindle. The forces that contribute to the static load are:
  - The body forces due to the weight of the spindle, work, and tool.
  - The components of the cutting forces. The feed results in the axial load and the depth of the cut results in bending and the tangential component results in torsional forces in the lathe.
  - Reaction forces at the supports.

The deflects due to static loads can be limited within a controlled value by proper design configuration and sufficient stiffness.

- ***Dynamic loads*:** Loads changing with time result in dynamic effects such as vibration and fatigue in machine spindles. Dynamic loads are caused by:
  - Intermittent cutting action during machining operation.
  - Unbalanced rotating masses.
  - Cyclic action of loads during spindle rotation.

- Bearing misalignments.
- Improper engagement of power transmitting systems such as gears.

The minimization of dynamic effects are possible by the following:

(*a*) Optimizing the structural design to reduce weight.

(*b*) Increasing the dynamic stiffness of the machine elements.

(*c*) Selecting materials for elements with higher endurance strength.

(*d*) Dynamic stability is improved by improving damping properties.

- ***Thermal loads:*** The localized heat generation mainly due to friction in the spindle and parts connected to the spindle results in temperature variations which lead to thermal expansion and thermal stresses in the spindles of machine tools. The major sources of heat generation are:
  - Friction in bearings and guide ways.
  - Friction in power transmitting systems such as gears and belt drives.
  - Metal-cutting processes.
  - Temperature variations in the environment.
  - Electrical and mechanical actuators.

To reduce inaccuracies due to thermal deformations in the spindle, the following measures can be adopted:

(*a*) Dissipating the frictional heat in the bearings by using a proper lubricating system.

(*b*) Heat generation in machining can be avoided by using a coolant system.

(*c*) Locating the drives and transmission systems away from the spindle.

(*d*) Geometric compensation in the design configuration reduces thermal stresses.

6.22 SELECTION OF SPINDLES

- ***Type A spindle nose:*** This type of spindle is provided with two rows of tapped holes on the face. The inner row provides a means for attaching a scroll chuck. The outer row of tapped holes are meant for securing the face plate and fixtures.

  Application: Lathe spindles, turrets.

- ***Cam-lock type spindle nose:*** This type is provided with a cam to attach the chuck with the spindle. The easy and fast assembly and removal of the chuck and attachment is possible with this type of spindle nose. The various standard sizes are available to suit various sizes of lathes.

Application: Lathe spindles, automatic lathes.

- ***Bayonet type spindle nose***: In this type of spindle nose, there are holes to fix the stud that gets enaged with the slots in the bayonet disc and gets clamped by the nut. The disc is attached to the chuck or the attachments. Easy engagement and removal is possible with this type.

  Application: Lathe spindles, single spindles, automatic and semiautomatic lathes.

- ***Self-release tapers:*** Self-release tapers are available in 7/24 tapers, morse tapers, and 1:3 tapers. They have tapering holes to locate tools and adapters, tennons to lock, and tapped holes to clamp.

  Application:

  - 7/24 tapers—used in milling machines.
  - Morse tapers—drilling and boring machines.
  - 1:3 tapers—spindles of grinding machines.

6.23 TYPES OF BEARINGS

The suitability of the types of spindle supports (bearings) depends on the deflection of the spindle nose under the action of the cutting forces. The compliance of the front and rear spindle bearings and rotational accuracy are the basic functional requirements considered in the choice of bearings. Diversified operating conditions form the basis for the selection of anti-friction, hydrostatic, hydrodynamic, or air-lubricated bearing types for the spindle supports of machine tools.

The general requirements of spindle supports, irrespective of their type, are:

- Accuracy of guidance for depth of cut.
- For varying speeds, the spindle should perform satisfactorily.
- For minimum end point deflection the stiffness should be high.
- Minimum heat generation and effective dissipation to minimize thermal loading and expansion.
- Dynamic stability by dominating damping effects.
- Broad classification of spindle supports.

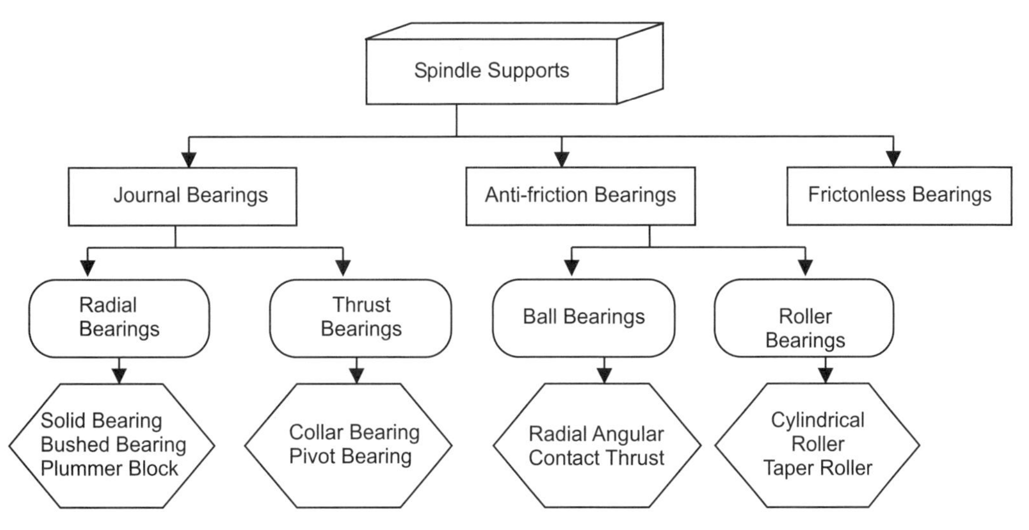

FIGURE 6.16 Classification of bearings.

6.24 SLIDING BEARINGS

The disadvantages of anti-friction bearings in applications requiring very high rotational speeds, high accuracy at high speeds, and damping needs when subjected to shock and vibration, favor the use of journal (sliding contact) bearings for machine tool spindles.

Based on the Hersey diagram shown in Figure 6.17 sliding bearings are classified as:

- Zero film bearings
- Thin film bearings
- Thick film bearings

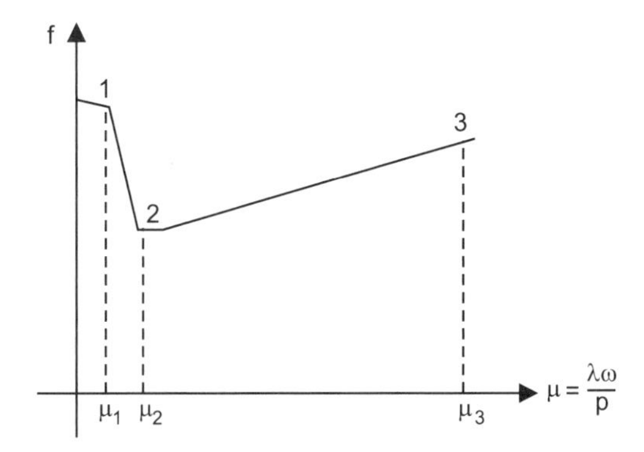

FIGURE 6.17 Hersey diagram.

The Hersey diagram is characterized by establishing the variation of friction coefficients with the parameter μ given by

$$\mu = \frac{\lambda\omega}{p},$$

where λ = absolute viscosity of the lubricant,

ω = angular velocity of rotation of journal, and

p = average pressure on the supporting surface;

between $0 - \mu_1$ = zero film lubrication zone,

$\mu_1 - \mu_2$ = thin film lubrication, and

$\mu_2 - \mu_3$ = thick film lubrication commonly observed in hydrostatic and hydrodynamic sliding bearings.

Sliding journal bearings can be classified by configuration as full bearings and partial bearings as shown in Figure 6.18. Partial bearings have more frictional losses than full bearings, but can be used in applications in which the load acts always in one direction.

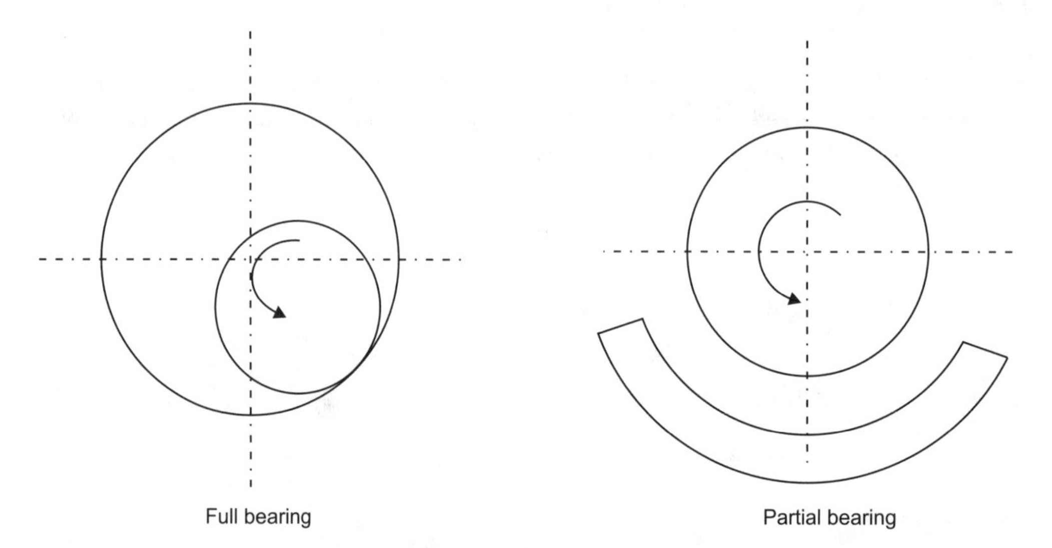

FIGURE 6.18 Full and partial journal bearings.

6.25 HYDRODYNAMIC JOURNAL BEARINGS

- At zero rotational speed the journal rests on the bearing with metal-to-metal contact at point A (see Figure 6.19A).

- As the journal starts rotating in the counter-clockwise direction, by the action of frictional force (static) the journal climbs to position B (see Figure 6.19B).
- As the speed increases the hydrodynamic force due to the oil wedge shifts the journal to contact with the bearing at point C. So for $\mu < \mu_2$ metal-to-metal contact persists.
- With the higher rotational speed $\mu < \mu_2$, a film of oil separates the journal from the bearing with minimum film thickness at point D.

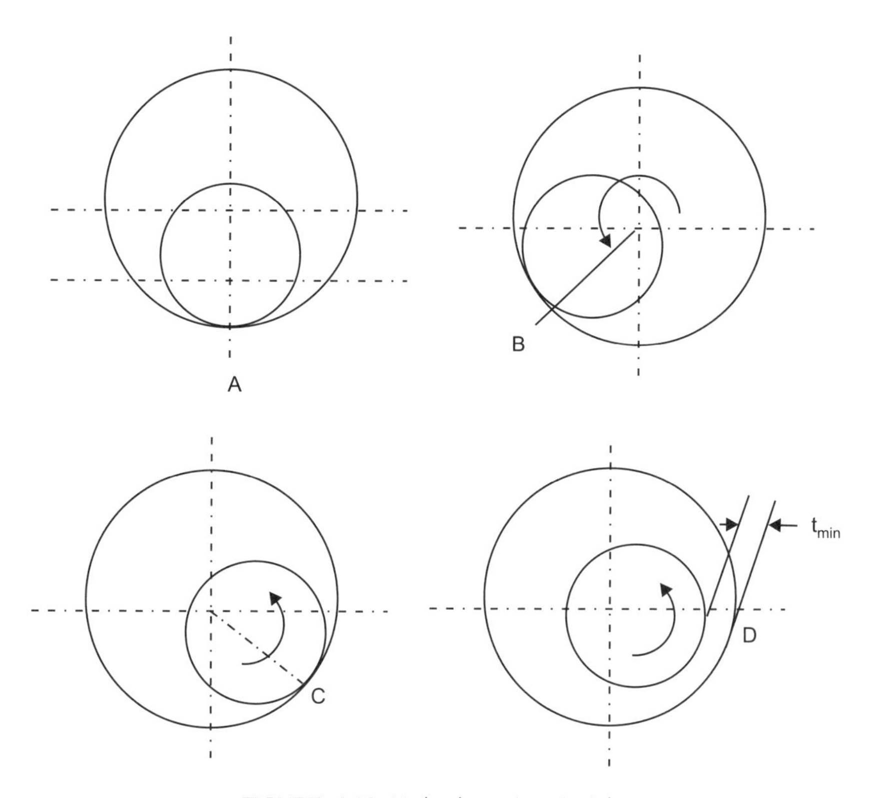

FIGURE 6.19 Hydrodynamic principle.

6.26 HYDROSTATIC JOURNAL BEARINGS

The shortcoming of a journal bearing with hydrodynamic effect lies in insufficient stiffness to produce a circular profile with high accuracy. The stiffness in hydrodynamic bearings changes with lubricant viscosity, temperature, and the rotational speed of the journal. This shortcoming is eliminated with the hydrostatic journal. The converging-diverging wedges of film in hydrodynamic bearings are replaced by uniform film in hydrostatic bearings with a supply of lubricating oil with constant high pressure. The hydrostatic bearings are single pad, multipad, or multirecess. The single-pad hydrostatic bearing is shown in Figure 6.20.

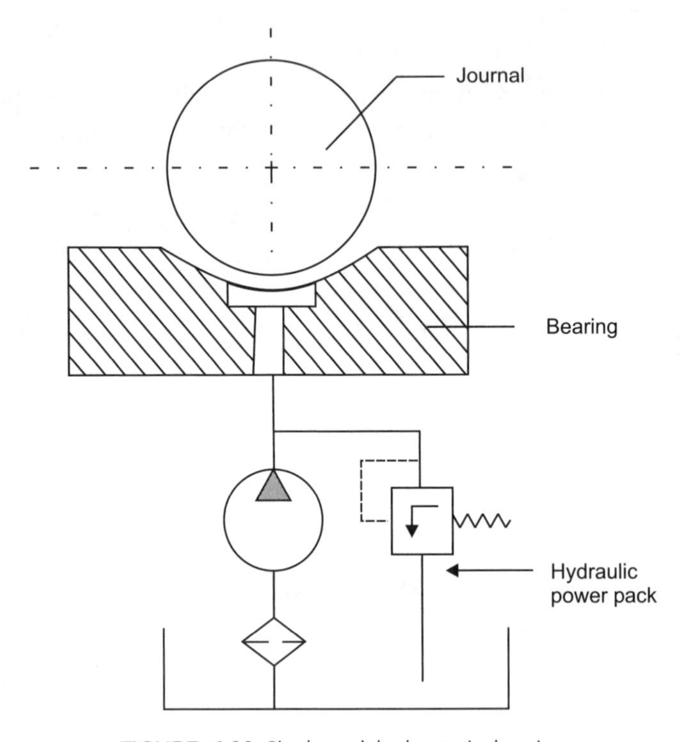

FIGURE 6.20 Single-pad hydrostatic bearing.

Single-pad hydrostatic bearings can support only unidirectional loads. Reversing the rotation of the journal requires multipadded journals. This type of journal is rarely used in machine tools with the spindle changing directions and with varying cutting forces.

6.27 BEARING MATERIAL SELECTION

Considerations in the selection of materials for sliding bearings include:

- High compressive strength.
- High fatigue strength.
- High thermal conductivity.
- High wear resistance.

- Low modulus of elasticity.
- High corrosion resistance.

6.28 ANTI-FRICTION BEARINGS

Anti-friction bearings have extensive applications in machine tools. They make use of rolling elements that have either a spherical ball shape or cylindrical and conical roller shape. Based on the shape of the rolling elements they are classified as ball or roller bearings. Machine tool spindles experience both radial and axial loads. The support may be designed to take these loads separately or in combination. There are anti-friction bearings that can take radial, axial, and combined loads. The various types of anti-friction bearings are given in Table 6.4.

TABLE 6.4 Anti-friction Bearings

Schematic Diagram	Features
1. Radial ball bearing 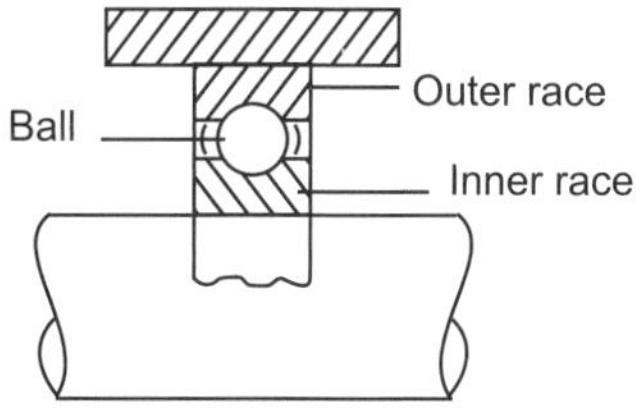 Outer race Ball Inner race	• The number of rows of balls can be one or two. • Balls located between the inner and outer race give radial point contact. • Basically meant for taking dial loads on the shaft.
2. Angular contact ball bearing 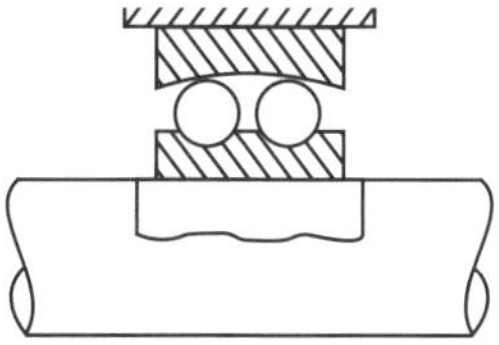	• The balls make contact with the outer race at an angle. • They can take radial and substantial axial loads.
3. Thrust ball bearing 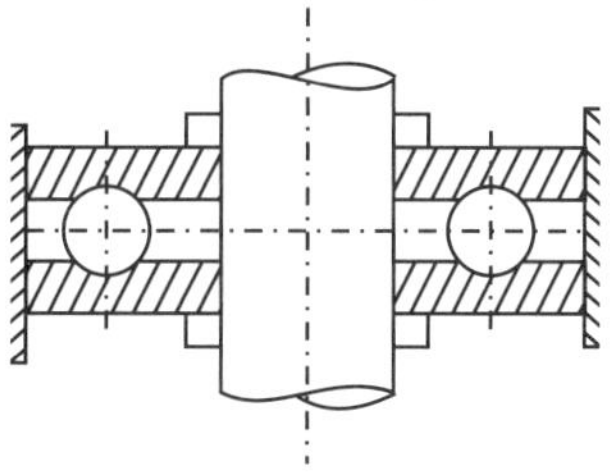	• Balls are located between top and bottom races in the circular grooves. • They take purely axial loads. • Rotational speed possible is relatively less.

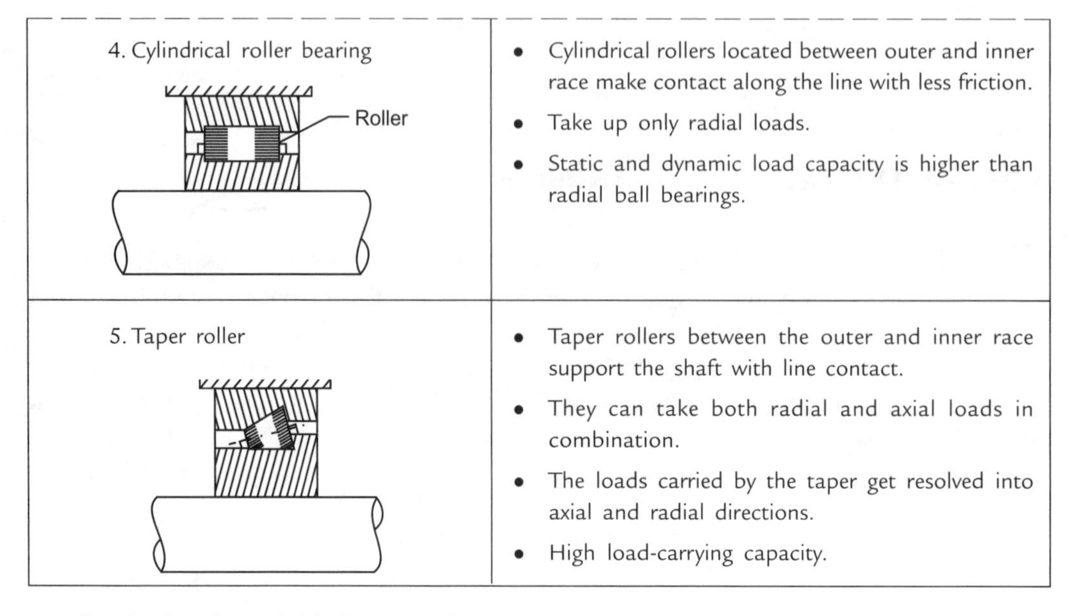

4. Cylindrical roller bearing	• Cylindrical rollers located between outer and inner race make contact along the line with less friction. • Take up only radial loads. • Static and dynamic load capacity is higher than radial ball bearings.
5. Taper roller	• Taper rollers between the outer and inner race support the shaft with line contact. • They can take both radial and axial loads in combination. • The loads carried by the taper get resolved into axial and radial directions. • High load-carrying capacity.

Applications of Anti-friction Bearings

- Use of taper roller bearings considerably increase the radial as well as axial run-outs.
- Thermal deformations are enhanced by using only cylindrical roller bearings.
- For lower heat generation and good stiffness at high rotational speeds angular contact ball bearings are recommended.

6.29 COMPARISON OF SLIDING BEARINGS AND ANTI-FRICTION BEARINGS

The following table compares sliding bearings and anti-friction bearings.

TABLE 6.5 Comparison of Sliding Bearings and Anti-friction Bearings

Feature	Sliding Bearing	Anti-friction Bearing
• Friction	High due to surface contact	Low due to point and line contact
• Heat generation	High heat generation	Low heat generation
• Starting resistance	High starting resistance, low running resistance	Low starting and running resistance

• Stick-slip phenomena	Exists due to variation in friction coefficient	Does not exist in operation
• Load capacity	Low per unit width of the bearing	High per unit width of bearing
• Space requirement	Low space and compact in size	Large space radially
• Maintenance	Relatively difficult due to wear	Easy and replacement is easily possible
• Lubricant consumption	High especially in hydrostatic bearings needing continuous supply	Low, does not require continuous supply
• Stiffness	Higher	Relatively low

6.29.1 Design Considerations of Bearing Supports

The following aspects should be considered in the design and selection of the right support for the spindles of machine tools:

- Radial stiffness of spindle assembly.
- Axial stiffness of spindle assembly.
- Radial run-outs of machine spindles.
- Axial run-outs of machine spindles.
- Lubrication system for dissipation of heat generated.
- Bearing wear restriction on maximum permissible rotational speed of spindles.
- Effect of thermal strains.
- Manufacturing ease and ease of assembly and disassembly.

6.30 PRE-LOADING OF ANTI-FRICTION BEARINGS

Pre-loading is the process of assembling the spindle supports with interference with induction of initial force so that the total nose deflection of the spindle under the radial load is reduced and the stiffness of the spindle support is enhanced.

If the bearing is assembled with clearance, the reversal of the radial load results in an abrupt change in deformation owing to the reduction in compliance of the support. This leads to a large deflection of the spindle nose. This is prevented by the assembly of the bearing with a certain amount of interference with the spindle. The interference requires initial pre-loading for which various methods exist.

Figure 6.21 shows the variation of spindle deformation due to radial load P on the bearing assembled with clearance and interference. With clearance in bearing support the variation is abrupt. In the bearing with pre-loading, variation is smooth, and at large loads the deflection becomes virtually constant which is desirable from the point-of-view of machining accuracy.

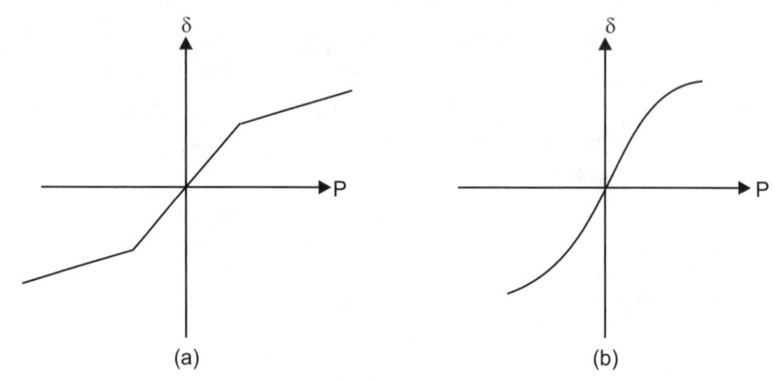

FIGURE 6.21 Variation of deflection. (*a*) Without pre-loading. (*b*) With pre-loading.

The interference due to pre-loading has to be optimum to prevent heating due to excessive interference which reduces bearing life on account of large contact deformations.

6.31 PRE-LOADING METHODS

Pre-loading methods are established by the relative displacement of the inner and outer races of anti-friction bearings. The relative displacement is achieved through various means such as grinding of the inner race using spaces of differential lengths, using adjustable helical compression springs, and using sleeves and split nuts. Figure 6.22 shows a ball bearing with the displacement of the inner race relative to the outer race. Table 6.6 gives the various pre-loading methods.

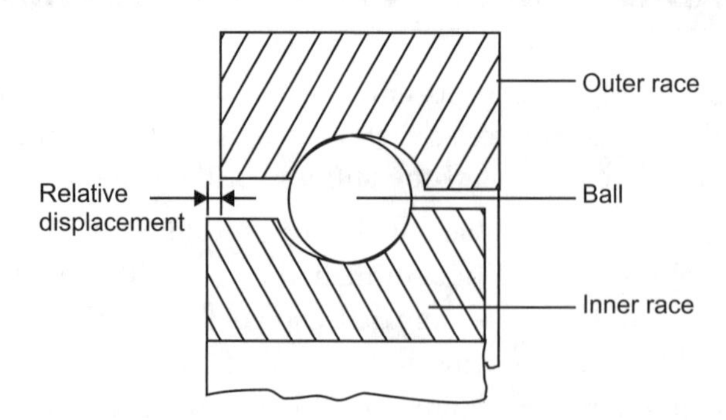

FIGURE 6.22 Pre-loaded bearing.

TABLE 6.6 Pre-loading Methods

Methods, Scheme	Process
• Grinding of Inner Race 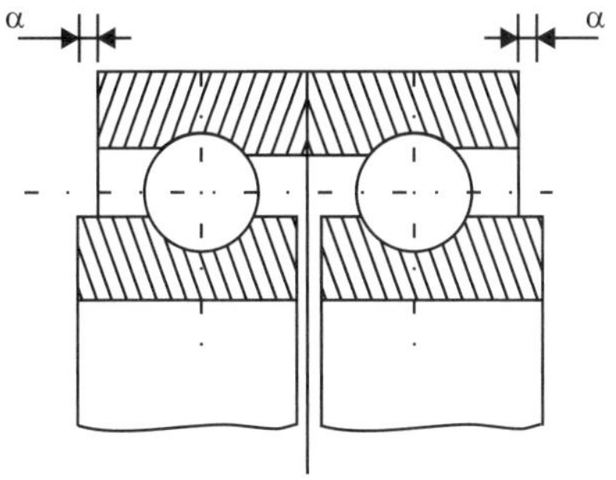 α α	• Pre-loading is achieved through grinding of the inner races of the bearings which results in a gap of 2α. This leads to interference of the balls with the races.
• Spacer Method 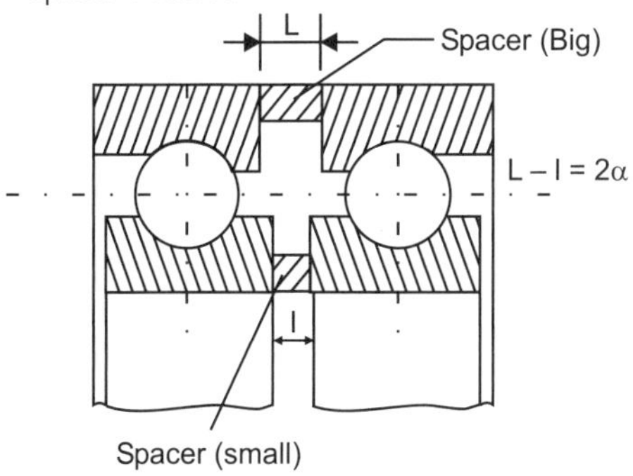 L Spacer (Big) $L - l = 2\alpha$ l Spacer (small)	• Between the outer races a spacer of length, L, and between the inner races a spacer of length, l, are assembled, where $L > l$ and $L - l = 2\alpha$ results in pre-loading.
• Spring Method 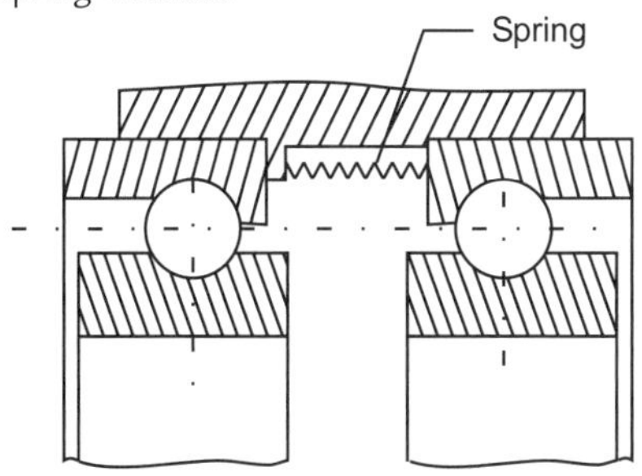 Spring	• The spring force acting on the outer sleeve exerted by the compression spring assembled between the housing and the outer race results in relative displacement leading to pre-loading. This method is adapted in precision bearings as accurate adjustment is possible.

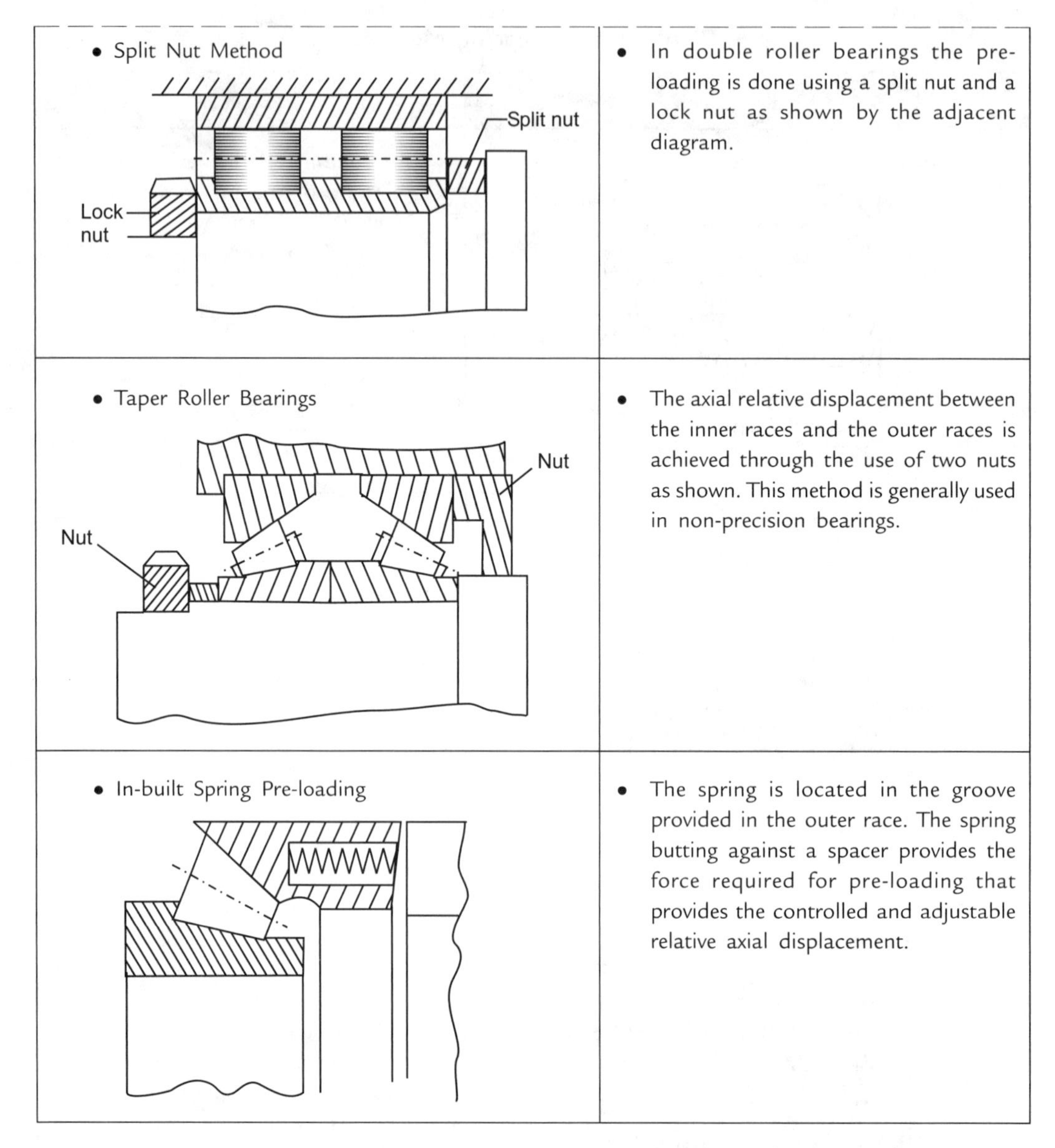

• Split Nut Method	• In double roller bearings the pre-loading is done using a split nut and a lock nut as shown by the adjacent diagram.
• Taper Roller Bearings	• The axial relative displacement between the inner races and the outer races is achieved through the use of two nuts as shown. This method is generally used in non-precision bearings.
• In-built Spring Pre-loading	• The spring is located in the groove provided in the outer race. The spring butting against a spacer provides the force required for pre-loading that provides the controlled and adjustable relative axial displacement.

The different types of pre-loading with changing interferences in accordance with the application exist in the assembly of spindles and bearings in machine tools. Applications include:

Lathes

- Small-sized and medium-sized — Pre-loading is of average importance
- Automatic — Pre-loading is of decisive importance

Grinding Machines

■ Universal grinding machines—small- and medium-sized	Pre-loading is of minor importance

Milling Machines

■ Universal milling machines	Pre-loading is of decisive importance

6.32 ADVANTAGES AND DISADVANTAGES OF ANTI-FRICTION BEARINGS

6.32.1 Advantages

- Maintain accurate shaft alignment for long periods.
- Can carry heavy momentary loads.
- Power loss due to friction is low.
- Particularly suitable for low speeds.
- Starting friction is low without stick-slip phenomena.
- Lubrication is simple and does not require due attention.
- Replacement of failed bearing is easy.

6.32.2 Disadvantages

- Design of shaft and housing is complicated.
- Initial cost is high.
- Larger radial dimension.
- Resistance to shock and vibration is high.
- Produces more noise at higher speeds.
- Are sensitive to dirt and grit.

6.33 SELECTION OF ANTI-FRICTION BEARINGS

- The static and dynamic capacities and the dimension of the selected series of bearings should suit the installation.
- The imposed load decides the type of bearing selected as radial, thrust, or combined.

- The length of service and reliability (life) are important in the selection of the type and size of the bearing to the installation in machine tools.
- Information in bearing catalogs of manufacturers makes the selection and application of anti-friction bearings easy.

6.34 FRICTIONLESS BEARINGS

The journal supports that create negligibly low or nil friction in the journal or the moving members are called *frictionless bearings*. Examples of frictionless bearings are:

- Conical pivots
- Knife edges
- Rolling edge bearings
- Flexure bearing pivots
- Twist flexures

These have low load capacity applications which are discussed in the following paragraphs.

1. ***Conical pivots:*** Figure 6.23 shows a conical pointed journal that rotates. A polished jewel bearing with a conical dip acts as a bearing that supports the journal. The journal tip may be turned with a small radius at the apex. These have negligibly low friction. This type of bearing is used in instruments such as voltmeters.

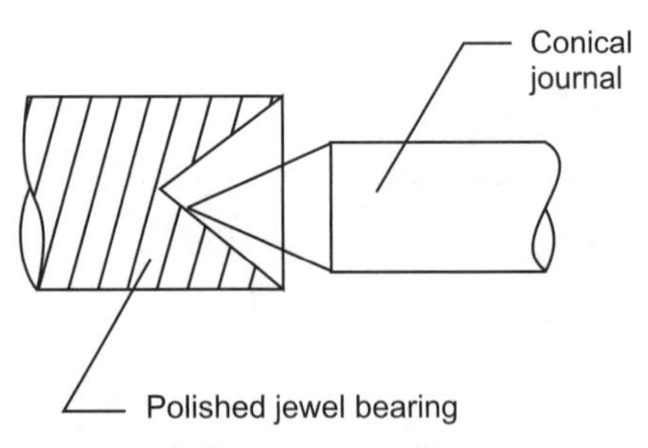

FIGURE 6.23 Conical pivots.

2. ***Knife-edge bearings:*** These are limited rotation bearings which consist of a tapering sharp edge of hardened steel that rotates or rolls on a hard, flat surface of stone or metal. There may be a sharp V groove on the surface for the location of the knife edge. They offer extremely low friction and are called frictionless bearings. They are used in laboratory analytical

balances and beam balances. The knife edge moves in a small arc to produce balancing action. (Refer to Figure 6.24.)

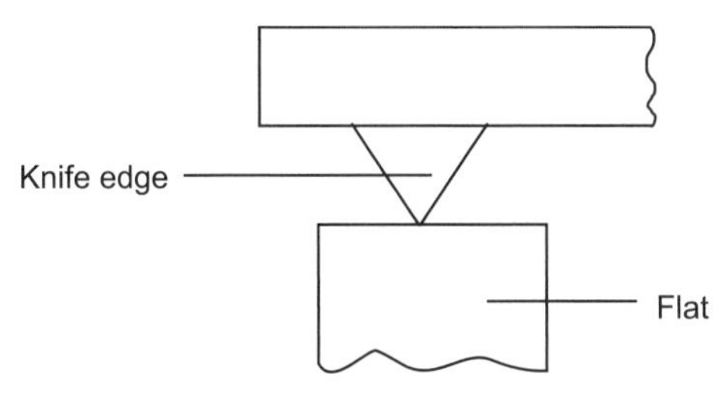

FIGURE 6.24 Knife edge bearing.

3. ***Rolling-edge bearings:*** These type of bearings are generally found in electromagnetic relays and contactors. A sharp-edged magnet armature with a wedge shape is supported by the corner formed by the magnetic core located on the flat part. Such a frictionless bearing is shown in Figure 6.25.

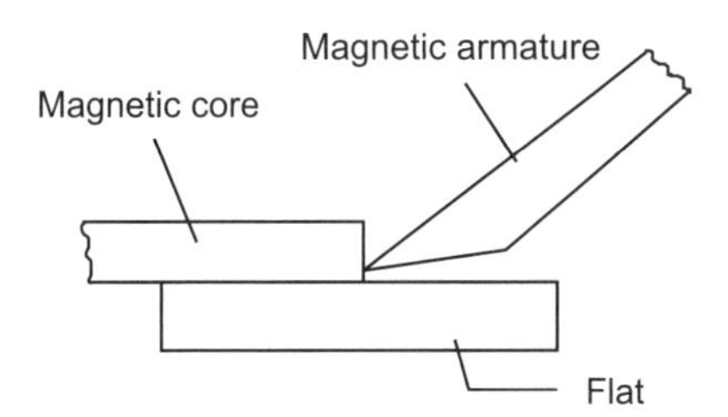

FIGURE 6.25 Rolling edge bearing.

4. ***Flexure pivots:*** A flexible strip called a flexure is clamped in between two supporting members leaving a gap in between where the mechanism is suspended. Such a frictionless and strictionless support is used in relay armatures, pendulum suspensions, and vibratory feeder suspensions. The flexures often act as electrical conductors and also as bearings. (Refer to Figure 6.26.)

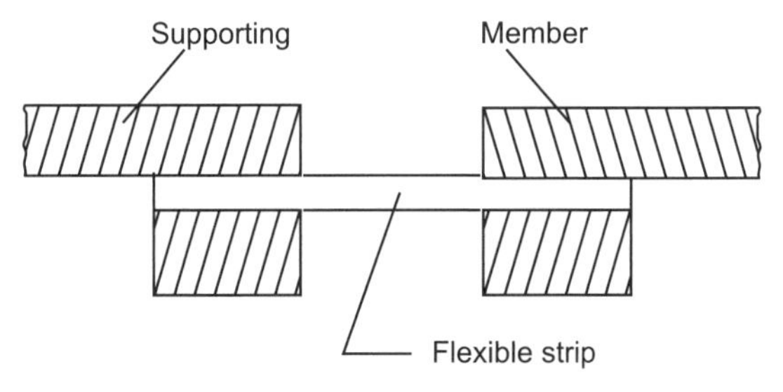

FIGURE 6.26 Flexure pivots.

5. ***Twist flexures:*** In this type the flexure is a wire, a quartz fiber, or a narrow ribbon. They twist about their own axes, have axial and torsional stiffness, and no transverse stiffness, hence, they can support hanging loads without friction and striction. They are used in instruments using twisting pendulums. (Refer to Figure 6.27.)

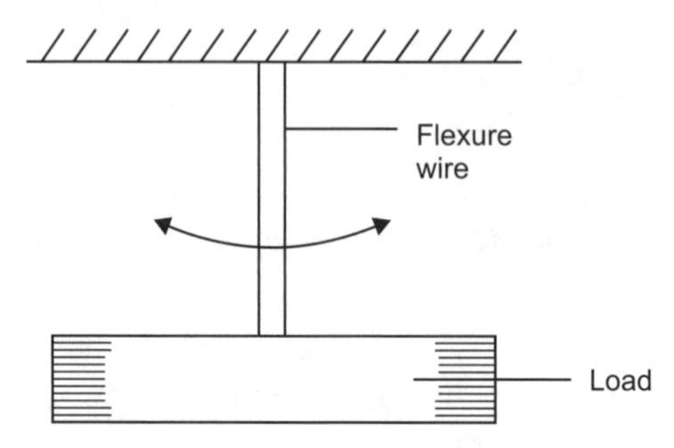

FIGURE 6.27 Twist flexures.

Advantages and Disadvantages of Frictionless Bearings

- They provide no or negligibly less frictional resistance to motion.
- They are used in precision instruments such as voltmeters and analytical balances.
- They are relatively costly.
- They are difficult to manufacture.
- They have size limitations.
- They cannot bear higher loads.
- They are used in limited motion applications.
- They cannot be used in high speed applications.
- They do not exhibit stick-slip phenomena.
- They can also be used as flexible suspensions and electrical conductors.
- They are sensitive to vibration.

EXERCISES

1. Sketch and explain the workings of a recirculating ball lead screw.
2. Explain the different types of lubrication conditions in linear and circular bearings.
3. Sketch the different guide ways used in machine tools.

4. Explain the workings of recirculating ball screws in CNC machines with a sketch.
5. Define the phenomenon of stick-slip.
6. Explain how stick-slip phenomena can be reduced.
7. Explain with a sketch the working principle of hydrodynamic bearings.
8. What are the requirements of a CNC machine spindle tool assembly?
9. How do you classify guide ways? Explain any one with a sketch.
10. Explain the working principle of hydrostatic and hydrodynamic bearings. Also ennumerate their advantages and disadvantages.
11. What are the requirements of a modern machine tool spindle assembly?
12. What is thermal displacement of ball screws and the measures taken to reduce the effect?
13. Categorize machine tool structures.
14. State the requirements of design and manufacture of machine tool structures.
15. List the design considerations of the structure of modern machine tools.
16. Explain the various loads on machine tool structures.
17. What is a guide way? Give examples.
18. What are the requirements of a good guide way?
19. Classify guide ways.
20. Explain the principle of a slide way.
21. Enumerate the principle of an anti-friction way.
22. Describe with applications the features of different shapes of slide ways.
23. Explain the features of ball and roller type anti-friction ways.
24. What is a recirculating type anti-friction way? Explain with a sketch.
25. What is the principle of function of a hydrostatic slide way?
26. Explain the principle of a hydrodynamic slide way.
27. List the advantageous features of hydrostatic and hydrodynamic slide ways.
28. Give a comparison between slide ways and anti-friction ways.
29. Explain with a sketch recirculating ball screws and nuts. Give the applications.
30. State the advantages and disadvantages of recirculating type ball screws and nuts over conventional screws and nuts.
31. Explain with a sketch two types of pre-loading in ball screws and nuts.
32. Explain the construction of a roller screw.
33. How does a planetary roller screw function?
34. List the advantages of roller screws over ball screws.
35. What are the functions of a machine tool spindle?
36. What are the requirements in the design of machine tool spindles?

37. Suggest suitable materials for spindles based on accuracy and size.
38. Give three different design configurations for spindle noses.
39. Discuss the types of loads on machine spindles.
40. How do you minimize dynamic effects in spindles?
41. How do you compensate for the inaccuracies due to thermal deformations in machine spindles?
42. Describe the selection of spindle noses.
43. State the general requirements of spindle supports.
44. Give the broad classification of spindle supports.
45. What are the considerations in the selection of suitable material for bearings?
46. Give the features of a ball bearing with a sketch.
47. Give the features of a roller bearing with a sketch.
48. Compare the features of a sliding bearing with an anti-friction bearing.
49. List the design considerations of bearing support.
50. What is pre-loading of bearings? Why is it done?
51. What are the different pre-loading methods? Explain any two with sketches.
52. Discuss the application of pre-loading in machine tool bearings.
53. Discuss the advantages and disadvantages of anti-friction bearings.
54. List the considerations in the selection of spindle bearings.
55. What is a frictionless bearing? List the types of frictionless bearings.
56. Explain with applications three types of frictionless bearings.
57. What are the advantages and disadvantages of frictionless bearings?

CHAPTER 7

SIGNAL CONDITIONING

The signals handled by microprocessors need to be conditioned for effective processing. The signals from sensors/transducers may be non-linear, noise-filled, or weak in magnitude. Such signals should be modified and magnified to suit the specification rating of the microprocessors and microcontrollers. There may be a need to convert the signal from one form to another. These signal conditioning functions are most commonly done using operational amplifiers of various types. The microprocessors should be protected from excessive voltage and reverse polarity of the signal using proper electronic protection circuits. The signal conditioning and protection are accomplished by the interface systems inserted between the sensors/actuators and the microprocessors. The following topics pertaining to signal conditioning are covered in this chapter:

- Concept of signal conditioning and its needs.
- Various types of operational amplifiers.
- Theory and application of protection.
- Types of filters and filtering.
- Wheatstone bridge and compensation.
- Digital signal processing and Shannon's theorem.

- Analog-to-digital converters.
- Digital-to-analog converters.
- Types and applications of multiplexers.
- Data acquisition.
- Amplifier errors and related terminology.

7.1 INTRODUCTION TO SIGNAL PROCESSING

In Chapter 2 we discussed that certain parameters such as speed, position, force, etc., are sensed before measurement to give some form of signal to be processed. Such a signal, before being input to the next stage of processing, may have to be conditioned for the following reasons:

- The signal may be weak and has to be magnified.
- It may have interference such as noise which has to be filtered.
- The non-linear signal has to be manipulated.
- The signal may have to be converted for a change in resistance and voltage.
- The analog form may be required in the digital form.
- The digital form may need conversion to the analog form.

Hence, the signal which may be weak, non-linear, noise-filled, and of a certain form may need transformed to an amplified signal, linear, noise-free, or some other required form through an amplifier, signal manipulator, filter, or a converter before being supplied to a processor in a process known as *signal conditioning*.

For example, the output from a strain gauge is a voltage change in a few millivolts which has to be converted to a current of reasonable size without noise and provided with linearization, and may have to be in a digital form before it is processed by a microprocessor. The processed signal also needs conditioned to a form needed by the output devices.

7.2 CONCEPT OF SIGNAL CONDITIONING

The following diagram illustrates signal processing.

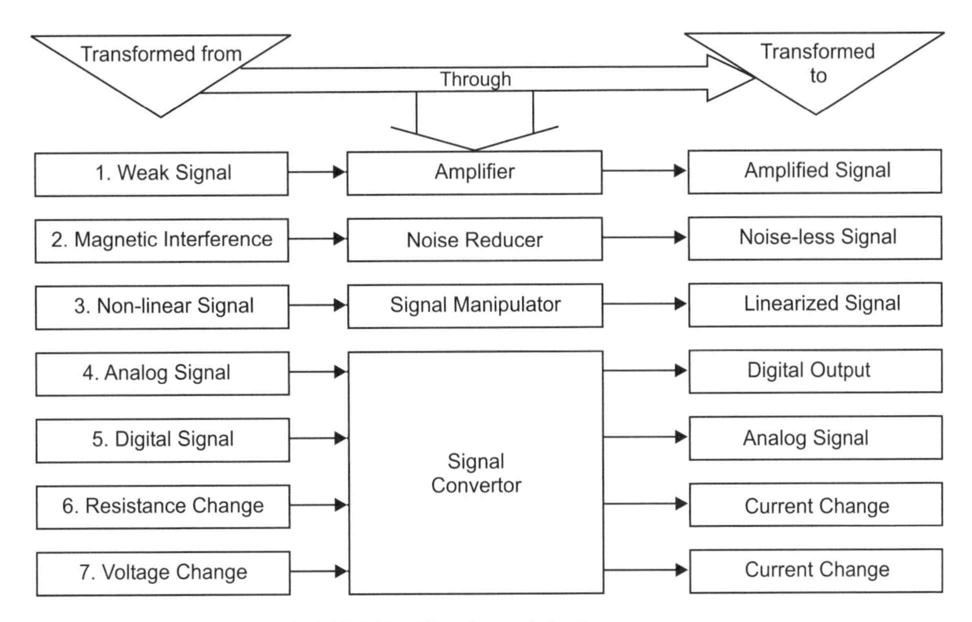

FIGURE 7.1 Signal conditioning concept.

7.3 NEED FOR SIGNAL CONDITIONING

The signal required by the microprocessor or the microcontroller cannot be in raw form from input devices such as sensors. The signals need transformation which is accomplished by the interface systems, connected between the input devices and processors, and the processor and output devices. The need for conditioning signals arises for the following reasons:

- The processor has to be protected from erratic input signals of excessive voltage and incorrect polarity.
- The processor needs protection from a sudden output signal.
- The processor can process the signal that is in a compatible form with the system characteristics.
- The processing system can receive signals that have ratings suitable to their specifications.
- The processor requires noise-free and disturbance-free signals to perform correctly.
- The non-linearity in the signal output from the input devices needs to be manipulated to transform it into a linear signal.

The interface system between the input devices such as the sensors, switches, keyboards, and the microprocessor is first amplified, then converted from analog form to digital form, and then protected from excessive voltage and incorrect polarity.

Again, the output from the processor should be protected from the back input signal from the output port. The signal is decoded and converted to analog form before being accepted by output devices such as actuators and displays.

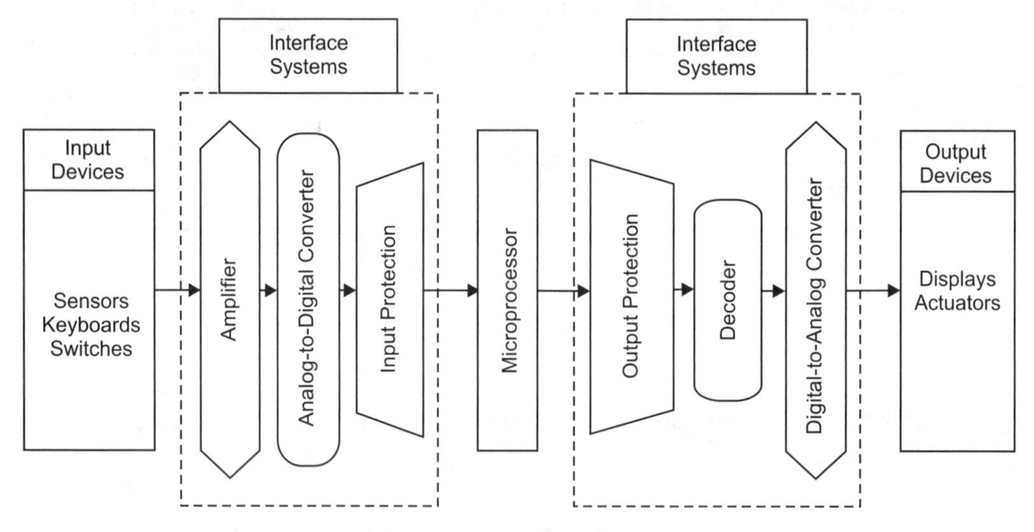

FIGURE 7.2 Interface diagram.

7.4 OPERATIONAL AMPLIFIERS

By definition, an operational amplifier is a high gain D.C. device that magnifies the input signal (current or voltage) up to an order of 10^6 or more. It is supplied as a silicon chip with an integrated circuit in it. A typical chip with op-amp is shown in Figure 7.3.

The types of inputs for an operational amplifier are:

- Negative inverting input.
- Positive non-inverting input.
- Negative voltage supply.
- Positive voltage supply.
- Two offset null inputs for extracting non-ideal behaviors from op-amps.

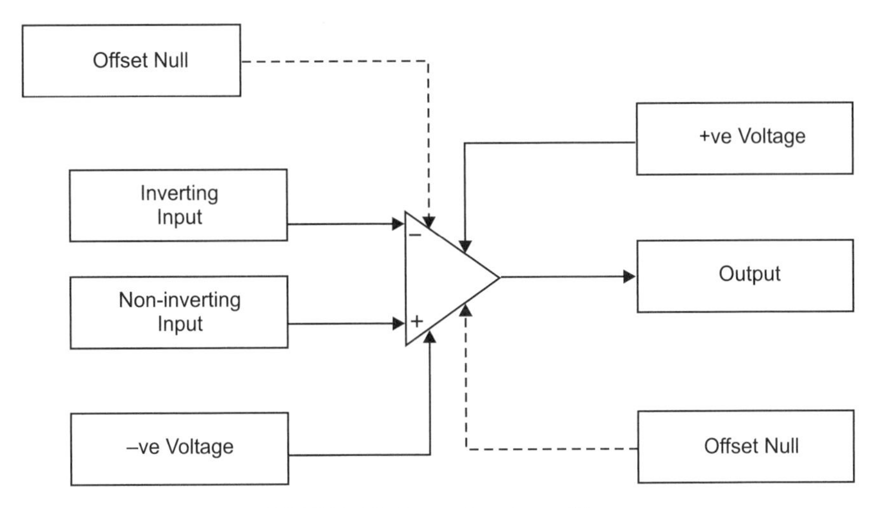

FIGURE 7.3 Operational amplifier.

Depending on the inputs and connection of impedance, the op-amps perform as:

- Voltage-to-current or current-to-voltage converter.
- Signal adder.
- Signal magnifier.
- Non-linear to linear manipulator.
- A filter for noise reduction.
- Analog-to-digital and digital-to-analog converter.
- Interface between sensors and microprocessor.
- Op-amplifier analysis.

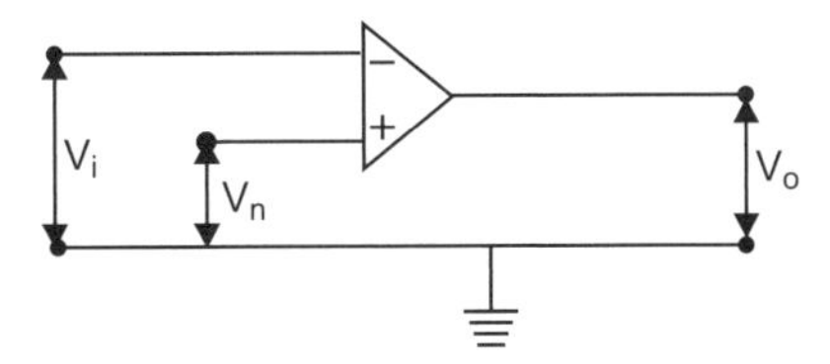

FIGURE 7.4 Op-amp configuration.

The output, V_o, of an operational amplifier depends on the difference between the voltage signals at its inverting (–ve) and non-inverting (+ve) inputs.

For example, $V_o = A(V_n - V_i)$, where A is the amplifier gain. (7.1)

7.5 VOLTAGE-TO-CURRENT CONVERTER: INVERTING TYPE OPERATIONAL AMPLIFIER

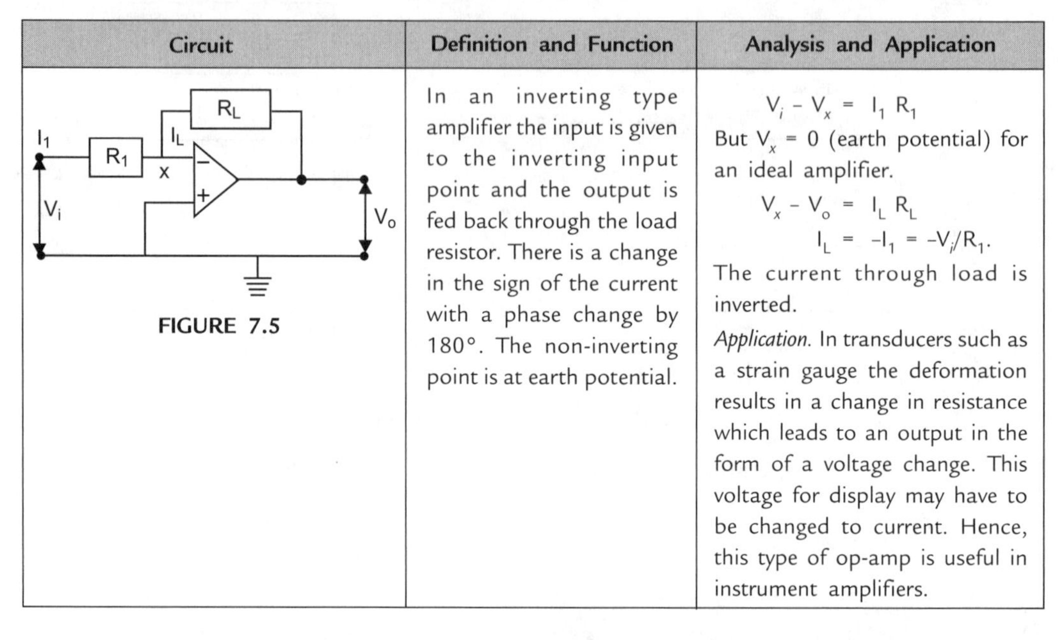

Circuit	Definition and Function	Analysis and Application
FIGURE 7.5	In an inverting type amplifier the input is given to the inverting input point and the output is fed back through the load resistor. There is a change in the sign of the current with a phase change by 180°. The non-inverting point is at earth potential.	$V_i - V_x = I_1 R_1$ But $V_x = 0$ (earth potential) for an ideal amplifier. $V_x - V_o = I_L R_L$ $I_L = -I_1 = -V_i/R_1$. The current through load is inverted. *Application.* In transducers such as a strain gauge the deformation results in a change in resistance which leads to an output in the form of a voltage change. This voltage for display may have to be changed to current. Hence, this type of op-amp is useful in instrument amplifiers.

7.6 CURRENT-TO-VOLTAGE CONVERTER: INVERTING TYPE OPERATIONAL AMPLIFIER

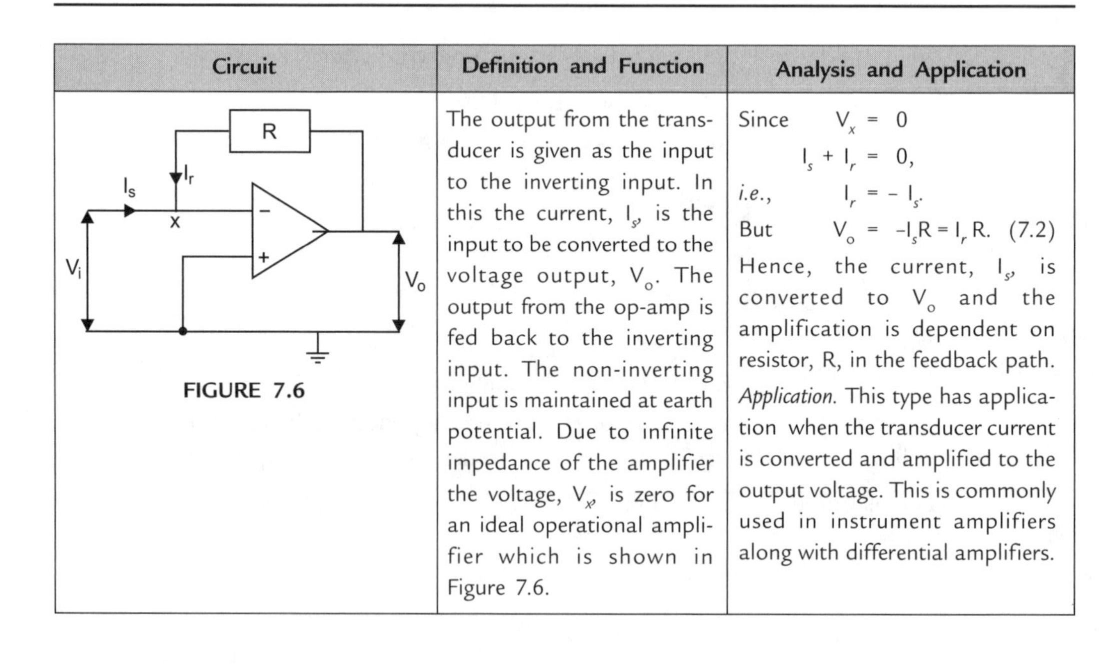

Circuit	Definition and Function	Analysis and Application
FIGURE 7.6	The output from the transducer is given as the input to the inverting input. In this the current, I_s, is the input to be converted to the voltage output, V_o. The output from the op-amp is fed back to the inverting input. The non-inverting input is maintained at earth potential. Due to infinite impedance of the amplifier the voltage, V_x, is zero for an ideal operational amplifier which is shown in Figure 7.6.	Since $V_x = 0$ $I_s + I_r = 0$, *i.e.*, $I_r = -I_s$. But $V_o = -I_s R = I_r R$. (7.2) Hence, the current, I_s, is converted to V_o and the amplification is dependent on resistor, R, in the feedback path. *Application.* This type has application when the transducer current is converted and amplified to the output voltage. This is commonly used in instrument amplifiers along with differential amplifiers.

7.7 NON-INVERTING OPERATIONAL AMPLIFIER

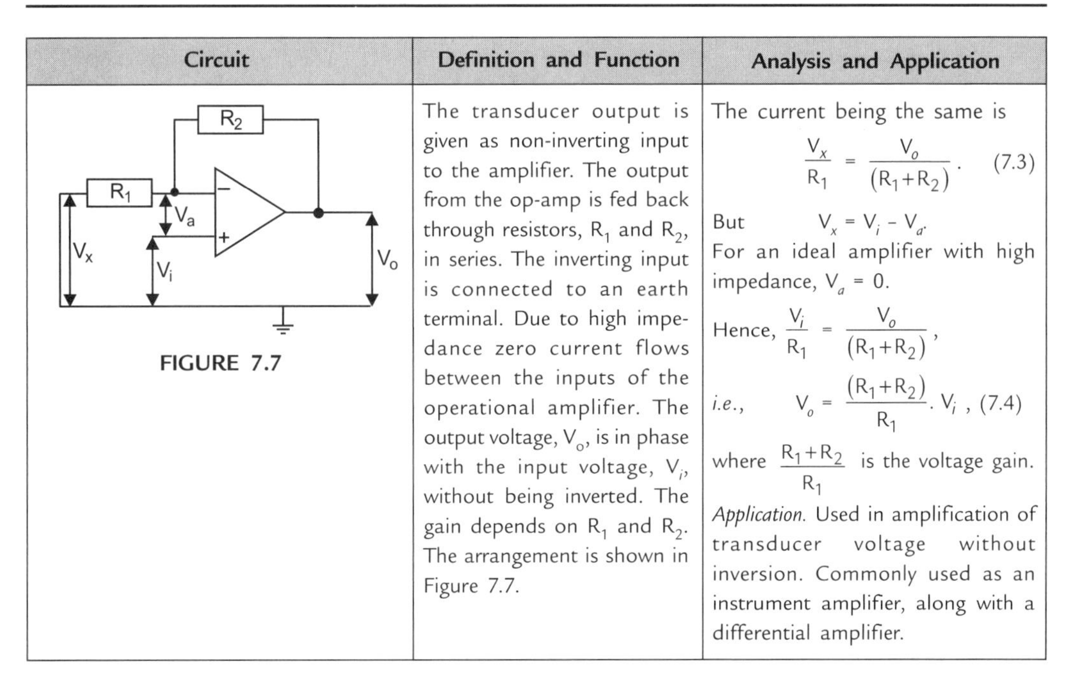

Circuit	Definition and Function	Analysis and Application
FIGURE 7.7	The transducer output is given as non-inverting input to the amplifier. The output from the op-amp is fed back through resistors, R_1 and R_2, in series. The inverting input is connected to an earth terminal. Due to high impedance zero current flows between the inputs of the operational amplifier. The output voltage, V_o, is in phase with the input voltage, V_i, without being inverted. The gain depends on R_1 and R_2. The arrangement is shown in Figure 7.7.	The current being the same is $\frac{V_x}{R_1} = \frac{V_o}{(R_1+R_2)}$. (7.3) But $V_x = V_i - V_a$. For an ideal amplifier with high impedance, $V_a = 0$. Hence, $\frac{V_i}{R_1} = \frac{V_o}{(R_1+R_2)}$, *i.e.*, $V_o = \frac{(R_1+R_2)}{R_1} . V_i$, (7.4) where $\frac{R_1+R_2}{R_1}$ is the voltage gain. *Application.* Used in amplification of transducer voltage without inversion. Commonly used as an instrument amplifier, along with a differential amplifier.

7.8 SUMMING AMPLIFIER

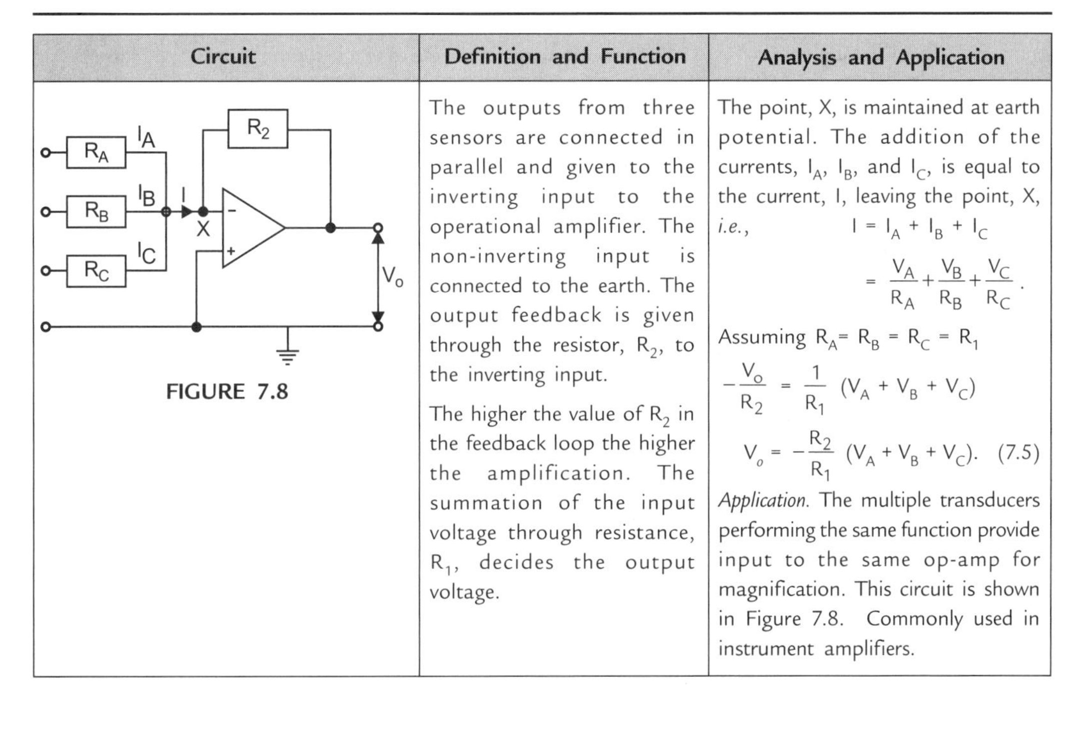

Circuit	Definition and Function	Analysis and Application
FIGURE 7.8	The outputs from three sensors are connected in parallel and given to the inverting input to the operational amplifier. The non-inverting input is connected to the earth. The output feedback is given through the resistor, R_2, to the inverting input. The higher the value of R_2 in the feedback loop the higher the amplification. The summation of the input voltage through resistance, R_1, decides the output voltage.	The point, X, is maintained at earth potential. The addition of the currents, I_A, I_B, and I_C, is equal to the current, I, leaving the point, X, *i.e.*, $I = I_A + I_B + I_C$ $= \frac{V_A}{R_A} + \frac{V_B}{R_B} + \frac{V_C}{R_C}$. Assuming $R_A = R_B = R_C = R_1$ $-\frac{V_o}{R_2} = \frac{1}{R_1}(V_A + V_B + V_C)$ $V_o = -\frac{R_2}{R_1}(V_A + V_B + V_C)$. (7.5) *Application.* The multiple transducers performing the same function provide input to the same op-amp for magnification. This circuit is shown in Figure 7.8. Commonly used in instrument amplifiers.

7.9 INTEGRATING AMPLIFIER

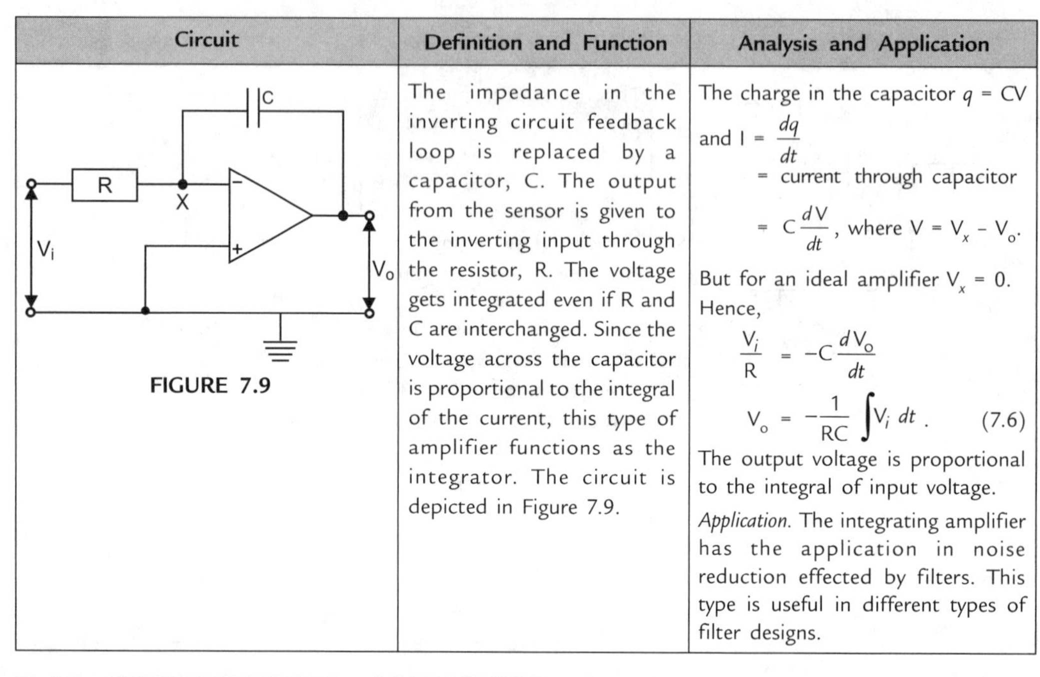

Circuit	Definition and Function	Analysis and Application
FIGURE 7.9	The impedance in the inverting circuit feedback loop is replaced by a capacitor, C. The output from the sensor is given to the inverting input through the resistor, R. The voltage gets integrated even if R and C are interchanged. Since the voltage across the capacitor is proportional to the integral of the current, this type of amplifier functions as the integrator. The circuit is depicted in Figure 7.9.	The charge in the capacitor $q = CV$ and $I = \frac{dq}{dt}$ = current through capacitor $= C\frac{dV}{dt}$, where $V = V_x - V_o$. But for an ideal amplifier $V_x = 0$. Hence, $\frac{V_i}{R} = -C\frac{dV_o}{dt}$ $V_o = -\frac{1}{RC}\int V_i\,dt$. (7.6) The output voltage is proportional to the integral of input voltage. *Application.* The integrating amplifier has the application in noise reduction effected by filters. This type is useful in different types of filter designs.

7.10 DIFFERENTIAL AMPLIFIER

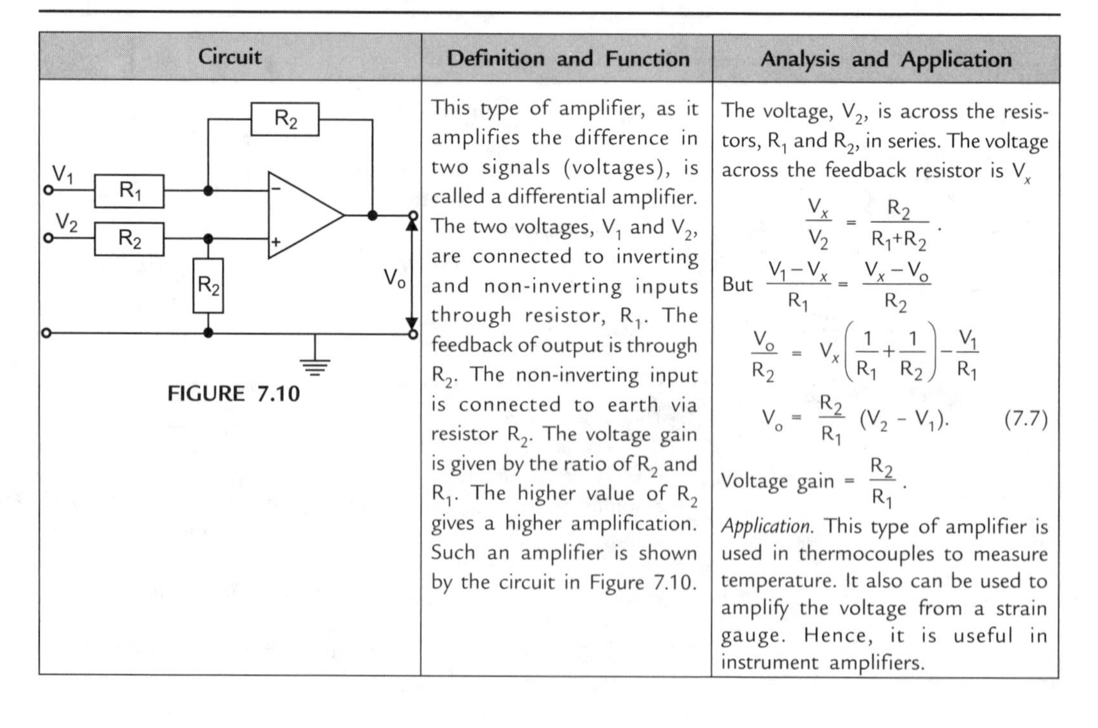

Circuit	Definition and Function	Analysis and Application
FIGURE 7.10	This type of amplifier, as it amplifies the difference in two signals (voltages), is called a differential amplifier. The two voltages, V_1 and V_2, are connected to inverting and non-inverting inputs through resistor, R_1. The feedback of output is through R_2. The non-inverting input is connected to earth via resistor R_2. The voltage gain is given by the ratio of R_2 and R_1. The higher value of R_2 gives a higher amplification. Such an amplifier is shown by the circuit in Figure 7.10.	The voltage, V_2, is across the resistors, R_1 and R_2, in series. The voltage across the feedback resistor is V_x $\frac{V_x}{V_2} = \frac{R_2}{R_1+R_2}$. But $\frac{V_1 - V_x}{R_1} = \frac{V_x - V_o}{R_2}$ $\frac{V_o}{R_2} = V_x\left(\frac{1}{R_1}+\frac{1}{R_2}\right) - \frac{V_1}{R_1}$ $V_o = \frac{R_2}{R_1}(V_2 - V_1)$. (7.7) Voltage gain $= \frac{R_2}{R_1}$. *Application.* This type of amplifier is used in thermocouples to measure temperature. It also can be used to amplify the voltage from a strain gauge. Hence, it is useful in instrument amplifiers.

7.11 LOGARITHMIC AMPLIFIER

Circuit	Definition and Function	Analysis and Application
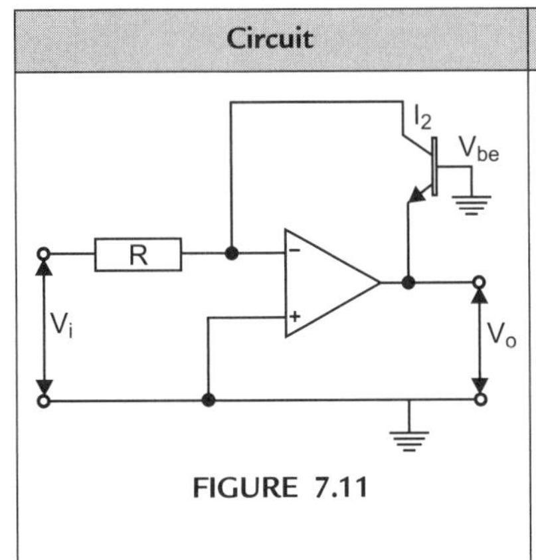 FIGURE 7.11	The impedance in the feedback loop of the inverting type amplifier is replaced either by a diode or by a transistor. The voltage, V_i, is given to the inverting input through resistor, R. The non-inverting input is connected to the earth. The non-linear input to the amplifier from the sensor would result in linearized output, V_o. The output, V_o, is logarithmically related to the non-linear input V_i, and the linearization takes place. The circuit is shown is Figure 7.11.	The I_2 and V_o are related by $I_2 = I_{be} \exp (q\, V_{be}/kT)$. q = electronic charge, I_{be} = saturation current of base-emitter junction, k = Boltzman constant, T = Temperature. The relation between V_o and V_i can be written as $V_o = -C_1 \log_e (C_2 V_i)$, (7.8) where $C_1 = \frac{kT}{q}$; $C_2 = \frac{1}{R\, I_{be}}$. The *npn*-transistor can handle +ve inputs, and the *pnp*-transistor can handle -ve inputs. *Application.* The outputs from sensors such as thermocouples are non-linear. But there has to be a linear relation between temperature and voltage for processing. This can be achieved by accommodating a logarithmic amplifier.

7.12 SCHMITT TRIGGER AMPLIFIER

Circuit	Definition and Function	Analysis and Application
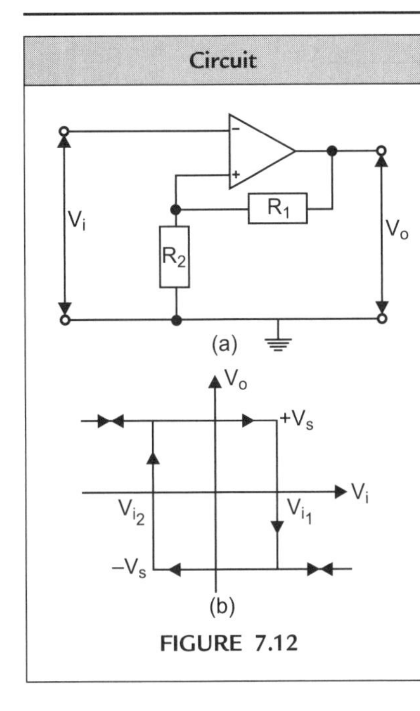 FIGURE 7.12	In this amplifier the non-inverting input receives the feedback of the output voltage, V_o. The non-inverting input is also connected to the earth through resistor, R_2. The Schmitt trigger circuit is as shown in Figure 7.12(*a*) and the output voltage is shown in Figure 7.12(*b*). The output voltage can be positive or negative, that depends on the supply, V_i.	The input and output relation in voltage is given by $V_{i_1} = V_s \frac{R_2}{(R_1+R_2)}$ V_s positive (7.9) $V_{i_2} = V_s \frac{R_2}{(R_1+R_2)}$ V_s negative. The voltage gain is given by $A = \frac{R_1+R_2}{R_2}$. The large value of R_1 (feedback loop resister) can give high voltage gain. This type of amplifier is used as a switch that eliminates the contact bounce that causes irregularity in the signal to be input to the processor.

7.13 AMPLIFIER ERRORS

The offset in the output voltage due to bias in the input current and the slow drift, the maximum change in the rate of the output signal due to frequency response and the slew rate, and the variation in the output gain are some of the errors in an amplifier's operation.

7.13.1 Input Bias Current

The bipolar transistor inputs to the operational amplifier can result in input bias current because of which bias voltages are produced in the external resistors connected to the inverting input of the amplifier. The effect of this bias voltage is the offset in the output voltage produced by the amplifier. By supplying bias voltages of equal magnitude to the inverting and non-inverting input, the result is the cancellation of the offset in the output.

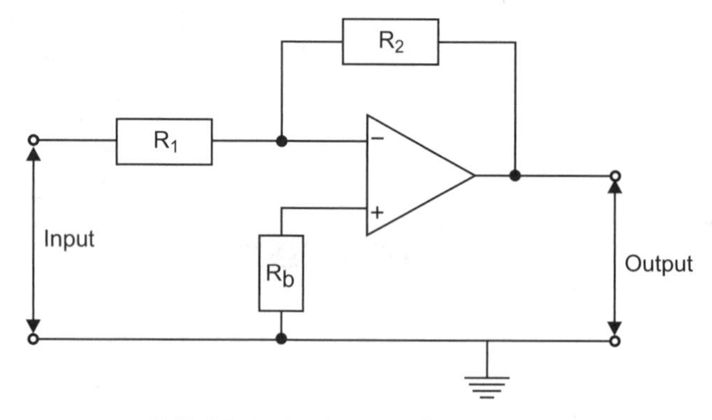

FIGURE 7.13 Bias cancellation circuit.

A bias resistor included in series with the non-inverting input to the amplifier minimizes the offset in the output voltage from the amplifier. The bias resistor, R_b, as shown in Figure 7.13, is given by

$$R_b = \frac{R_1 R_2}{(R_1 + R_2)}. \tag{7.10}$$

7.13.2 Drift

The variation in output is observed in the long run of the amplifier. The long-term effects are due to the following reasons:

- Temperature variations due to changing seasons.
- Supply voltage variations due to changing input to source.
- Aging of the component with decreased reliability.

The results of all these possibilities lead to variation in the output of a D.C. coupled amplifier, which is known as *drift*. Since the variation of the output voltage to the input signal is very slow, the effect of drift cannot be identified and distinguished on a short-term basis.

7.13.3 Frequency Response

The frequency response limits the frequency of the signals that an operational amplifier can accommodate. The effective gain depends on the feedback fraction, which is always less than the D.C. gain. The decrease in effective gain results in a decrease in the frequency response. This leads to an increase in the bandwidth. This can be observed by Eqs. (7.11) and (7.12)

$$A_{eff} = A/(1 + \beta A) \tag{7.11}$$

$$B_{eff} = B\,(1 + \beta A). \tag{7.12}$$

A typical frequency response curve of an operational amplifier is shown in Figure 7.14.

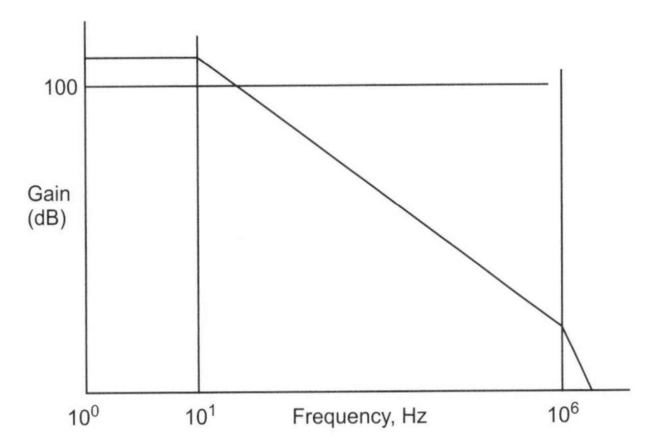

FIGURE 7.14 Frequency response.

7.13.4 Slew Rate

The rate of change of an input signal to which an operational amplifier responds also limits the frequency of the input signal. This is referred to as the *slew rate*. The slew rate gives the maximum rate of change of the output signal to unit step change in the input signal. The unit of slew rate is V/S. The slew rate is related to the full power bandwidth of the operational amplifier. Hence, slew rate is the criterion used to select the highest frequency and the full voltage sine wave for the input signal.

7.13.5 Gain Variation

In situations when the operational amplifier has to function as a differential amplifier, error is introduced in the output as gain varies for non-inverting and inverting inputs (A_1 and A_2, respectively).

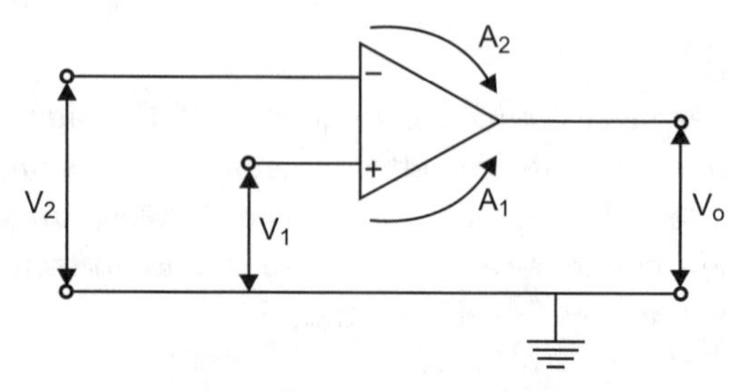

FIGURE 7.15 Variation in gain.

The voltage output of the op-amp is given by

$$V_o = A_1V_1 - A_2V_2. \quad (7.13)$$

By rearrangement,

$$V_o = G_d (V_1 - V_2)/2 + G_c (V_1 + V_2)/2, \quad (7.14)$$

where $G_d = (A_1 + A_2)$

= {different mode voltage gain}

$G_c = (A_1 - A_2)$

= {common mode voltage gain}.

The ratio of different mode voltage gain (G_d) to common mode voltage gain (G_c) is called the Common Mode Rejection Rate (CMRR).

$$\text{CMRR} = \frac{G_d}{G_c} = \frac{(A_1 + A_2)}{(A_1 - A_2)}. \quad (7.15)$$

An increase in the common mode voltage gain decreases the CMRR. A higher CMRR results in less error in the output voltage of an differential amplifier.

7.13.6 Comparator

The manner in which an operational amplifier compares two signals as required during conversion is categorized into a comparator as shown in Figure 7.16. Because the gain of the amplifier is high, the signals with small differences in voltage are enough for a large swing in the output voltage. The comparator integrated circuits give rapid response with minimum error compared to an operational amplifier. Reversal of polarity is also possible in this arrangement of comparison by the comparator.

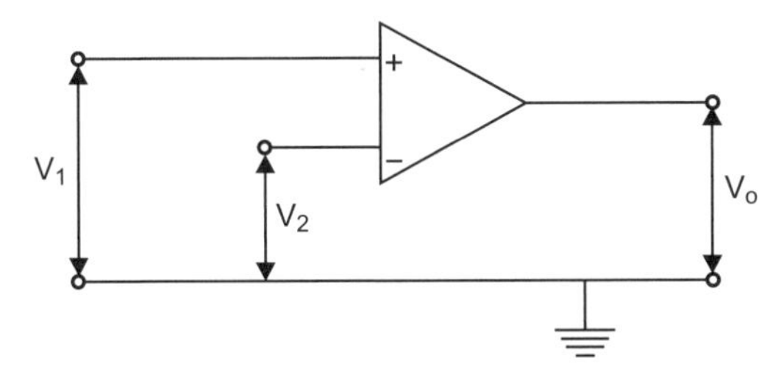

FIGURE 7.16 Comparator.

7.14 PROTECTION

The output from sensors such as thermocouples or strain gauges or LEDs cannot be given to the next stage, *i.e.*, to the microprocessor directly. The microprocessor or the microcontroller may be damaged by irregular and erratic signal from the sensors. The undue signal may be:

- High current beyond the acceptable limit of the microprocessor.
- High voltage exceeding the rating of the microprocessor.
- Incorrect polarity leading to malfunctioning of the microprocessor.

To safeguard the microprocessor from these irregularities *protection* is provided in the interface between the sensor or actuator and the microprocessor.

7.14.1 Zener Diode Protection

- To limit the high current a resistor may be connected in the input line and a fuse to break the circuit in case the current exceeds the safe level.
- A Zener diode connected in parallel with the input and output protects from excessive voltage and wrong polarity. The Zener diode conducts up to a certain breakdown voltage and beyond which drops the resistance to a low value and the output to the next stage also drops. As the Zener diode has low resistance in one direction and high resistance in the other direction, it provides protection to the next stage receiving output against the wrong polarity.

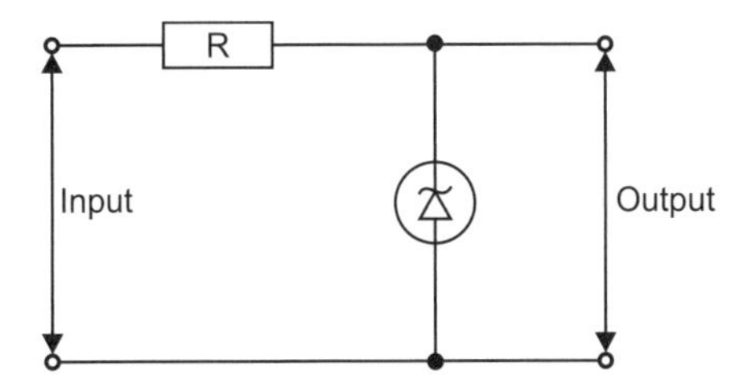

FIGURE 7.17 Protection circuit.

7.15 FILTERING

By definition, *filtering* is the process of removing a certain band of frequencies from a signal and permitting the remaining. The levels of filtering include:

- *Pass band.* Range of frequency passed by the filter.
- *Stop band.* Range of frequency not passed by the filter.
- *Cut-off frequency.* The boundary between the stop band and pass band.

7.15.1 Types of Filters

Depending on the range of frequency transmitted and rejected filters are classified as:

1. ***Low-pass filters:*** These filters have a pass band from zero to a certain value as shown in Figure 7.18.

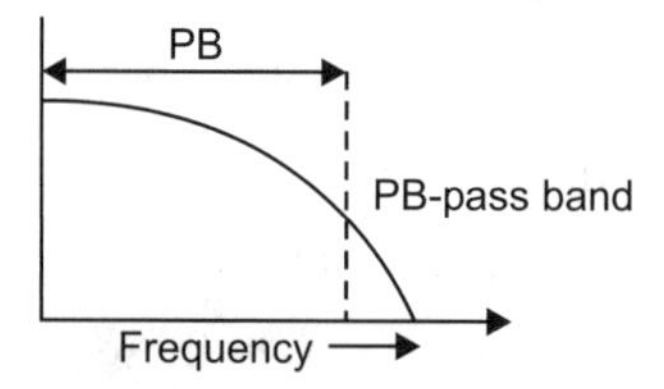

FIGURE 7.18 Low-pass filter.

2. ***High-pass filters:*** These filters transmit frequencies between a certain value to infinity. The high-pass filters transmit only high frequencies as shown in Figure 7.19.

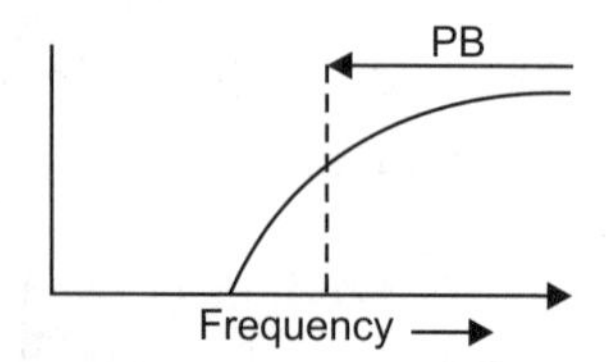

FIGURE 7.19 High-pass filters.

3. ***Band-pass filters:*** These type of filters transmit frequencies within a specified range. With any beyond the range, it stops transmitting. (Refer to Figure 7.20.)

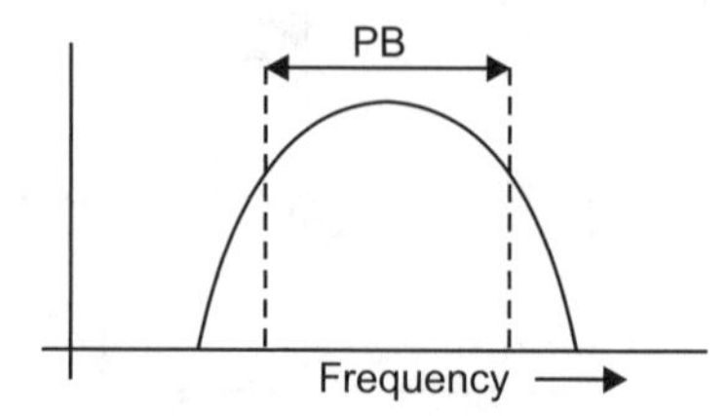

FIGURE 7.20 Band-pass filters.

4. ***Band-stop filters:*** Within a certain range of frequency, the filter stops transmitting. It operates from zero to a certain value, stops in a range, and transmits beyond the range. (Refer to Figure 7.21.)

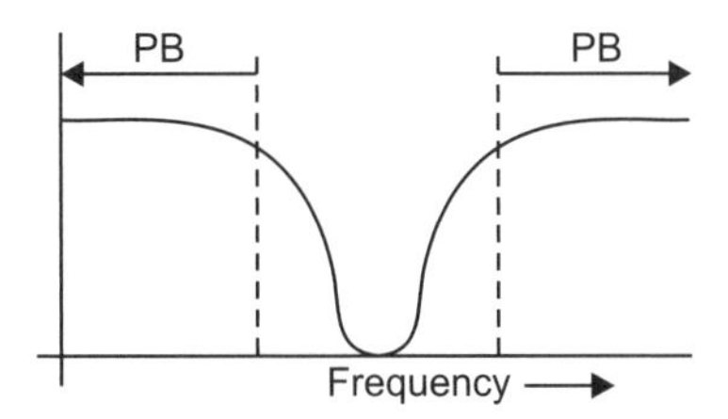

FIGURE 7.21 Band-stop filter.

5. ***Low-pass passive filter:*** They operate in the lower frequencies and are generally used to filter noise occuring in the higher frequencies as in the case of interferences in an A.C. main supply. They are called passive because they make use of only resistors, inductors, and capacitors to make the filters. (Refer to Figure 7.22.)

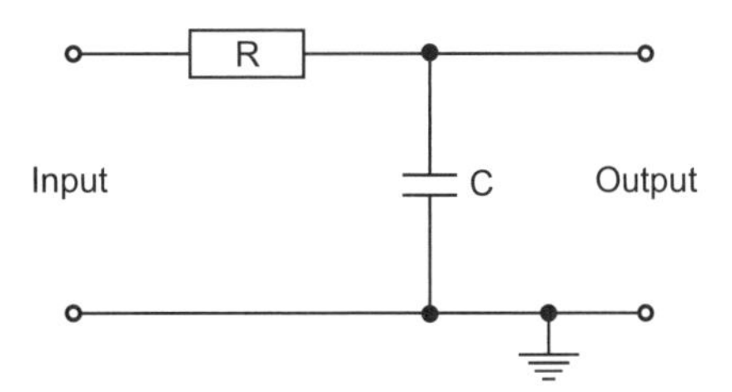

FIGURE 7.22 Low-pass passive filter.

6. ***Low-pass active filters:*** An integrating operational amplifier using a capacitor in its feedback loop acts as a low-pass active filter that removes noise in the high frequency range. Figure 7.23 shows the basic form of a low-pass active filter.

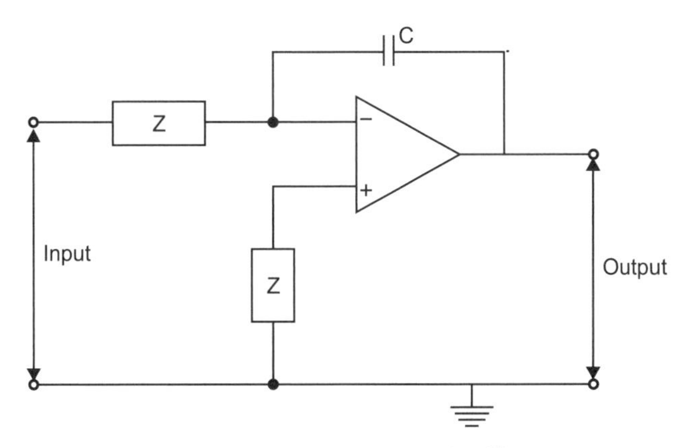

FIGURE 7.23 Low-pass active filter.

7.15.2 High-pass Filters

Passive high-pass filters filter noise at lower frequencies and transmit the higher frequencies as shown in Figure 7.24. Two capacitors arranged in series between the input and output with a parallel inductor function as a high-pass passive filter. A high-pass active filter using an operational amplifier is shown in Figure 7.25. The feedback output is given to both inverting and non-inverting inputs through a resistor.

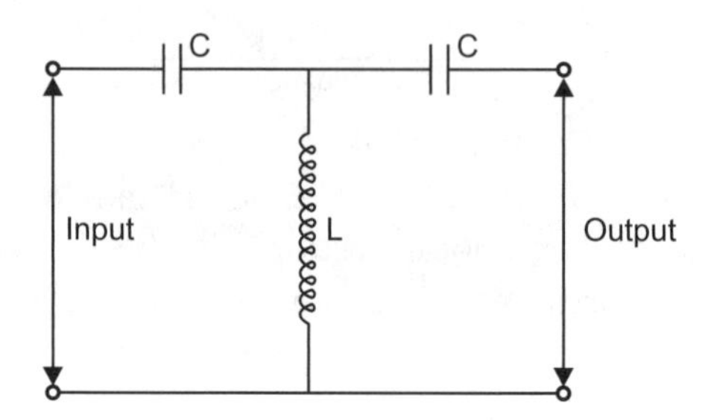

FIGURE 7.24 Passive high-pass filter.

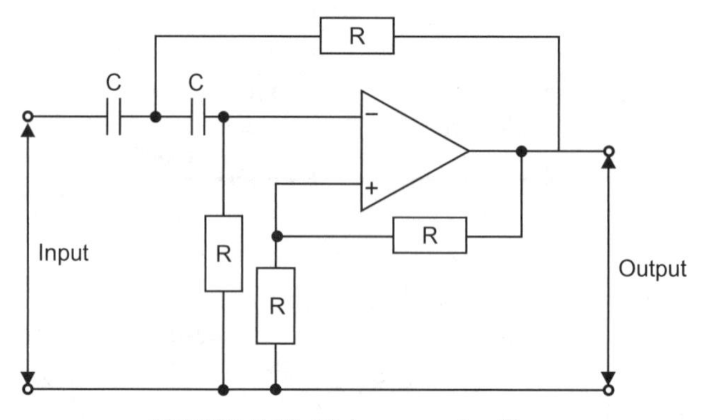

FIGURE 7.25 High-pass active filter.

Two capacitors in series with inverting input do the filtering of noise at high frequencies.

7.16 MULTIPLEXERS

Sometimes a processor has to receive and process in sequence the data from multiple sources. A multiplexer enables a processor to share a single data channel between multiple input sources. By selecting the required channel the output from the selected source is processed. A block diagram representation is shown in Figure 7.26. Hence, a multiplexer is basically an electronic switching device.

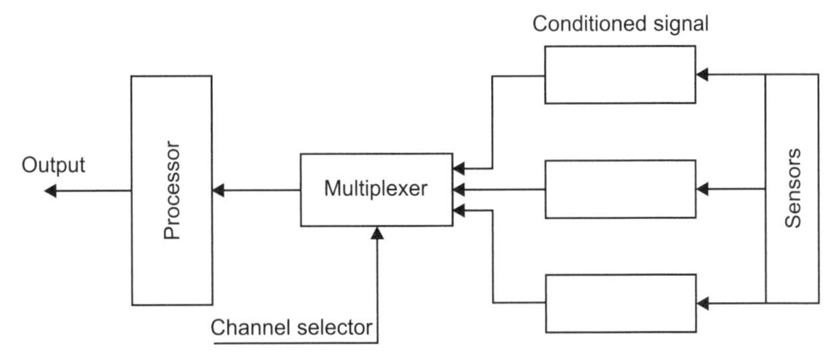

FIGURE 7.26 Multiplexer function.

There are two types of multiplexers:

- Time Division Multiplexer (TDM)
- Frequency Division Multiplexer (FDM)

1. In a Time Division Multiplexer (TDM), each of the data sources is connected to the data channel transmitting the data. The data is then made available to the processor at different times. When large data slowly varies with time the TDM is useful. Examples of data sources for measurement are pressure, temperature, and static strain. These data have discrete values in series useful to the TDM. Figure 7.27 shows time division multiplexing.

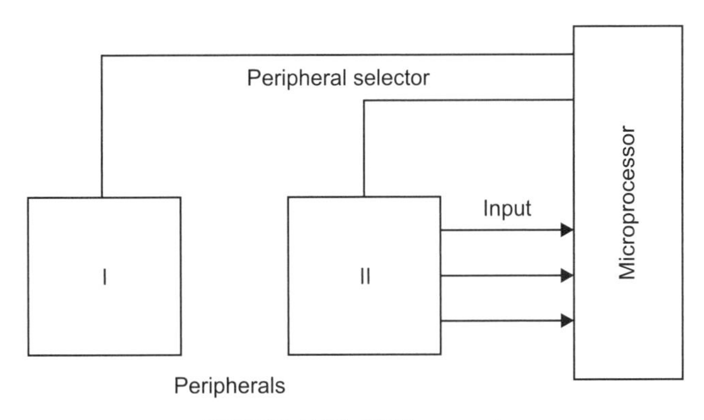

FIGURE 7.27 TDM concept.

2. The Frequency Division Multiplexer (FDM) is particularly suited for analog signals. Each data source is modulated to a subcarrier frequency. All the subcarriers are combined in a mixer and modulated to higher frequency carriers. Then the information is transmitted to the receiver. The receiver, after demodulation, separates into such carriers and is filtered to contain the individual information data. This uses both amplitude modulation systems. A block diagram of the FDM is shown in Figure 7.28.

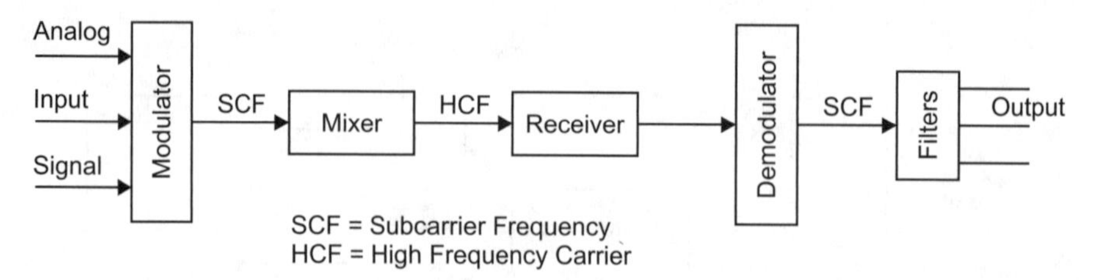

FIGURE 7.28 FDM concept.

7.17 WHEATSTONE BRIDGE

A Wheatstone bridge (Figure 7.29) is formed by four resistors forming the sides of the quadrilateral. The diagonally opposite nodes form pairs. One pair is connected to the source (input) with voltage V_i, and the output, V_o, is measured from the other pair.

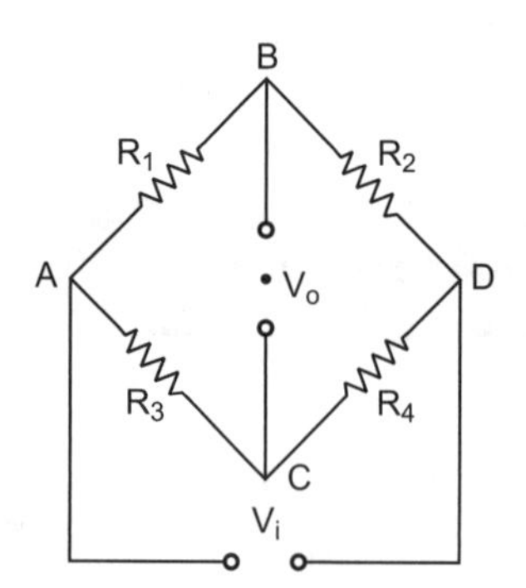

FIGURE 7.29 Wheatstone bridge.

Since the potentials between AB and AC are equal and also between BD and CD, the potential differences are the same, and it holds that

$$\frac{R_1}{R_2} = \frac{R_3}{R_4}. \tag{7.16}$$

This is the condition for the balanced bridge. The output voltage, V_o, is given by

$$V_o = V_{AB} - V_{AC}$$

$$= V_i\left(\frac{R_1}{R_1+R_2} - \frac{R_3}{R_3+R_4}\right). \tag{7.17}$$

For a small change in resistance R_1 to $(R_1 + \delta R_1)$ the change in output voltage, δV_o, can be

$$\delta V_o \approx V_i\left(\frac{\delta R_1}{R_1+R_2}\right). \tag{7.18}$$

Eq. (7.18) shows that a change in one of the resistances proportionately reveals in the output. By replacing, say R_1, by the strain gauge (a sensor) the bridge provides a means to measure the strain in terms of the output voltage change which is converted to indicate the change. Such a bridge circuit with a sensor in one of its legs is shown in Figure 7.30.

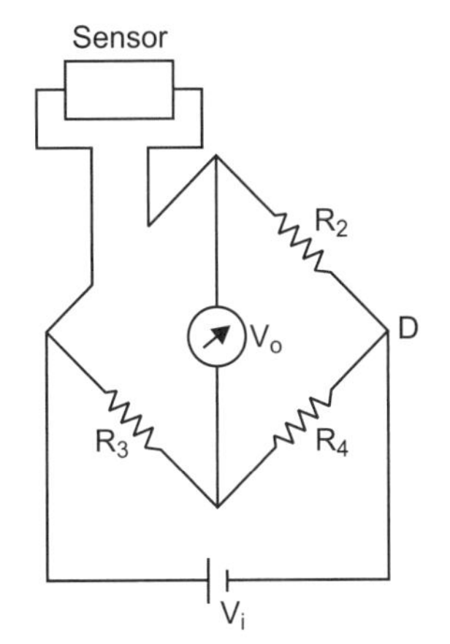

FIGURE 7.30 Bridge with sensor.

7.17.1 Temperature Compensation

The expansion or contraction of the elements of a sensor due to variations in temperature results in a change in resistance. This change results in an offset in the output leading to erroneous measurements. This problem can be tackled by providing a dummy sensor replacing R_3 so that the offset is nullified and the error is eliminated. Hence, a dummy strain gauge which changes due to temperature change compensates the change in an active strain gauge. This effect is known as *temperature compensation*. The bridge circuit showing the arrangement is given in Figure 7.31.

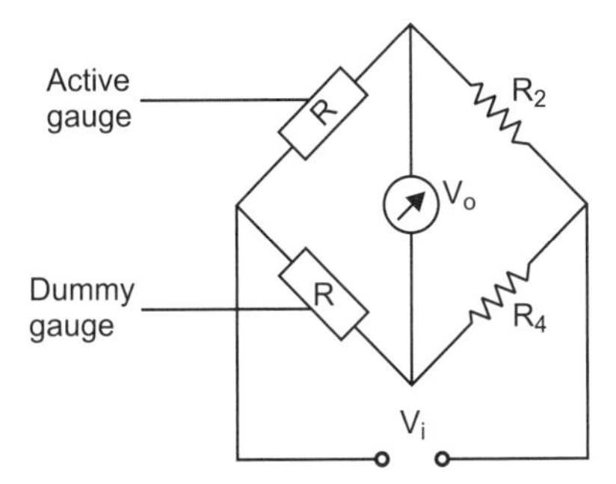

FIGURE 7.31 Temperature compensation using a dummy gauge.

7.17.2 Load Cell

The temperature compensation in a load cell is achieved by using four strain gauges forming the four legs of the Wheatstone bridge. The arrangement of strain gauges in the load cell is shown in Figure 7.32. The orientation of the mounting of strain gauges is such that two are in tension and the other two are in compression. As all are affected by temperature changes, equally, the change in resistance is compensated automatically.

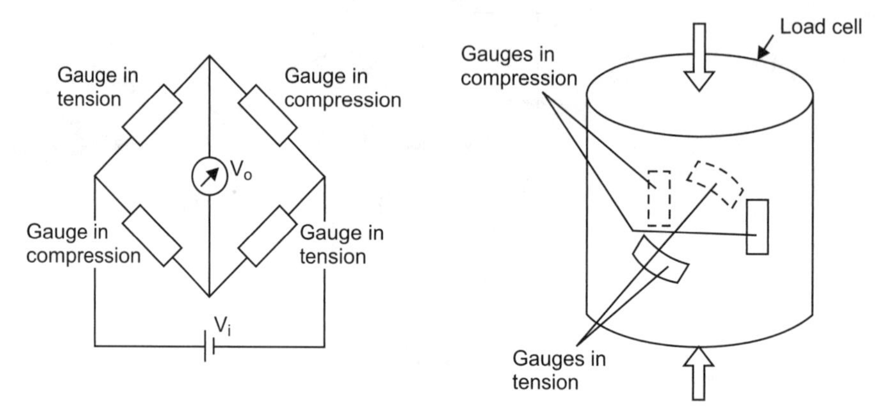

FIGURE 7.32 Bridge in load cell.

7.18 SIGNAL PROCESSING

A microprocessor with a substanitially rated processing capability allows the performing of operations on a wide range of data provided by the transducers. The processing capabilities of microprocessors are:

- Linearization
- Compensation
- Signal Averaging
- Fourier Analysis
- Self-tests
- Self-calibration

7.18.1 Linearization

In practical situations the output of the transducer or sensor bears a complex relation with the parameter to be measured. Analog signal processing cannot handle such non-linear relations effectively. The lookup tables and the built-in computing capability of microprocessors favor the use of non-linear transducers. Hence, in digital signal processing non-linear signals can be accomodated easily.

7.18.2 Compensation

The use of active material such as silicon in transducers make them sensitive to environmental conditions such as variations in temperature. This distorts the output signal from the sensors. But in digital signal processing the microprocessor separately senses the changing conditions and provides suitable compensation so that the output is realizable.

7.18.3 Signal Averaging

In digital signal processing the noise in the signal of random nature is removed by signal averaging to recover the original signal from the noise. This is done by mixing the noise ridden signal with a Gaussian noise signal. If r_i signals with noise have N successive sets of data then the average value r_{ave} of the sample is given by

$$r_{ave} = \frac{1}{N}\sum_{i=1}^{N} r_i \,. \tag{7.19}$$

The r_{ave} closely matches with the actual value of the signal.

7.18.4 Fourier Analysis

The system behavior analysis in the time domain and frequency domain to extract information is widely accomplished using the powerful analytical tool of Fourier analysis based on forward and reverse Fourier transform equations. This is used to obtain information about power spectral density in the frequency domain to which analytical tools such as correlation and cepstrum analysis are applied.

In digital signal analysis, it is the Discrete Fourier Transform (DFT) that is extensively applied to time and frequency domain signals of periodic and discrete form. In the operation of DFT the frequency is defined by Nyquist frequency which is half the sampling frequency. Figure 7.33 shows the DFT in the time domain and frequency domain.

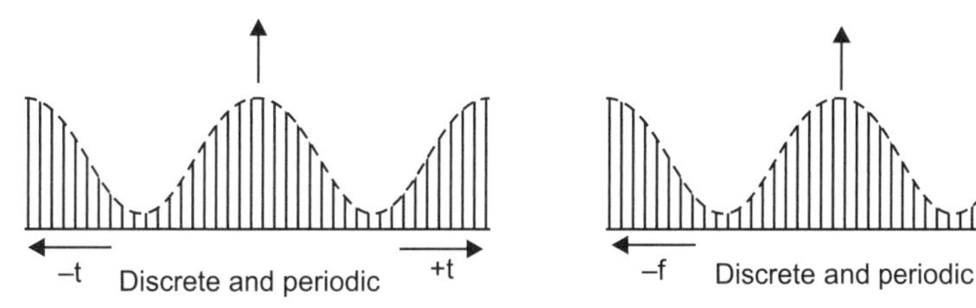

FIGURE 7.33 DFT in time and frequency domain.

7.19 DIGITAL-TO-ANALOG CONVERTER (DAC)

The conversion of binary digital signals to analog signals using an operational amplifier is accomplished by Digital-to-Analog Converters (DACs). The output of the microprocessor is in digital form which has to be converted to analog signal before being given to the displays. There are mainly two types of DACs based on the arrangement of the resistors connected to the operational amplifier:

1. Graded resister DAC
2. Ladder type DAC

7.19.1 Graded Resistor DAC

This is the simplest form of DAC (as shown by Figure 7.34) which consists of a series of graded resistors connected to the inverting input of the operational amplifier. The reference voltage is connected to the resistor through switches which respond to binary '1' or binary '0'. The binary 1 indicates the normally closed (NC) status and binary '0' is the normally open (NO) condition. The output voltage can be given by

$$V_o = \left(\frac{R_f}{R}\right) (S_0 + 2S_1 + 4S_2 + 8S_3 + \ldots + 2^{n-1} S_{n-1})V_{ref}. \quad (7.20)$$

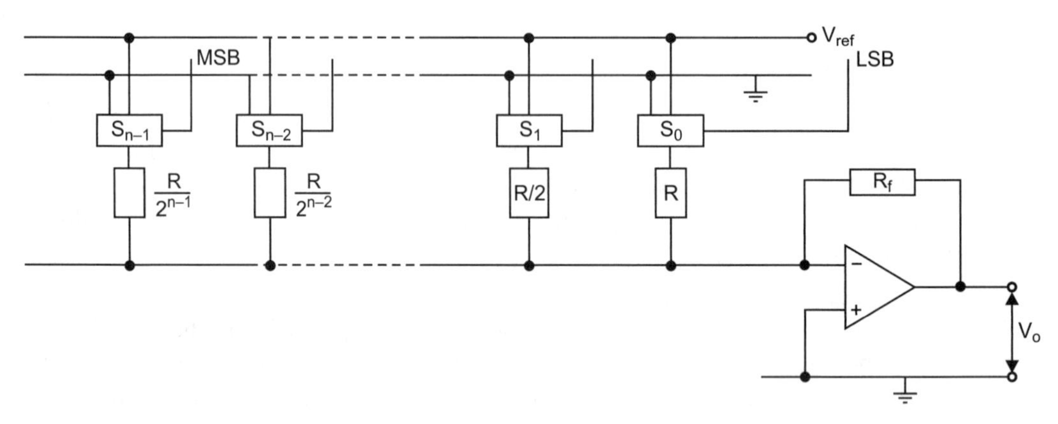

FIGURE 7.34 Graded resistor DAC.

S_0 represents the Least Significant Bit (LSB) and S_{n-1} represents the Most Significant Bit (MSB).

7.19.2 Ladder Type DAC

This is a high-performance device which is shown in Figure 7.35. The ladder network uses only two types of resistors connected to the operational amplifier. One is the feedback resistor R_f. The other is a series of the same value resistor connected in ladder form to the inverting input. This type of connection presents the same value of impedance at each node.

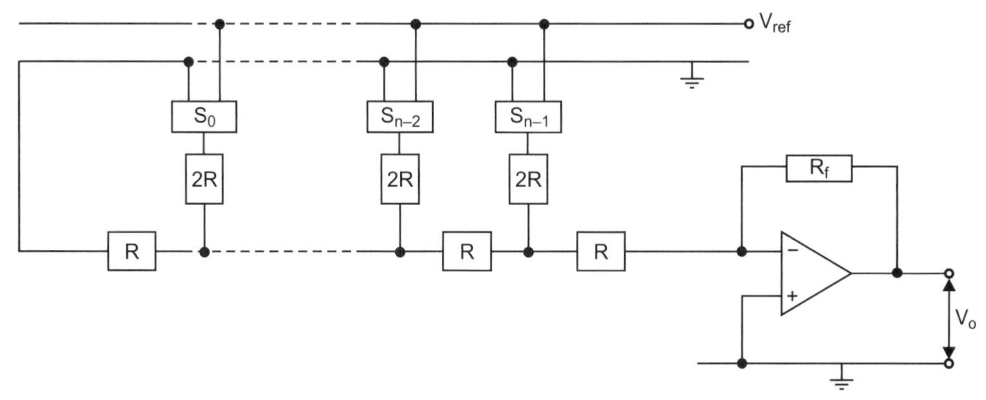

FIGURE 7.35 Ladder type DAC.

In the graded resistor DAC, to be the high performance type, the resistors have to be carefully matched within close tolerances. Such requirements create problems in the manufacturing of DACs. The resistors are required in a large range with the use of high bit numbers. Furthermore, low input impedance that varies with switching introduces offset errors in the output voltage.

The better solution is the use of a ladder type DAC. In this type, by the laser trimming technique, the individual resistances are adjusted within close tolerances, and it maintains the same impedance value at each node eliminating the offset in the output signal.

7.19.3 Specifications of DACs

- *Full scale output.* Producing the output for all 1s in the input is known as full scale output.
- *Resolution.* The number of bits in the memory in a DAC decides the resolution. The higher the bits the higher the resolution.
- *Settling time.* The time taken by the DAC to reach a steady state within ½ LSB of the new voltage signal.
- *Linearity.* The maximum deviation from the straight line joining zero and a full range of output in the ramp.

7.20 ANALOG-TO-DIGITAL CONVERTER (ADC)

The conversion of analog signal to digital signal is required as the microprocessor can process digital signal only. This is accomplished by using an Analog-to-

Digital Converter (ADC). Different types of ADCs are available to suit the application requirements of performance characteristics. Of the many types of ADCs the following are two simple forms:

- Successive Approximation ADC
- Integrating ADC

7.20.1 Successive Approximation ADC

The output of a digital counter is given input to a DAC that converts the signal to an analog signal which is given to the non-inverting input of an operational amplifier. The analog signal to be converted to digital signal is given as inverting input to the op-amp that acts as a comparator. When the output of the DAC equals or exceeds the analog signal, which indicates the completion of conversion, the count is stopped. This value of the count is the actual required digital value. If the count starts from zero and increases sequentially, the process of conversion is slow, which also depends on the value of the analog signal. The successive approximation counter incorporates a switching strategy that reduces the range in which the solution is reached.

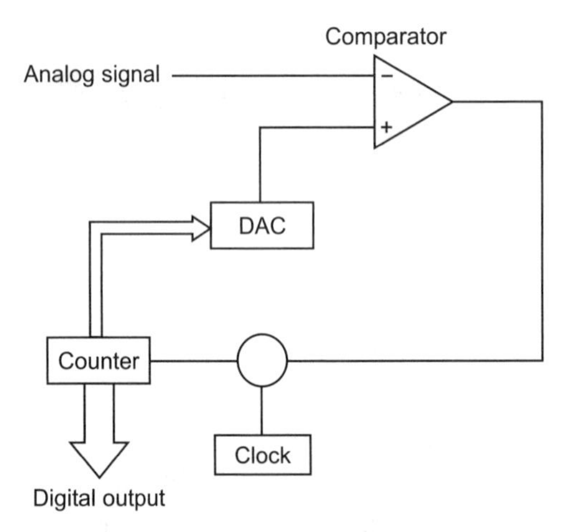

FIGURE 7.36 Successive approximation ADC.

7.20.2 Integrating ADC

The integrating A/D converter is shown by the block diagram in Figure 7.37. The reference voltage, by using an integrator, is converted to an analog value. The output of the integrator and the measured analog signal are input to a comparator (operational amplifier) which compares the two signals. When the two signals equal the comparator gives the digital solution in the binary form.

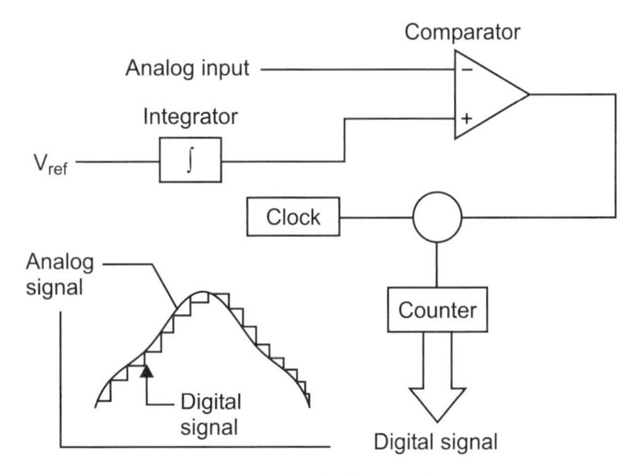

FIGURE 7.37 ADC, integrating type.

7.20.3 Sampling Theorem

The sampling theorem is also known as Shannon's sampling theorem by Nyquist criteria. The theorem states that "In the reconstruction of the analog signals from the samples, it is at the sampling rate at least twice the highest frequency of the analog input signal that the original form of the signal is sampled to the digital

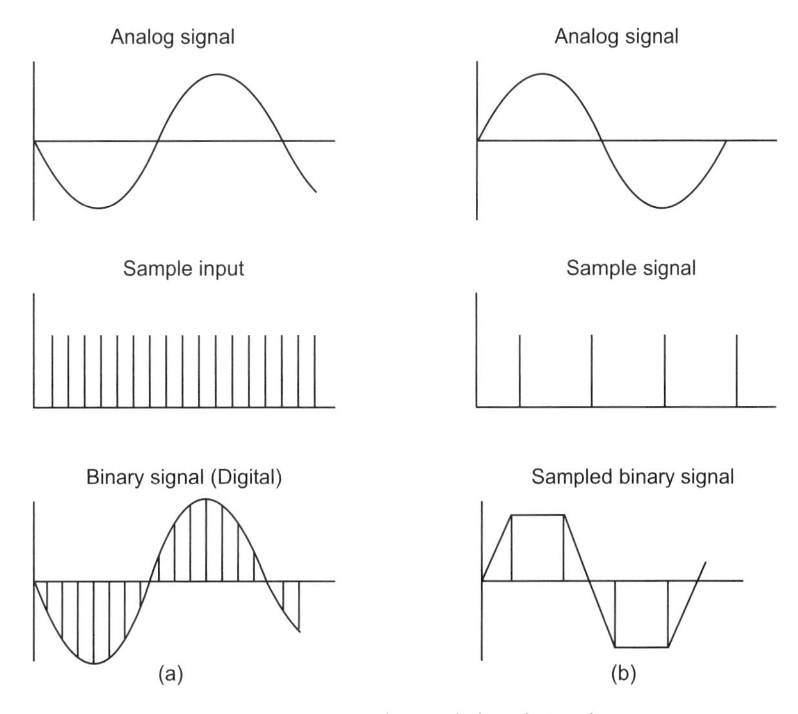

FIGURE 7.38 Analog and digital signals.

form." At lower frequencies a false representation of the original signal is produced which is known as *aliasing*. Figure 7.38 shows the conversion of analog signal into digital signal at a high sampling rate and a low sampling rate. The errors due to aliasing and high frequency noise are minimized by passing the signal through low-pass filters before being given to the A/D converter for conversion into digital signal.

7.21 DATA ACQUISITION (DAQ)

The data to be processed by the computer has to be acquired from the sensors or transducers with the aid of the data acquisition system. The process is known as Data Acquisition (DAQ). The sensors and the data acquisition systems are interfaced by the signal conditioning elements before being connected to computers for processing. For an analog input signal, DAQ is a printed circuit board that consists of a multiplexer, amplifier, A/D converter register, and controllers. The output of DAQ is plugged into the computer to transfer data for processing.

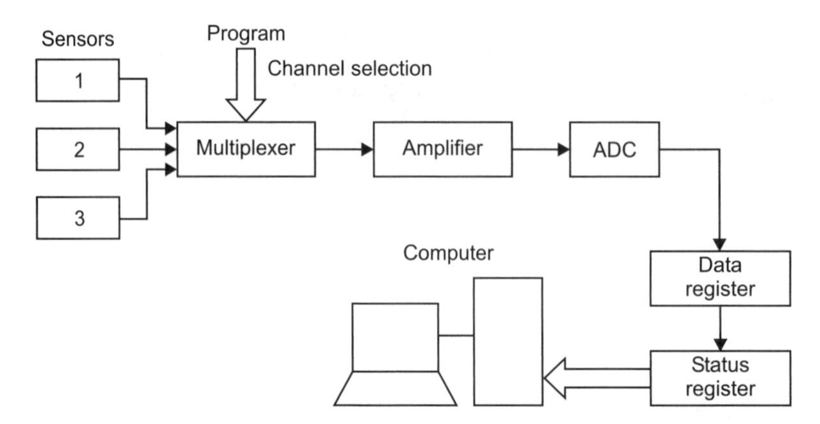

FIGURE 7.39 DAQ function diagram.

The output signals from the multiple sensors are connected to the multiplexer which selects the particular signal by a channel selection controlled by the program. The selected signal is amplified and given to the analog-to-digital converter. The digital signal is input to the computer through data registers and status registers. The block diagram of DAQ is shown in Figure 7.39. Figure 7.40 shows the interface diagram of DAQ with the computer.

7.21.1 Specifications of DAQ Board

- The maximum frequency of the input signals is fixed by Nyquist criteria.
- The sampling rate for analog inputs is determined by doubling the rate from the maximum frequency component.

- The DAQ outputs are supplied as analog signals.
- DAQ uses trigger systems such as timers and counters for the sensor elements.

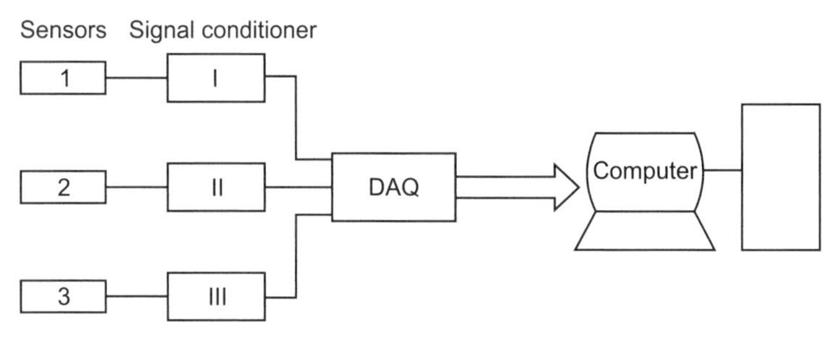

FIGURE 7.40 DAQ as an interface system.

EXERCISES

1. Give the basic process of an operational amplifier.
2. Explain the Wheatstone bridge circuit used for strain measurement.
3. Differentiate between analog and digital signals.
4. What is the significance of an operational amplifier? How is it used as an inverting amplifier?
5. What is a multiplexer? State the basic principle of a two-channel multiplexer.
6. Define data acquisition.
7. Draw the circuit for the following amplifiers and write the voltage gain of the circuit:
   (*a*) Inverting (*b*) Non-inverting
   (*c*) Summing (*d*) Integrating
8. Explain with an illustration Shannon's sampling theorem and the term aliasing.
9. Explain the salient features of the following:
   (*a*) Multiplexers
   (*b*) Data acquisition system
10. Explain the following terms related to operational amplifiers:
    (*a*) Input offset voltage (*b*) CMRR
    (*c*) Slew rate (*d*) Input offset current
11. Explain the various types of filters with a sketch.
12. What is signal conditioning? Explain the processes in signal conditioning.
13. Draw a block diagram showing the concept of signal conditioning.
14. What are the needs of signal conditioning?

15. Explain the interface elements between sensors or actuators and the processor.
16. What are the various possible inputs to an operational amplifier?
17. What are the functions that an operational amplifier can perform?
18. Explain a voltage-to-current converter.
19. How does a current-to-voltage converter function?
20. Explain a non-inverting type amplifier with voltage gain.
21. Describe the features of a differential amplifier.
22. What is the function of a logarithmic amplifier in signal conditioning? Explain the transistor type.
23. Explain the Schmitt trigger amplifier with its application.
24. What are the possible types of amplifier errors? Explain any three.
25. How does frequency response play a role in the selection of an amplifier?
26. Explain gain variation in amplifiers.
27. Explain why protection is needed. How is protection provided?
28. Explain the different types of filters.
29. Differentiate between active filters and passive filters.
30. Differentiate between high-pass and low-pass filters.
31. Define Time Division Multiplexer (TDM).
32. How does a Frequency Division Multiplexer (FDM) work?
33. What are the capabilities of digital signal processing?
34. Explain the following:
    (*a*) Linearization (*b*) Compensation
    (*c*) Signal averaging (*d*) Discrete Fourier Analysis
35. What is the function of a DAC? What are its types?
36. Explain a graded resistor DAC.
37. How does the ladder type of DAC function? Why is it superior to the graded type?
38. Give two types of ADCs and their functions.
39. Explain the successive approximation type ADC.
40. Discuss the features of the integrating type ADC.

CHAPTER 8

MICROPROCESSORS AND MICROCONTROLLERS

Mechatronic systems need the application of microprocessors and microcontrollers to process information and control operations. The logic functions for digital processing of signals should be studied. Boolean created binary mathematics to handle data in the form of binary codes.

This chapter on microprocessors and microcontrollers covers the following topics:

- Microprocessor-based digital control.
- Digital number systems—binary and hexadecimal.
- Logic functions and logic gates.
- 8085A microprocessor architecture, with related terminologies such as CPU, ALU, registers, bus, address, data, state, interrupt, read, write, and fetch cycles.
- Microcontroller definition.
- Differences between microprocessors and microcontrollers.
- Requirements of microcontrol action.

- Implementation and classifications, with explanations of different types of microcontrollers.

8.1 INTRODUCTION TO MICROPROCESSORS

The elements of control and information processing need, typically, the application of a microprocessor in a mechatronics system. The role of the microprocessor is predominant over other methods of control and processing for the following reasons:

8.1.1 Stored Program Control

A program stored in the memory, which is the sequence of operations based on mathematical and logical algorithms, can configure the microprocessor to function at a very high speed. The instructions in the program are deterministic and repeatable during the implementation of control and signal processing.

8.1.2 Digital Processing

Microprocessors process the information represented in binary form that is not disturbed by analog noise. For the binary form of signal, the variable resolution can be chosen according to situational requirements.

8.1.3 Speed of Operations

The response time of a machine, device, or a system which is controlled by a microprocessor is much higher than the time taken by a microprocessor to execute instructions even though sequential in order.

8.1.4 Design Flexibility

Within the hardware limitations of input and output ranges of memory, the microprocessor offers very flexible system solutions by changing the execution program. The conditional execution of a program, and the communication with other microprocessors and the operators, renders mechatronic systems using microprocessors to be called *smart* or *intelligent* systems.

8.1.5 Integration

The restriction on physical space availability and power supply to the mechatronics system and the incorporation of a microprocessor on a single chip provides a higher degree of integration, benefiting designers.

8.1.6 Cost

The cost component in commercial applications is made favorable by the flexibility and space of occupation by the microprocessor in a mechatronics system. With an increase in viable applications exponentially, the cost of the microprocessor is reduced drastically.

8.1.7 Definition

"A self-starting and self-contained device that works as a control system, and processes instructions to carry out specific functions with the use of its own program, without the intervention of a human operator, is known as a microprocessor."

The three basic parts of a microprocessor system are:

1. ***Input and output port.*** The communication between the computer system and the surrounding systems are handled by input and output interfaces (ports).
2. ***Central processing unit (CPU).*** The CPU identifies and executes the sequence of instructions contained in the program with the aid of a microprocessor chip.
3. ***Memory.*** The program instructions are processed and the processed data are stored in a space called memory.

8.2 MICROPROCESSOR-BASED DIGITAL CONTROL

If the control action is exercised by incorporating a microprocessor or a computer in the feedback control circuit that reacts to the error signals digitally, the action is known as *digital control*.

The microprocessors for digital control can handle only digital signals and give outputs only in digital form. Hence, the input signals that are in analog form from the sensors are transformed by passing through the Analog-to-Digital Converter (ADC). The desired values are set in the programs of the microprocessor that compare the input signal with the set value to generate an error signal. The error signals are processed to give digital signals. Since the correction system can accept only analog signals, the digital signal from the microprocessor has to pass through the Digital-to-Analog Converter (DAC) before being used for process control. A computer in the feedback loop can provide digital feedback of the output from the process or the system. The block diagram of digital control is shown in Figure 8.1.

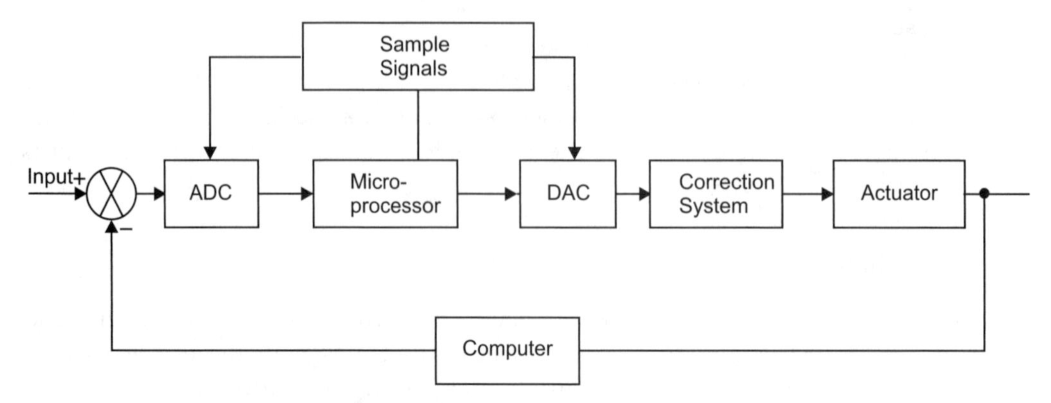

FIGURE 8.1 Digital control.

The functions of microprocessor-based digital control include:

- Sampling of the measured value of signal.
- Establishing the error signal by comparing it with the set value.
- Obtaining the output signal based on calculations carried out on error values and stored input/output history.
- Communicating the digital output with the DAC.
- Repeating the cycle for the next sample inputs.

8.2.1 Advantages of Digital Control

- The control action can be altered by altering the computer software.
- The alteration can be exercised online (during the execution of the process) without altering hardware.
- Separate controllers are not required for each process, not as in analog controllers.
- Digital control actions are more accurate than analog control actions.
- Digital control signals do not suffer from drift.
- The change in characteristics with time and varying environmental conditions are compensated in the microprocessor.
- By using a multiplexer, multiple signals can be handled by a single processor on the required channel selection.

The disadvantages of analog control over digital control have enabled more versatile use of digital controllers in mechatronic systems rendering them fast and accurate in response and they are able to carry out different functions.

8.3 DIGITAL NUMBER SYSTEMS

Consider a number with several digits in it. The weight attached to each digit is representative of its particular number system. For a decimal system the weight is 10 and each digit from the left to right increases by a factor $10^{(n-1)}$, where n is the digit position. Hence, the decimal system uses ten digits (symbols) in its representation. The digits are 0, 1, 2, 3, 4, 5, 6, 7, 8, 9.

For example, in the number 257, the digit 2 has the weight $10^{(3-1)} = 100$. The number 3689 can be written in the expanded form as $(3 \times 10^3 + 6 \times 10^2 + 8 \times 10^1 + 9 \times 10^0)$.

8.3.1 Binary System

The specialty of this system as indicated by the name is to use only two digits, 0, 1, in the representation of its numbers. The weight factor is given by $2^{(n-1)}$, where n represents the digit position.

For example, in the binary number 100, the weight of 1 is $2^{(3-1)} = 4$. The number (binary) 1101 can be expanded as $(1 \times 2^3 + 1 \times 2^2 + 0 \times 2^1 + 1 \times 2^0)$. Consider the decimal number 14. To convert it to a binary number,

$$\begin{array}{r|l} 2 & 14 \\ \hline 2 & 7 - 0 \\ \hline 2 & 3 - 1 \\ \hline & 1 - 1. \end{array}$$

Hence, $(14)_2 = 1110 = 1 \times 2^3 + 1 \times 2^2 + 1 \times 2^1 + 0 \times 2^0 = 14$.

To convert a binary number 10101 to a decimal number, for example, $(10101)_{10} = 1 \times 2^4 + 0 \times 2^3 + 1 \times 2^2 + 0 \times 2^1 + 1 \times 2^0 = 21$.

8.3.2 Binary Addition

The addition of 14 and 21 in the decimal system yields 35. In the binary system the addition is carried out like the following:

$$\begin{array}{rl} 01110 & - \text{ Augend} \\ 10101 & - \text{ Addend} \\ \hline 100011 & - \text{ Sum} \end{array}$$

It may be verified that

$$1 \times 2^5 + 0 \times 2^4 + 0 \times 2^3 + 0 \times 2^2 + 1 \times 2^1 + 1 \times 2^0 = 35.$$

8.3.3 Binary Subtraction

In the decimal system subtracting 14 from 21 results in 7. To do it in the binary way

$$\begin{array}{rl} 10101 & - \text{ Minuend} \\ 01110 & - \text{ Subtrahend} \\ \hline 1\bar{1}0\bar{1}1 & - \text{ Difference} \end{array}$$

$$\text{To verify this } (1\bar{1}0\bar{1}1) = 1 \times 2^4 - 1 \times 2^3 + 0 \times 2^2 - 1 \times 2^1 + 1 \times 2^0$$

$$= 16 - 8 - 2 + 1 = 7.$$

The binary system, as it indicates two states (ON or OFF), is widely used in computers.

8.3.4 Hexadecimal System

This system uses 16 symbols to represent a number. The numbers used are 0, 1, 2, 3, 4, 5, 6, 7, 8, 9, A, B, C, D, E, F. In this system the decimal number 10 is A and the decimal number 15 is F. The representative of the number system is 16. The digits from the right to left in a hexadecimal number increase by the weight factor $16^{(n-1)}$.

For example, the number B9C in a hexadecimal system can be converted to a decimal number as

$$11 \times 16^2 + 9 \times 16 + 12 \times 16^0 = 2962.$$

Since the data becomes compact on conversion to hexadecimal numbers, this system is used to represent numbers in microprocessors.

To convert a decimal number to a hexadecimal number, consider 333,

$$\begin{array}{r|l} 16 & 333 \\ \hline 16 & 20 \;-\; 13 \\ \hline & 1 \;-\; 4 \end{array}$$

Hence, $(333)_{16} = 14D.$

To verify $- 1 \times 16^2 + 4 \times 16^1 + 13 \times 16^0 = 256 + 64 + 13 = 333.$

TABLE 8.1 Number Systems

Decimal	Binary	Binary Coded Digit	Hexadecimal
0	0000	0000 0000	0
3	0011	0000 0011	3
5	0101	0000 0101	5
7	0111	0000 0111	7
9	1001	0000 1001	9
11	1011	0001 0001	B
13	1101	0001 0011	D
15	1111	0001 0101	F
18	10010	0001 1000	12

8.4 LOGIC FUNCTIONS

In the beginning logic networks were designed using a hit-or-miss approach of guesswork. Nowadays the procedure calls for a systematic technique based mainly on the formal apparatus of mathematical logic. Let the logic network be described by the following pair of equations:

$$\left.\begin{aligned} Y_i &= f_i\,(x_1, \ldots, x_n) \qquad (a) \\ Y_i &= T\,(x_1, \ldots, x_n) \qquad (b) \end{aligned}\right\}. \qquad (8.1)$$

A logic network receives the item of input information designated as $\{x_1, \ldots, x_n\}$. These parameters can assume either of the two switching values which are labeled 1 and 0 in binary code. The network yields the output information marked $\{y_1, \ldots, y_n\}$. The functioning of a logic network is described by Eq. (8.1).

FIGURE 8.2 Logic network representation.

The functioning of the logic network is based on the logic relationship correlating the output information with the input information using binary code. This is established in Eq. (8.1*a*). The transformation of input symbols into output symbols is described by Eq. (8.1*b*). The physical structure of the network elements and the kind of energy carrying medium have relevance in the nature of transformation of the symbols. The possible logic functions are AND, OR, NAND, NOR.

Logic functions are explained in detail in the tables on the following pages.

8.5 MICROPROCESSOR ARCHITECTURE TERMINOLOGY

8.5.1 Central Processing Unit (CPU)

The CPU in a microprocessor forms the heart of the system and is responsible for the overall control of all devices connected to it.

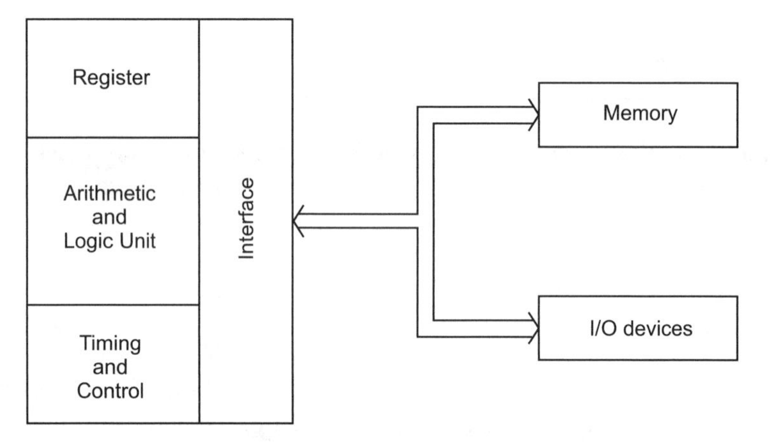

FIGURE 8.3 Block diagram of CPU.

Certain functional components associated with the CPU are:

- *Register section.* A set of registers to store instructions, data, and addresses temporarily.
- *Arithmetic and logic unit.* This part of the CPU is the hardware for performing primitive arithmetic and logical operations.
- *Interface section.* This forms the input and output lines through which the microprocessor unit (MPU) communicates with the outside world.
- *Timing and control section.* This is the hardware for coordinating and controlling the activities of the various sections within the MPU and other devices connected to the interface section.

TABLE 8.2 AND Logic Gate

Physical Representation	Graphical Symbol	Digital Input and Output	Truth Table
A B	A Inputs B Output A . B	Inputs A B Output Q	(see below)
The physical visualization of the AND gate is realized by an electrical circuit with two switches in series. Only when A and B are in the normally closed (NC) positions will the current flow. A gate will give high output when both inputs A and B are high and for all others, it gives low output and is called the AND gate. In the circuit shown if any one or both switches are closed, the current will not flow.		In the digital waveforms shown two digital inputs that vary with time are given to the AND gate. The output is high in a situation where both A and B are high. The output is low at all other instances of time. The truth table reveals that Q = 1 only when A = 1 and B = 1. In all other combinations Q = 0. The Boolean equation for the AND gate can be written as Q = A.B. (8.2)	

Truth Table:

Input		*Output*
A	B	Q
0	0	0
0	1	0
1	0	0
1	1	1

TABLE 8.3 OR Gate

Physical Representation	Graphical Symbol	Digital Input and Output	Truth Table		
Input, A, B, Output	A, B, Inputs, Output, A + B	A, B, A + B	*Inputs*		*Output*
			A	B	A + B
			0	0	0
			1	0	1
			0	1	1
			1	1	1

An electrical circuit with two switches arranged in parallel provides the physical visualization for the OR function. In the normally open (NO) position of the switches the current will not flow through the circuit. In all other states the current will flow through the circuit.

The gate represented by the graphical symbol has the lowest output when both A and B are low. In other conditions the output of the OR gate is higher.

The OR gate is provided with two digital signals, A and B, which are different and vary with time. At instances when both the inputs are low the gate output will also be low. In all other cases the digital output is high for the OR gate.

The output $Q = 0$, when $A = 0$ and $B = 0$. In all other combinations the output $Q = 1$.

The Boolean expression for OR gate is

$$Q = A + B. \tag{8.3}$$

TABLE 8.4 NOT Gate

Physical Representation	Graphical Symbol	Digital Input and Output	Truth Table	
R R	Input A Output $\overline{A}$	Input A Output $\overline{A}$	*Input*	*Output*
			A	$\overline{A}$
			0	1
			1	0
The NOT gate physically is an inverting type operational amplifier with the feedback resistor equal to the input resistance. The output obtained in this gate is the inverse of the input signal. When the input is high the output is low and when the input is low the output is high. Hence, the output is the inverted form of the input in the NOT gate.		The digital signal input to the NOT gate results in an output that is complementary to the input. The 0 state of input is converted to the 1 state and the 1 state is changed to the 0 state in the output. The output digital wave is the inverse of the input digital signal in the time domain. The Boolean equation is $Q = \overline{A}$, where bar (–) indicates the inversion. (8.4)		

TABLE 8.5 NAND Gate

Physical Representation	Graphical Symbol	Digital Input and Output	Truth Table

Truth Table

Inputs		Output
A	B	$\overline{A . B}$
0	0	1
1	0	1
0	1	1
1	1	0

Physically it is the combination of the AND gate and the NOT gate. The output of the AND gate is given as the input to the NOT gate. When the output of the AND gate is low the output of the NAND gate is high. At high output of the AND gate the NAND gate produces the low output. Two switches in series connected to the inverting input of the op-amp represent physically the NAND gate.

Two digital inputs that vary with time, as shown, produce an output to represent the functioning of the NAND gate. When A = 0 and B = 0, the output Q = 1. When A = 1 and B = 1 the output Q = 0. In the other two intermediate states in which the AND gate is zero (0), Q = 1.

The Boolean equation for NAND gate is

$$Q = \overline{A . B} . \qquad (8.5)$$

TABLE 8.6 NOR Gate

Physical Representation	Graphical Symbol	Digital Input and Output	Truth Table

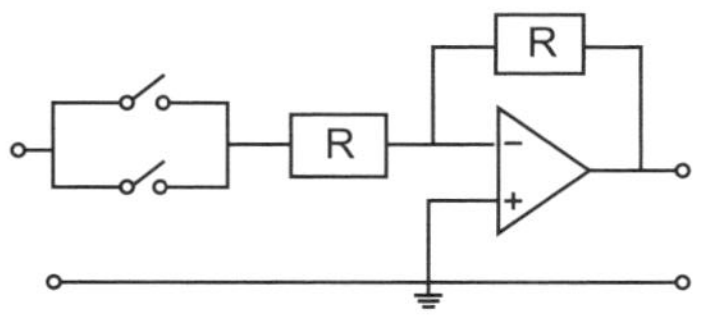

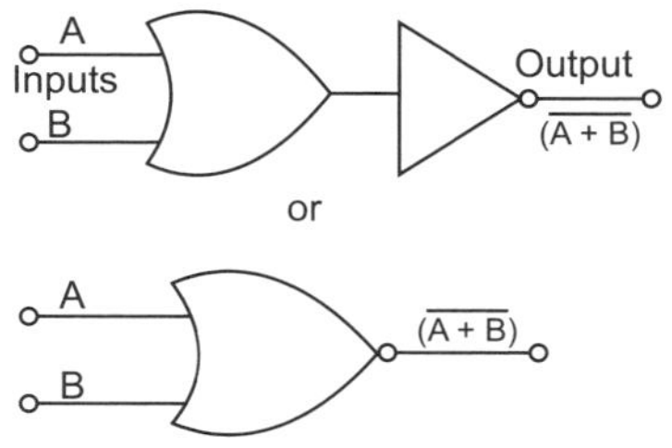

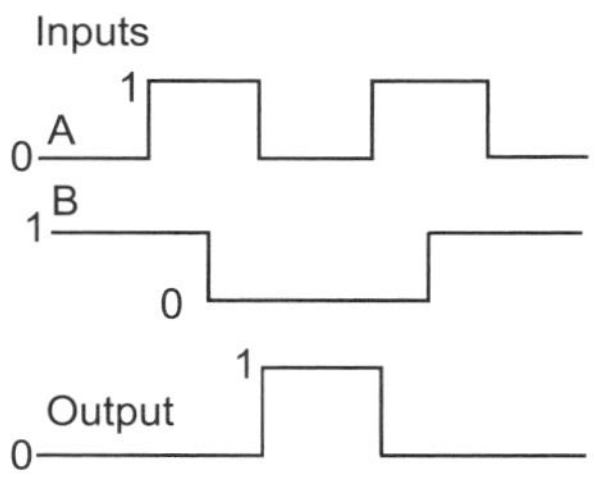

Truth Table

Inputs		Output
A	B	$\overline{A + B}$
0	0	1
1	0	0
0	1	0
1	1	0

An OR gate is connected to the NOT gate serially to form the NOR gate. Two switches are arranged in parallel whose output is connected to the inverting input of the op-amp. This combination produces a high output when the input is low. In other cases when the output of the OR gate is high the output of the NOR gate is low. Hence, the NOR gate adds the signals and inverts them to give the output.

Two digital signals, A and B, vary with time when the input to the NOR gate produces the output signal varying with time as shown. When A and B are both zero Q is equal to one. When any one of A and B or both A + B are 1 the output is zero (Q = 0).

The Boolean equation takes the form

$$Q = \overline{A + B}. \quad (8.6)$$

TABLE 8.7 XOR Gate

Physical Representation	Graphical Symbol	Digital Input and Output	Truth Table		
A Inputs B $A + \overline{B}$ Output	A B $A + \overline{B}$	1 0 A 1 0 B 0 1 $\overline{B}$ 1 0 $A + \overline{B}$	*Inputs*		*Output*
			A	B	$A + \overline{B}$
			0	0	1
			1	0	1
			0	1	0
			1	1	1

The exclusive OR gate (XOR) is the combination of a NOT gate and an OR gate. The output of a NOT gate is given as input to the OR gate. One of the inputs of the OR gate is inverted by the NOT gate so that of the two inputs one is non-inverted and the other is inverted. The combination of NOT and OR gates has the symbol shown by the figure.

The digital waveform of the signals, A and B, vary with time. B is inverted to $\overline{B}$ before being added to A. When the input A is low and input B is high, *i.e.*, A = 0 and B = 1, the output Q is low, *i.e.*, Q = 0.

In all other cases the output is high. The Boolean representation is

$$Q = A + \overline{B}. \tag{8.7}$$

8.5.2 Registers

The internal architecture of a microprocessor is used to determine how and what operations can be performed with the given data. These operations are:

- Store 8-bit data.
- Perform arithmetic and logical operations.
- Test for conditions.
- Sequence the execution of instructions.
- Store the data temporarily in the stack.

One of the elements needed by the microprocessor is the *register*, which is programmable for data manipulation by writing instructions.

TABLE 8.8 Registers

Type of Register	Bit Capacity	Function
1. Accumulator	8-bit	• Part of ALU and performs arithmetic and logic operations. • Stores results and data.
2. Flag	8-bit	• Five bit positions out of 8 bits are used to store outputs from five flip-flops. • Participates in the decision-making process of the microprocessor. • Tells data condition.
3. Program Counter (PC)	16-bit	• Performs sequencing of the execution of instructions. • Points to the memory address from which the next byte is to be fetched.
4. Stack Pointer (SP)	16-bit	• Used as a memory pointer. • Points to the memory location in the R/W memory called the stack. • Address in SP marks the beginning of the stack.

8.5.3 Memory

An essential component of a microcomputer is the *memory* which stores binary instructions and data for the microprocessor. If a microprocessor stores 8-bits of information as a group, the memory word length should be 8-bits.

The communication with the memory by the MPU is accomplished by:

1. Selecting the chip.
2. Identifying the register.
3. Reading from and writing into the register.

The executing and storing is done by the primary memory or main memory. Examples of main memory are R/W M and ROM. The other type of memory is storage memory. This includes hard disks. Several types of primary memory are as follows:

R/W M. This type of memory is also known as Random Access Memory (RAM). This memory is volatile, *i.e.,* the memory contents are destroyed when the power is turned off. Two types of RAM are static and dynamic RAM. Static memory is made of flip-flops that store bits as voltage. Dynamic memory is made of transistor gates and stores the bits as a charge. Because a large number of gates can be placed on the chip, they are high in density and faster than static memory.

ROM. Read-only memory is non-volatile memory. It retains the data or the stored information even after the power is switched off. The diodes arranged in the matrix format constitute the ROM structure. The presence of the diode stores 1 and the absence 0. The diode representation is the simplified version of an actual MOSFET memory cell. Four types of ROM are presently available. They are: Masked ROM; PROM; EPROM; and EEPROM.

1. ***Masked ROM.*** In this memory the bit pattern is permanently recorded by the masking and metalization process which is an expensive and specialized process.
2. ***PROM (Programmable Read-only Memory).*** This memory has nichrome or polysilicon wires arranged in a matrix arrangement. The wires function as diodes or fuses. A special PROM programmer that selectively burns the fuses according to the bit pattern required to be stored is used to store information in this type of memory. The information stored is permanent by the process of *burning the PROM.*
3. ***EPROM (Erasable PROM).*** The information stored in this memory is semipermanent. The exposure of the memory to ultraviolet light through a quartz window installed on the chip erases the information stored in a PROM. The regaining of the memory chip favors the reprogramming repeatedly. This is used in product development and experimental projects.
4. ***EEPROM (Electrically Erasable PROM).*** The functions are very similar to EPROM. The information can be altered using electrical signals at the register level rather than erasing all the information. The complex

manufacturing process renders these PROMS very costly, and cannot be applied to common use.

In a mechatronics system using a microprocessor, the programs are written in ROM, and varying data information is stored in the R/W M during processing.

8.5.4 ALU

ALU stands for Arithmetic and Logic Unit and is the unit which performs computing functions. The primitive arithmetic and logical operations are performed by the ALU and the realization of complex functions are also done using it. The architecture of the ALU includes:

- the accumulator,
- the temporary register,
- the arithmetic and logic circuits, and
- five flags.

The temporary register holds the data during arithmetic and logic operations. The accumulator stores the result. The flags (flip-flops) are set or reset according to the result of the operation. The data condition of the accumulator is reflected by the flags in most cases.

The most common operations performed by the ALU of an MPU are:

8.5.5 Arithmetic

Addition
Subtraction
Increment
Decrement
Comparison

8.5.6 Logical

AND
OR
EXOR
NOT
SHIFT/ROTATE
CLEAR

In earlier MPUs, because of limited chip area, complex operations such as multiplication and division were not available. But in recent MPUs with Intel 8086 processors complex operations are incorporated.

8.5.7 Address

The first step in the communication by the microprocessor functions is to "identify the peripheral or the memory location." Each peripheral and memory location is identified by a binary number called the *address*. This is similar to the mailing address of a house. A house is identified with a numbering scheme. The number of address lines of the MPU determines its capacity to identify different memory locations or peripherals. If the address lines with an MPU are 16 it is capable of addressing $2^{16} = 65536$ memory locations (generally known as 64K). In most 8-bit MPUs the address lines are 16 and their memory capacity is 64K.

8.5.8 Buses

MPU chips are equipped with a number of pins for communication with the outside world. This is known as the *system bus*. The number of pins for the system bus are limited by both technological and economical considerations. For an 8-bit MPU the optimum pin number is estimated to be 40 pins. To make use of a limited number of pins the information communicating lines are designed and classified into four groups as follows:

1. Address Bus
2. Data Bus
3. Control Bus
4. Utility Lines (Bus)

Address Bus. The memory and the I/O devices are external to the MPU. Hence, for communication of information, the MPU has to be equipped with a mechanism to connect memory and I/O devices with the MPU. These lines of communication are called the address bus, which carries the memory address and the I/O device address. Most 8-bit MPUs have a 16-bit address bus providing an address facility up to $2^{16} = 65536$ memory locations.

Hence, the address bus is a group of 16 lines that are unidirectional: The flow of bit occurs in one direction. This allows the external devices to gain control of the address bus. The unidirectional flow of bit is accomplished by tristate buffers.

Data Bus. The information, instruction, data, or address is communicated with the outside world by transfer in both directions by the same pins known as the data bus. In an 8-bit processor the data bus is a group of eight lines. The width of the data bus is normally equal to the data or word size of the ALU. The bidirectional tristate buffers provided to the data bus help the external devices gain control of the data transfer. If the data bus has 8 lines it can manipulate 8-bit data numbering $2^8 = 256$. The largest number conveyed by the data bus is 11111111 (*i.e.*, 255_{10}). Microprocessor architecture is designated by the size of the data bus. A microprocessor with 8 lines in the data bus is designated as an 8-bit processor.

Control Bus. The generation of various signals and the receiving of signals for coordination and control of different operations pertaining to external devices is the responsibility of the control bus. The control lines are categorized as:

(*a*) Memory and I/O control lines and

(*b*) CPU and bus control lines.

The read/write lines, memory, ready/wait lines, address latch enable lines, and status lines fall under the first category, and the interrupt lines, RESET lines, bus request, and bus grant lines are included in the second category.

Utility Lines. The MPU requires 2 pins for the power supply lines. The MPU requires a clock as it has a sequential circuit. The clock requires a bus to communicate when situated external to the MPU. For communication in bit-serial form the MPU should be provided with serial I/O lines. The MPU requires one control line for controlling the refresh operation of the dynamic RAM known as the *dynamic refresh control line.*

The bus structure of an 8-bit microprocessor unit is shown in Figure 8.4.

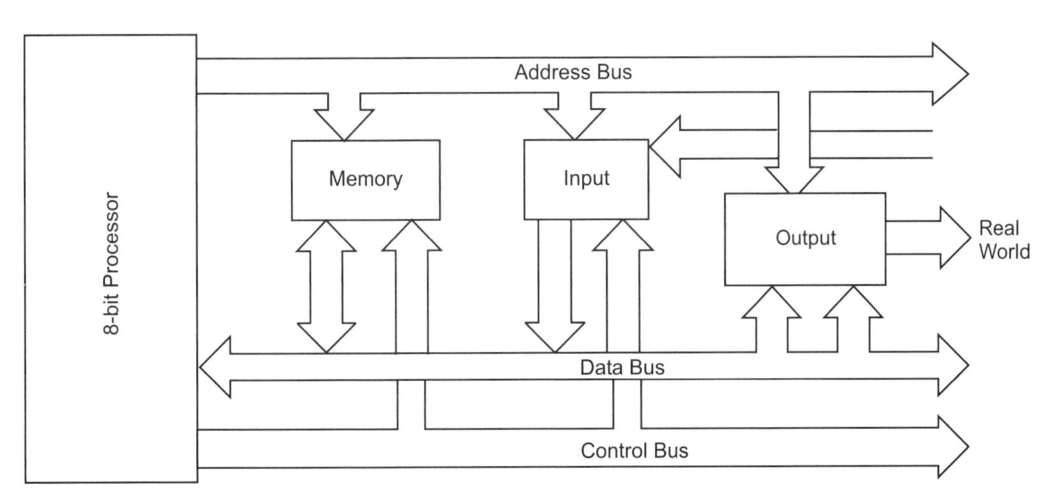

FIGURE 8.4 Bus architecture of an 8-bit MPU.

State. The data grouped in fields and displayed in octal or hex code along with the binary form for better user interaction describes the circuit (logic) under test and is considered the *state*. This data is presented in tabular form and is known as the *state table*. For digital circuits the state of the signal at various points can be visually displayed by a Logic State Analyzer (LSA) which is similar to the oscilloscope for analog circuits. The LSA has six subsections which are shown by the block diagram in Figure 8.5. The subsections are input, real-time processor, memory, clock and triggering, display, and control.

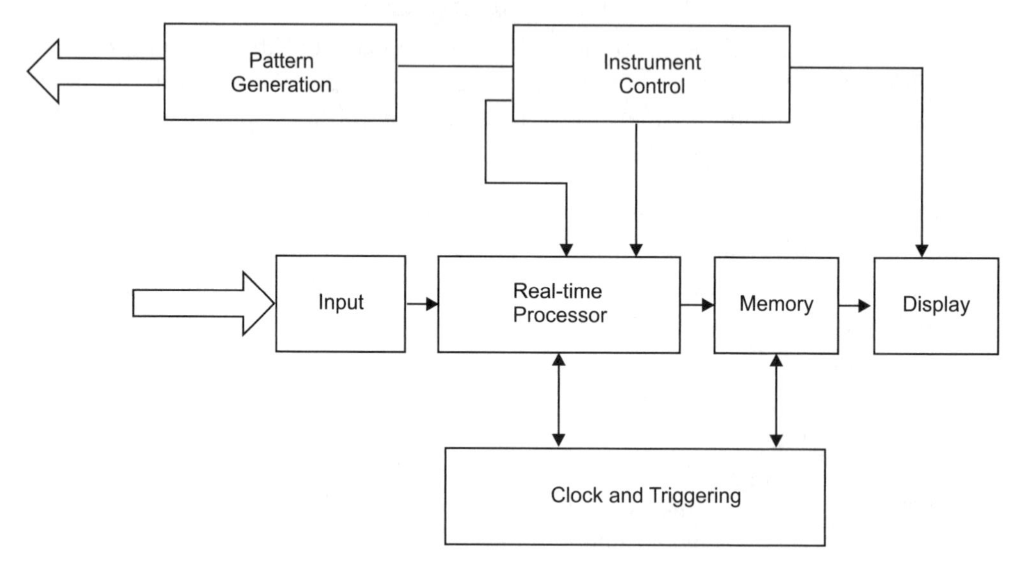

FIGURE 8.5 Block diagram of LSA.

8.5.9 Data

The signed integers, BCD, bits, bytes, words, long words in integer form, packed BCD, and strings are known as *data*. The majority of 8-bit MPUs support one or two types of data—namely, signed integers and BCDs. But 16-bit and 32-bit microprocessors can support a wide variety of data types such as bits, bytes, words, long words in integer form, packed BCDs, and strings. The format for representing a decimal number is the Binary Coded Decimal (BCD) format. There are two ways of storing BCD data: packed BCD and unpacked BCD.

Packed BCD. A string of decimal digits is stored in the sequence of a 4-bit group, *e.g.*, 56 is stored in a byte as 0101 0110.

Unpacked BCD. Each digit is stored in the lower order of a 4-bit of a byte and the higher 4-bit group is left unused, *e.g.*, 56 is stored in two bytes as XXXX0101 XXXX0110.

8.5.10 Interrupts

The definition of an *interrupt* follows the literal meaning of the word interrupt. That is, forcing the MPU to suspend the execution of the current program and jump to a subroutine at a predetermined location. After servicing the interrupt subroutine the processor returns control to the interrupted program. MPUs are provided with two or more control lines which the external devices can use. Interrupts are one of two types: non-maskable interrupts and software interrupts.

Non-maskable interrupt (NMI). This interrupt cannot be disabled or masked by any means. An NMI forces the MPU to interrupt the signal unconditionally. For example, emergency situations such as power failure are compulsorily communicated to the MPU by this NMI.

Software interrupt. The interrupt inputs are disabled or masked by the software. This type of interrupt is commonly used by I/O devices for data communication.

Most MPUs have a minimum of one NMI and one software interrupt. In the second type the selective masking or total masking is possible by software means. The I/O interrupt from an external device or peripheral is a data transfer process of informing the MPU that it is ready for communication and demands attention. The process is asynchronous and can be initiated any time without reference to the system clock. For example, in the keyboard routine the processor stays in the loop until a key is pressed and cannot perform any other tasks.

8.5.11 Assembler

The instructions are entered by the keyboard in the mnemonic language which is very friendly to the user. This assembly language program is translated into a binary machine code understandable to the MPU by means of a program set known as the *assembler*. Each microprocessor has its own assembler. Each has certain rules to be learned by the programmer. The steps to be followed in writing the assembler are:

1. Write the instructions in mnemonics using the manufacturer's instruction set.
2. Find the hexadecimal machine code for each instruction with the help of the instruction set.
3. Load the program in the user memory in sequential order.
4. The pressing of the 'Execute' key executes the program and the answer is displayed in LED.

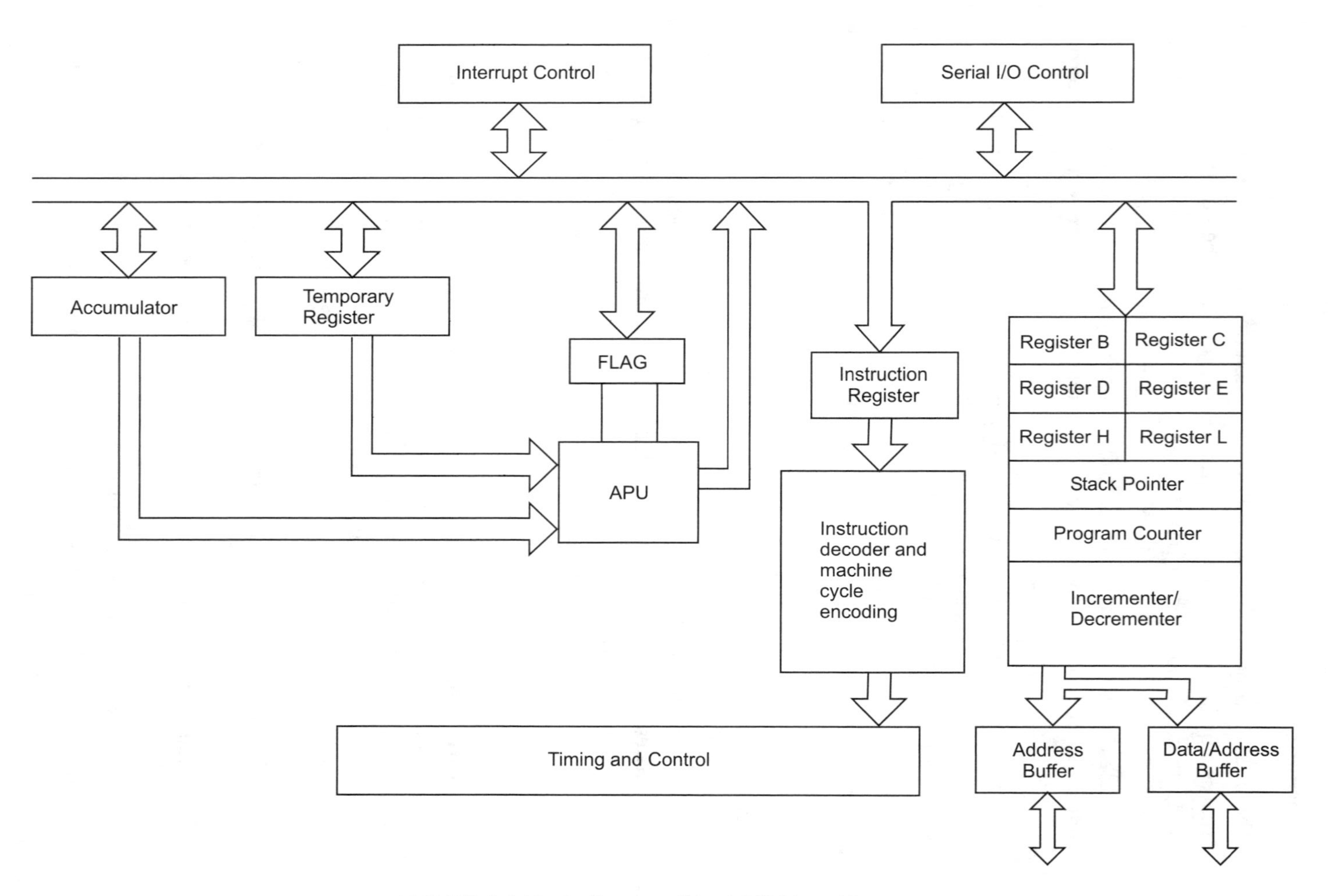

FIGURE 8.6 Block diagram of Intel 8085A architecture.

8.5.12 Read Cycle

The read cycle consists of three clock periods as far as the external logic is concerned. Figure 8.7 shows the timing of how a data byte is transferred from the memory to the MPU. It shows five different groups of signals in relation to the system clock. The MPU has to perform the following three steps in the read cycle:

Step 1. The program counter allows the address bus to place a 16-bit memory address.

Figure 8.7 reveals that at T_1 the high order memory address is placed on the address lines A_{15}–A_8. The low order memory address is placed on the data bus AD_7–AD_0 and the Address Latch Enable (ALE) signal goes high. Similarly, the status signal $IO/\overline{M}$ goes low. This indicates a memory related operation.

Step 2. The control unit sends the control signal $\overline{RD}$ (Read) enabling the memory chip.

The control signal $\overline{RD}$ is initiated during the clock period, T_2, and remains active for two clock periods, T_2 and T_3.

Step 3. The data bus gets placed with the byte from the memory location.

The instruction byte is placed on the bus AD_7–AD_0 when the memory is enabled, transferring the byte to the microprocessor. The $\overline{RD}$ signal is responsible for this operation. With a high $\overline{RD}$, the bus impedance is high. Figure 8.7 shows only the clock periods of three states of the read cycle.

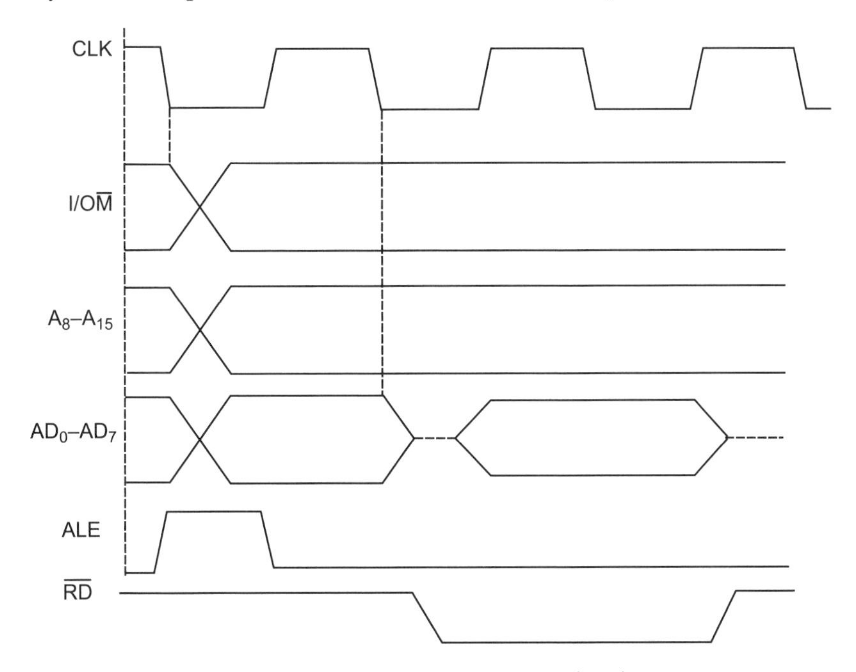

FIGURE 8.7 Memory read cycle.

8.5.13 Write Cycle

The write cycle is similar to a read cycle and consists of three clock periods, (T_1–T_3), concerned with external logic. The timing sequence of the memory write cycle is shown in Figure 8.8. The microprocessor places the address of the memory location A_{15}–A_8 during the clock period, T_1. This forms the first step of the write cycle. In the second step, the possible logic level combinations are examined and assumed by the address lines A_7–A_0. In the third step at the period, T_2, the microprocessor places the data on the data bus and sends the $\overline{WR}$ (Write) signal. During periods, T_2–T_3, the memory locations are identified. With this information the data is written into the memory location. Hence, the memory write cycle is similar to the memory read cycle which is revealed when Figure 8.8 and Figure 8.7 are compared.

8.5.14 Explanation of Terminology

1. ***ALE (Address Latch Enable).*** Every time the microprocessor begins an operation a positive pulse ALE is generated. This is the indicator of address bits on the AD_7–AD_0 lines. The latching of the low order address from the multiplexed bus using the ALE signal generates eight separate address lines (A_7–A_0).

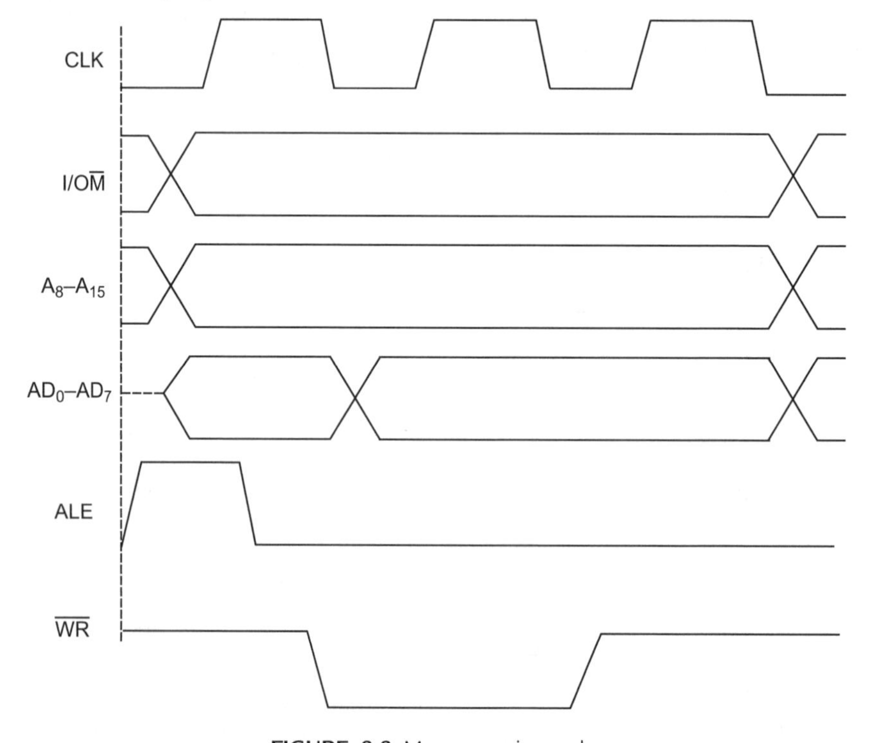

FIGURE 8.8 Memory write cycle.

2. ***IO/$\overline{M}$***. This is a status signal which distinguishes the I/O from memory operations. With a high status signal, the I/O operation is indicated. The low status signal indicates a memory operation.
3. ***$\overline{RD}$* (Read).** This is a read control signal indicating that the selected I/O or memory device is to be read from the data available on the data bus.
4. ***$\overline{WR}$* (Write).** This indicates the write control signal which is active when low. The data on the data bus is to be written onto the I/O device or memory location when this signal is indicated.

8.5.15 Fetch Cycle

Depending on the type of instruction the fetch machine cycle of the microprocessor consists of four or six clock periods. The fetch cycle is classified into the following three steps of activation:

Step 1. The low order address is made available on the data bus, AD_7–AD_0.

The address in the external buffers is latched by the generated ALE signal. This is operational during the clock period, T_1.

Step 2. Activation of the $\overline{RD}$ signal.

At the falling edge of T_1 the $\overline{RD}$ signal is activated, which is withdrawn at the leading edge of T_3. This signifies the completion of the read cycle of the op-code.

Step 3. Decoding of the op-code.

During the fourth clock period, T_4, the op-code that is fetched is decoded. If the subsequent memory access is not involved in the instructions, the fetch cycle is extended to two more clock periods, T_5 and T_6. This is needed to complete the internal operation specified by some of the fetched instructions. During T_5 and T_6, the buses AD_0–AD_7 and A_8–A_{15} contain unspecified data. The signals in the op-code fetch cycle are shown by Figure 8.9.

The MPU performs the following operations during the fetch cycle:

1. The address available on the PC point is generated on the address bus along with other memory control signals for read in the op-code from the memory.
2. After a certain amount of delay, the information available on the data bus is placed on the instruction register.
3. The instruction is decoded to determine the operations to be performed to complete the execution of the instructions.

All three operations have to be completed in one or few clock period(s).

8.6 MICROCONTROLLERS

Microcontrollers are single-chip microcomputers that are primarily meant to perform dedicated functions such as signal processing, guidance and control, speech processing, acquiring and processing analog signals, and industrial automation. They can be independent as in machines of mechatronics or dependent as in distributed processing.

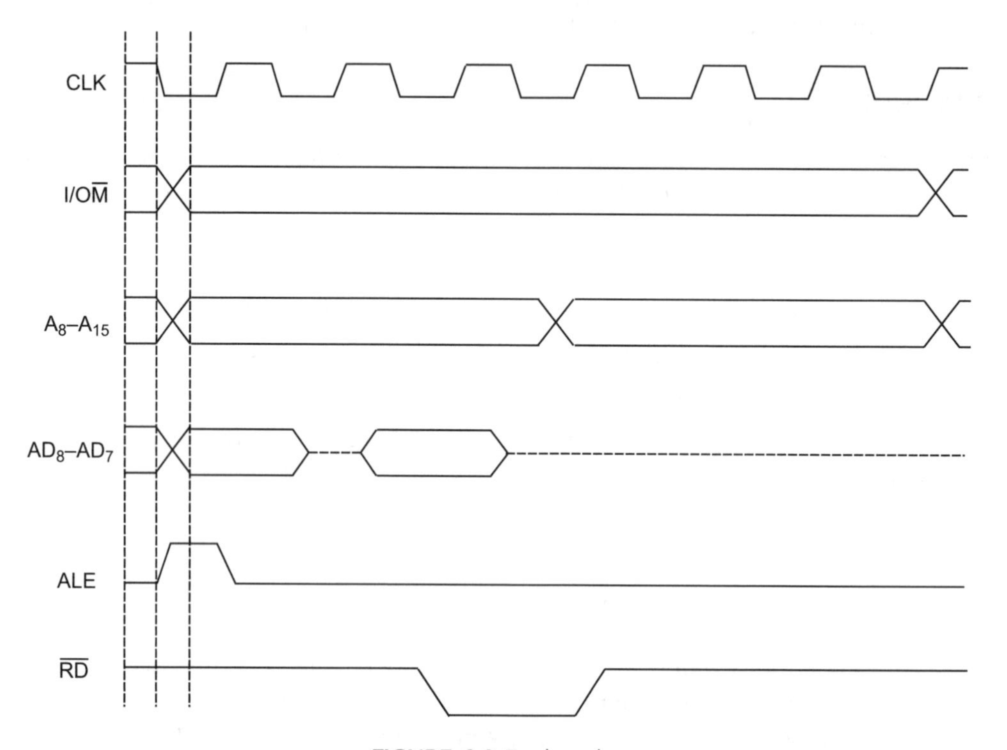

FIGURE 8.9 Fetch cycle.

The essential elements that constitute microcontrollers are the MPU, read/write memory, ROM, and the input/output lines. The specifications for microcontroller elements are as follows:

Elements	Specifications
Word size	8-bit
R/W memory	64 bytes
ROM	1 K bytes
I/O Lines	16 to 32 lines

Use of EPROM on the chip facilitates easy programming of microcontrollers. An unlimited variety of microcontrollers are available for diversified industrial needs.

I/O Functions. On a microcontroller the I/O functions generally include parallel input and output ports and serial input and output ports for communication with other devices or terminals.

Timer/Counter. Frequency and period measurements are made using a timer/ counter which generates fixed or variable pulse waveforms.

Interrupt Controller. The interrupt sources in a microcontroller are internal peripherals, I/O devices, timers, and general purpose interrupt outputs and inputs. All these sources are managed by one chip interrupt controller.

Special purpose microcontrollers also include the following:

- Analog-to-digital converter
- Voltage comparator
- Interface circuitry
- Display drivers
- Stepper motor drivers

8.6.1 Instruction Sets

The instructions, in a single-chip microprocessor (the microcontroller), are designed to generate dense code. Bit addressing is commonly used with instructions operating on individual bits within the memory or register. The bit is modified with a logic operation rather than using a separate instruction to read a byte. Furthermore, the complete byte is written back to the same address.

In larger microcontrollers, the read/write memory size is as large as 512 bytes. The program memory is 16 kilobytes in size. Microcontrollers with 4-, 8-, and 16-bit architectures are also available today. Since microcontrollers are integrated with a small amount of memory, provisions are made generally to expand the memory capacity with off-chip memory.

8.7 REQUIREMENTS FOR CONTROL

The following table lists some applications and requirements of different microcontrollers.

Sl. No.	Applications	Microcontroller	Feasible Requirements
1.	Washing machine	4- or 8-bit type	• The microcontroller must be cheap • It should be immune to electrical noise • The response time should be low • It can have low resolution
2.	Autofocus camera	8-bit type	• Requires a high level of integration of sensors and actuators • Low response time is needed • High level functional accuracy
3.	Intelligent sensor or transducer	8-bit type	• Input and outputs are analog • Faster and correct communication • The system should consume low power
4.	Engine-management system	8- or 16-bit type	• Analog inputs and outputs are processed • Signals need to be processed • System should be immune to electrical noises (Requires proper filtering) • Characterized by high current output

8.8 DIFFERENCES BETWEEN MICROPROCESSORS AND MICROCONTROLLERS

The following table lists the differences between microprocessors and microcontrollers.

Feature	Microprocessor	Microcontroller
Application-wise	• More suitable for general-purpose applications	• More suitable for special purpose and custom-built applications
Complexity of task	• These can handle more complex tasks with better performance	• Tasks that are easy and simple suit the application of microcontrollers
Cost effectiveness	• Requires matching of performance with the requirements for these to be cost effective	• Tailor made to suit the requirements, hence, are cheap in cost
Speed and reliability	• High speed calculation and communications. Highly reliable results	• Speed is limited by memory capacity. Reasonably good reliability
Real-time processing	• With the induction of a fast interrupt response, they are capable of real-time processing of data	• They do not give results in real time as they do not have a fast interrupt response
Context switching	• Switching of CPU between two tasks is possible	• These are not featured by contex switching
Examples	• Used in speech processing, flight controllers, and robot controllers	• Find uses in washing machines, autofocus cameras, and engine management

8.9 CLASSIFICATION OF MICROCONTROLLERS

Single-chip microcontrollers that suit a variety of domestic and industrial applications are available in various types. The classification of microcontrollers from the popular Intel family is given as:

- Intel 8048 Family
- Intel 8051 Single-chip Family
- Universal Peripheral interface (UPI) 8041 Family
- Analog Signal Processor 2920 Family

The details of the architecture of these classifications of specifications, functioning, and applications are given in the following paragraphs.

8.10 IMPLEMENTATION OF CONTROL REQUIREMENTS

8.10.1 Intel 8048 Family

Figure 8.10 shows the block diagram of the 8048 single-chip microcontroller.

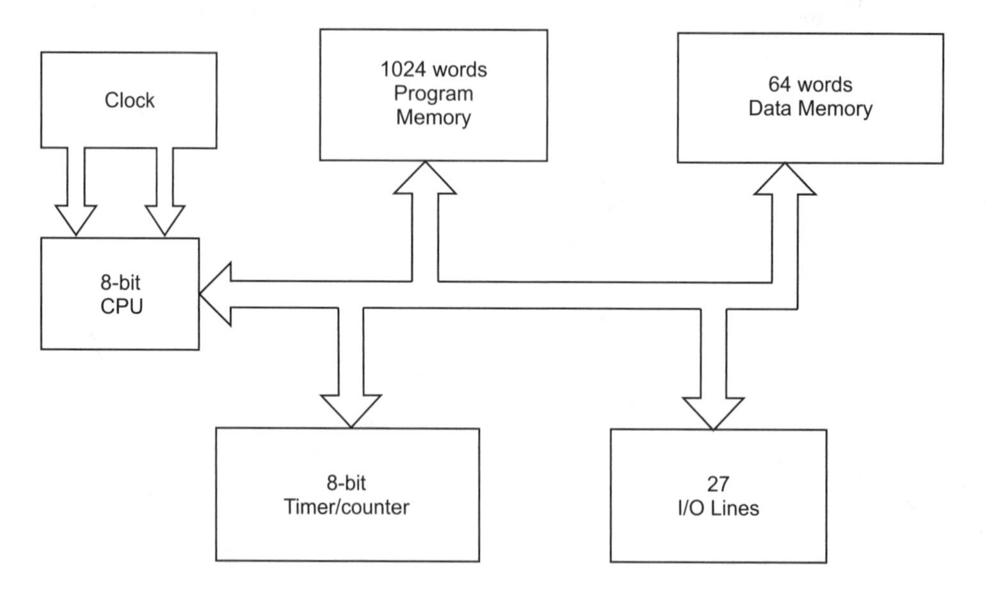

FIGURE 8.10 Block diagram of 8048 Intel family.

Specifications. The components of this family of Intel microcontrollers are:

	Elements	Specifications
(*i*)	I/O Lines	27
(*ii*)	ROM or EPROM	1 K byte
(*iii*)	Read/Write memory	64 bytes
(*iv*)	Timer/counter	8-bit
(*v*)	Accumulator memory	8-bit
(*vi*)	Scratchpad memory	64 bytes
(*vii*)	Program memory	1024 words

Functions. They perform these functions:

- Manipulation of signals
- BCD operations
- Conditional branching
- Table look-up
- External interrupt
- Single-step troubleshooting

Applications. The primary use of the Intel 8048 is found in low-cost, high-volume control applications. Additional features such as an 8-bit A/D converter and a serial I/O port make it versatile as a microcomputer.

8.10.2 Intel 8051 Single-chip Family

Figure 8.11 shows the block diagram of the 8051 chip. It includes the following features:

- 4 K bytes of ROM or EPROM
- 128 bytes of data memory and 21 special function registers
- Four programmable I/O ports
- Two 16-bit timer/counters
- Serial I/O port
- Five interrupt lines—two for external signals and three for internal operations

Functions. The tasks performed by the Intel 8051 are:

- Binary and BCD arithmetic operations
- Bit set/reset functions

- Logical functions
- Functions such as set, clear, complement
- Jump if set or not set
- Jump if set then clear
- Carry flag logical function

Applications. This microcontroller has higher capability and versatility from the design point-of-view. The Intel 8051 is commonly used in sophisticated real-time instrumentation and industrial control. It has a 12 MHz clock and a very powerful instruction set to make it more useful. This is the latest in the single-chip microcomputer family.

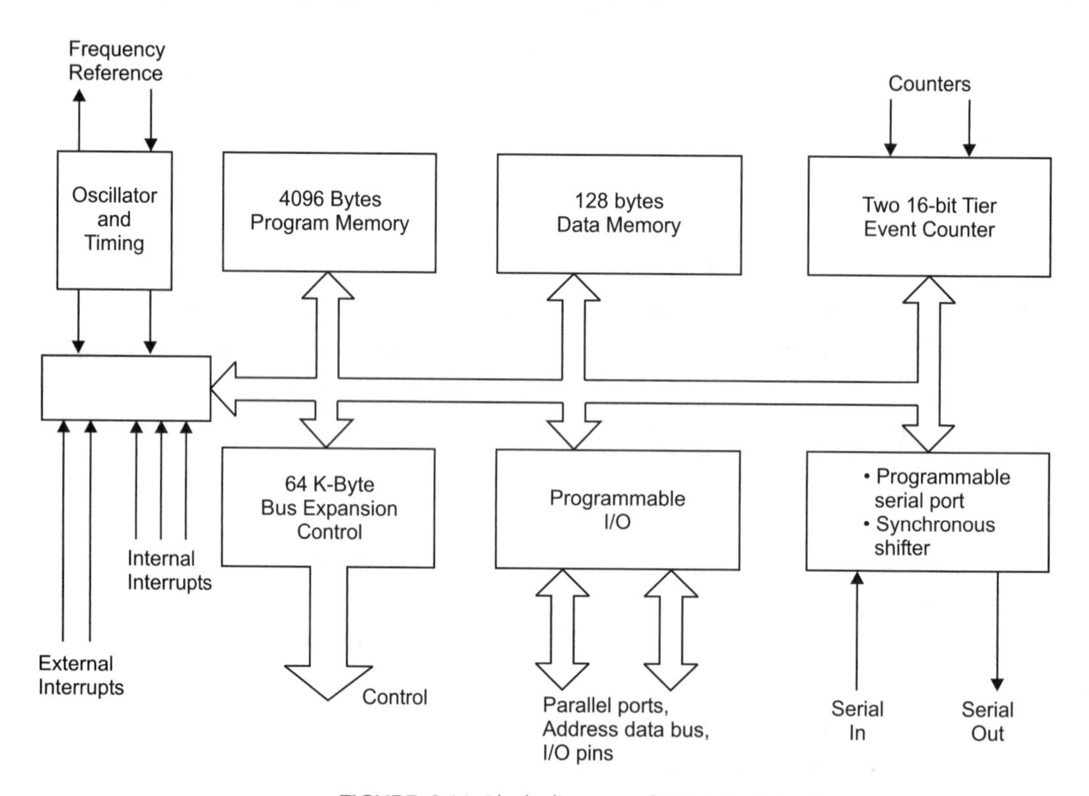

FIGURE 8.11 Block diagram of 8051 Intel family.

8.10.3 Universal Peripheral Interface (UPI): 8041 Family

The UPI is a single-chip microcomputer that is in many ways similar to the 8048 family. Apart from the elements of the 8048 it has other inclusions such as a data bus, input/output register, and a status register. The data byte written in the register by the master generates an interrupt to inform the 8041, and the data byte is read with a special instruction. The status register checks for the handshaking signal before writing the data byte and then the data byte is transferred from the slave to the master.

Functions. They perform functions such as handshaking and synchronizing the data transfer between the UPI and the master.

Applications. The chips are primarily meant for performing the functions of the slave MPU in distributed processing.

8.10.4 Analog Signal Processor: 2920 Family

The 2920 is a single-chip microcomputer specially designed to carry out the processing of analog signals. The inability of most microprocessors to process high frequency analog signals because of slow response led to the design of a microcontroller with a special architecture and instruction set for handling high speed signals.

The features of the 2920 family are:

- MPU and memory
- ADC and DAC
- Multiplexer to handle four different inputs
- Sample and hold circuit
- Demultiplexer unit

Applications. The 2920 microcontroller has wide application in acquiring and processing analog signals. Common applications include:

- Telecommunication
- Signal processing
- Guidance and control
- Speech processing
- Industrial processes and operation automation
- Mechatronic systems

EXERCISES

1. Write the truth tables for OR, NOR, AND, and NAND gates for three inputs.
2. Explain the working principle of digital-to-analog converters.
3. Explain the structure of a microprocessor. Draw the pin configuration of an 8085 processor.

4. Write the architecture of an 8085 microprocessor.
5. What is a microcontroller? Explain the use of a microcontroller with an example.
6. Give the classification of microcontrollers.
7. Give five application areas of microprocessors.
8. Describe with an example a microprocessor-based digital control system.
9. Explain the following terms:

   (*a*) Fetch cycle (*b*) State (*c*) Bus
10. Discuss memory and address as related to microprocessors.
11. State in detail the difference between a microprocessor and microcontroller.
12. Briefly give the classifications of microcontrollers.
13. With the help of a block diagram example briefly give the form of a microprocessor system.
14. Explain in brief the common types of registers used in microprocessors.
15. What are buses? Explain the functions of three forms of buses in a microprocessor system.
16. Explain in detail with a block diagram the architecture of an Intel 8085 A microprocessor.
17. What are microcontrollers? Explain the general from of a microcontroller with a sketch.

INDEX

A

A.C. Commutator Motors 129
A.C. Servomotor 133, 134, 135, 136
Acceleration 141, 142, 147, 148, 160, 161, 162, 164, 169
Accumulator 86, 87, 89
Actuation Module 13, 14
Actuation System 19, 20
Actuator 59, 60, 61, 63, 64, 80, 81, 82, 83, 84, 87, 89, 91
Address Bus 260, 265
Acceleration Feedback 148
ALE (Address Latch Enable) 266
ALU 243, 257, 259, 261
Amplifier Errors 216
Analog-to-Digital Converter (ADC) 245, 275
Annular Gear Motor 75
Anti-friction Bearings 176, 204, 209
Anti-friction Ways 184
Aperture Control 28
Assembler 263
Assembly Module 13
Association Diagram 37
Asynchronous Motors 124, 126, 128
Auto Focusing 26
Automatic Camera 26, 28
Automatic Washing Machine 29
Automation Objective 11
Automation Systems 7

B

Ball Recirculating Power Screws 189
Band-pass Filters 228
Band-stop Filters 229
Bathroom Scale 1, 12, 31
Bearings 175, 176, 193, 199, 203, 204, 209, 210
Bearing Material 202
Bent Axis Motor 70, 71
Binary Addition 247
Binary Coded Decimal 262
Binary Subtraction 248
Binary System 247, 248
Bipolar Transistor 109, 110
BLDC Motor 123, 124, 130, 134
BLDC Servomotor 135
Bridge Circuit 233
Bus 243, 260, 261, 265, 266, 267

C

Capacitance 141, 163, 169, 170, 171, 173
Check Valve 83, 85, 87, 89, 91, 92, 93, 94
CMRR 226
Common-emitter Circuit 110
Communication Module 13, 14
Communication Objective 11
Comparator 226, 227, 238
Compression Pre-loading 191
Concept of Signal Conditioning 215
Concurrent Integration 3, 5, 6
Conical Pivots 210
Contact Bounce 104, 105
Control Bus 260, 261
Control Objective 11
Control Program 12, 15, 21
Control System 11, 16, 19, 20, 21

Control Valves 141, 165, 166, 169
Controllers 1, 2, 3, 6, 11, 22
Counter-balance Circuit 60, 90, 92
CPU 243, 245, 250
Current-to-Voltage Converter 220
Cylinder Style 63, 66

D

D-Flip-Flop 105
Dampers 151, 152
Damping Element 144
Data Acquisition (DAQ) 216, 240, 241
DAQ Board 240
Darlington Pairs 111
Data Bus 260, 261, 265, 266, 267
Data Extraction 10
Data Integration 11, 18
D.C. Motor 117, 119, 122, 129, 133
D.C. Servomotor 158
Debounce 104
Deformation 162, 163
Design 6, 10, 12, 15, 16
Design Considerations 177, 205
Design Flexibility 244
Design of Mechatronic Systems 12
DFT 235
Differential Amplifier 220, 221, 222, 226
Digital Control 243, 245, 246
Digital Number System 243
Digital Processing 244
Digital Signals 236, 239
Digital-to-Analog Converter (DAC) 236, 237, 238, 245, 246, 275
Diodes 99, 106
Direction Control Circuit 60, 95
Direction Control Valve 83, 85, 168
Display Objective 11
Drift 224, 225
Dynamic Characteristics 38, 39
Dynamic Loads 178, 196

E

EEPROM 258
Electric Motors 115
Electrical Actuators 99
Electrical System 154
Energy Domain 35, 36, 38
Engine-management System 24
Engineering Skills 16
Environment Module 13
EPROM 258, 269, 273

F

FDM 231, 232
Fetch Cycle 267
Filtering 228
Flexure Pivots 211
Floating Body 44
Flow Control 85
Flow Control Valve 85
Flow Sensors 35, 43, 44
Four-pole D.C. Motors 117
Fourier Analysis 234, 235
Frequency Response 224, 225
Frictionless Bearings 176, 210, 212
Fuel Level Indicator 49, 50

G

Gain Variation 226
Gear Motor 68, 74, 75
Gear Pump 43, 44
Graded Resistor 236, 237
Guide Ways 175, 178, 179, 180, 182, 185, 197

H

Hall Effect 35, 46, 47, 48, 49, 50
Hall Effect Sensor 48, 49
Hall Sensor 50
Hersey Diagram 199, 200
Hexadecimal System 248
High-pass Filters 230

Hydraulic Actuators 18, 20
Hydraulic Circuit 87, 89
Hydraulic Cylinders 63, 64, 67, 68
Hydraulic Motors 59, 68, 69
Hydraulic Pump 88
Hydraulic System 59, 84, 90, 141, 160
Hydrostatic Journal Bearing 201
Hydrodynamic Journal Bearing 200, 201
Hydrostatic Bearing 202
Hydrostatic Slide Way 181, 186
Hydrostatic Transmission 60, 90, 96

I

Inductance 141, 162
Induction Motor 99, 125, 127, 134
inductive Flow Sensor 43, 45
Inductive Proximity Sensor 53
Inductive Sensor 53
Input Bias Current 224
Instruction Set 263, 269, 274, 275
Integrating Amplifier 222
Intel 8048 Family 272
Intel 8051 Single-chip Family 272, 273
Intelligent Control 7
Interface Diagram 240
Interface Module 13, 14, 15
Interrupts 263
Inverting Type Operational Amplifier 220

K

Knife Edge Bearing 210

L

Ladder Type DAC 236, 237
Light Sensors 35, 50, 51
Limit Switch 102, 103
Linear Actuator 163, 164
Liquid Bath 172
Liquid Flow Switch 102
Liquid Level Switch 101
Load Cell 234
Loads 175, 176, 178, 190, 192, 193, 194, 196, 197, 202, 203, 204, 206
Loads on Spindles 196
Logarithmic Amplifier 223
Logic Functions 243, 249, 250
Logic Gate 243
Logic State Analyzer 262
Low-pass Active Filters 229
Low-pass Filters 228, 240
Low-pass Passive Filter 229

M

Machine Tool 175, 208
Main Wash Cycle 29, 30
Masked ROM 258
Mechanical Design 6, 12, 16
Mechanical Elements 141, 142, 151
Mechanical Switches 99, 100, 101
Mechanical Systems 141, 142, 152, 169
Mechatronics Design 5, 12, 13
Mechatronics in Industry 8, 18
Mechatronics Technology 2, 15, 16
Mechatronics 1, 2, 4, 5, 7, 8, 9, 10, 11, 12, 15, 17, 18, 23, 33
Memory 245, 257, 258, 261, 262, 265, 266
Meter-out Circuit 90, 94, 95
Microdevices 7
Microcontrollers 11, 14, 19, 243, 244, 268, 269, 272, 273, 275
Microprocessors 243, 244, 245, 246, 248, 250, 257, 260, 261, 262, 263, 265
Microprocessor-based Digital Control 245
Microprocessor Controller 22, 28, 32
Modules in Mechatronics System 13
Morse Taper 196, 198
MOSFET 47, 48
Motion Control 7, 12, 15
Mounting Style 67
Multiplexer 216, 230, 231, 240

N

Need for Mechatronics in Industry 8

Needle Valve 80, 85
Number System 243, 247, 248, 249

O

Objectives of Mechatronics 10, 11
Operational Amplifier 215, 218, 219, 220, 221, 224, 225, 226, 229, 230, 236, 238
Optical Sensor 55, 56
Optoelectronic Systems 7
Output Characteristic 40

P

Parallel Arrangement 152
Permanent Magnet D.C. Motor 119, 120
Photo Cell 51
Photo Conductors 52
Photo Diodes 51
Photo Transistors 51, 52
Piston Style 64
Planetary Roller Screws 176, 192, 193
Planetary Rollers 192
Pneumatic Actuators 10, 14
Position Feedback 144
Pre-loading of Anti-friction Bearings 176
Pressure Control 59, 76, 90
Pressure Control Valves 76, 77
Pressure Relief Valve 78
Pressure Sensors 35, 41
Pressure Switch 101
Pre-wash Cycle 30
Principle of Motors 115
Principle of Slide Ways 180
Processor Module 13, 14, 15
PROM 258
Proximity Sensors 35, 53
Proximity Switch 102
Pulse-latching Solenoid 113
Pumps 84, 88

R

R-L-C Circuit 156, 157
R/W M 258, 259
Rack-and-Pinion Arrangement 152, 153
Radial Piston Motor 71, 72
Range and Span 39
Read Cycle 265
Recirculating Ball Screws 176
Recirculating Type of Anti-friction Ways 185
Reducing/Regulating Circuit 90
Registers 243, 250, 257, 273
Relays 99, 114, 115, 128
Relief Valve 77, 78, 79, 85, 87, 88, 91, 92, 93, 165, 167
Relieving Circuit 90, 91
Reservoir 70, 78, 86, 87, 88, 89, 91
Resistance 141, 142, 145, 160, 161, 162, 163, 164, 166, 167, 168, 169, 170, 172, 173
Resistor 154, 155, 156, 157
Rinse Cycle 29, 31
Rolling Edge Bearing 210, 211
Rotary Actuators 60, 63, 80

S

Sampling Theorem 239
Schmitt Trigger 105, 223
Screw-and-Nut Arrangement 189
Selector Switch 101
Self-commutated D.C. Motors 118
Sensing Parameters 38
Sensor 6, 14, 15, 22, 23, 24, 28, 36
Sensor Energy 37
Sensor Function 41
Sequence Circuit 93
Sequence Valve 79, 80
Sequential Integration 3, 4, 5
Series Arrangement 151
Servomotors 133
Servo-drive Control 134
Signal Averaging 234, 235
Signal Conditioning 216, 217
Signal Processing 234
Single-phase Induction Motors 125
Single-phase Stepper Motor 131

Slew Rate 225
Slide Ways 180, 186
Sliding Bearings 176, 199, 200
Software Module 13, 14
Solenoid 99, 111, 112, 113, 114, 115
Solid-state Switches 99, 106
Spark Timing 25
Speed Measurement 37
Speed of Operations 244
Speed Switch 102
Spin Cycle 29, 31
Spindle Bearings 176, 193, 198
Spindle Noses 194
Spindles 175, 176, 193, 194, 196, 197, 198, 199, 203, 205
Spring-mass-damper 148, 150
SR-Flip-Flop 105
Static Loads 178, 196
Steep Taper Nose 195
Stepper Motor 28, 129, 132
Stick-slip Phenomena 179, 181, 182
Stored Program Control 244
Summing Amplifier 221
Swash Plate Motors 69
Symbols for Hydraulic Systems 84, 85, 86, 87
Synchronous Motors 115, 127, 128, 130
Synergistic Integration 3
System Integration 7, 11, 14
System Models 141

T

Task Program 23
Temperature Compensation 233
Temperature Controller 172
Temperature Switch 102
Tension Pre-loading 191
Thermal Capacitance 169, 170, 171
Thermal Loads 175, 178, 194, 197
Thermal Resistance 141, 169, 170, 171
Thermometer System 171
Throttle Valve 166
Thyristor 99, 106, 107, 108, 109, 111, 135, 136
Time Division Multiplexer (TDM) 231
Transducers 35, 45, 55
Transfer Function 143, 144, 146, 169, 172
Triac 99, 106, 109, 136
Turbine Sensor 44
Twist Flexures 210, 212
Type A Spindle Nose 194
Types of Filters 215, 228

U

Ultrasonic Proximity Sensor 53, 54
Ultrasonic Sensors 53
Utility Lines 260, 261

V

Valve Actuation 24
Valves 59, 76, 77, 81, 85
Vane Motors 72, 73
Velocity Feedback 146, 150, 154
Voltage-to-Current Converter 220

W

Weighing Machine 12
Wheatstone Bridge 215, 232, 234
Write Cycle 266

Z

Zener Diode Protection 227